Coal and Coal Products: Analytical Characterization Techniques

Coal and Coal Products: Analytical Characterization Techniques

E. L. Fuller, Jr., EDITOR
Union Carbide Nuclear Division

Sponsored by the ACS
Divisions of Analytical,
Fuel, and Colloid
and Surface Chemistry

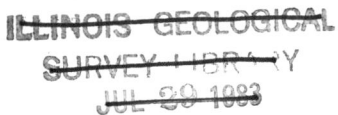

ACS SYMPOSIUM SERIES **205**

AMERICAN CHEMICAL SOCIETY
WASHINGTON, D. C. 1982

Library of Congress Cataloging in Publication Data

Coal and coal products.

(ACS symposium series, ISSN 0097–6156; 205)
Includes bibliographies and index.

1. Coal—Analysis—Congresses.
I. Fuller, E. L. II. American Chemical Society. Division of Analytical Chemistry. III. American Chemical Society. Division of Fuel Chemistry. IV. American Chemical Society. Division of Colloid and Surface Chemistry. V. Series.

TP325.C5145 1948   622.6'22   82–18442
ISBN 0–8412–0748–8   ACSMC8 205   1–326
                                    1982

Copyright © 1982

American Chemical Society

All Rights Reserved. The appearance of the code at the bottom of the first page of each article in this volume indicates the copyright owner's consent that reprographic copies of the article may be made for personal or internal use or for the personal or internal use of specific clients. This consent is given on the condition, however, that the copier pay the stated per copy fee through the Copyright Clearance Center, Inc. for copying beyond that permitted by Sections 107 or 108 of the U.S. Copyright Law. This consent does not extend to copying or transmission by any means—graphic or electronic—for any other purpose, such as for general distribution, for advertising or promotional purposes, for creating new collective work, for resale, or for information storage and retrieval systems. The copying fee for each chapter is indicated in the code at the bottom of the first page of the chapter.

The citation of trade names and/or names of manufacturers in this publication is not to be construed as an endorsement or as approval by ACS of the commercial products or services referenced herein; nor should the mere reference herein to any drawing, specification, chemical process, or other data be regarded as a license or as a conveyance of any right or permission, to the holder, reader, or any other person or corporation, to manufacture, reproduce, use, or sell any patented invention or copyrighted work that may in any way be related thereto.

PRINTED IN THE UNITED STATES OF AMERICA

ACS Symposium Series

M. Joan Comstock, *Series Editor*

Advisory Board

David L. Allara	Marvin Margoshes
Robert Baker	Robert Ory
Donald D. Dollberg	Leon Petrakis
Robert E. Feeney	Theodore Provder
Brian M. Harney	Charles N. Satterfield
W. Jeffrey Howe	Dennis Schuetzle
James D. Idol, Jr.	Davis L. Temple, Jr.
Herbert D. Kaesz	Gunter Zweig

FOREWORD

The ACS SYMPOSIUM SERIES was founded in 1974 to provide a medium for publishing symposia quickly in book form. The format of the Series parallels that of the continuing ADVANCES IN CHEMISTRY SERIES except that in order to save time the papers are not typeset but are reproduced as they are submitted by the authors in camera-ready form. Papers are reviewed under the supervision of the Editors with the assistance of the Series Advisory Board and are selected to maintain the integrity of the symposia; however, verbatim reproductions of previously published papers are not accepted. Both reviews and reports of research are acceptable since symposia may embrace both types of presentation.

CONTENTS

Preface ix

1. Theoretical and Experimental Approaches to the Carbonization of Coal and Coal Blends 1
   Maggi Forrest and Harry Marsh

2. Characterization of Alkanes in Extracts of Coals, Lignites, and Related Fuels 27
   K. D. Bartle, D. W. Jones, and H. Pakdel

3. Fourier Transform IR Spectroscopy: Application to the Quantitative Determination of Functional Groups in Coal 47
   Paul C. Painter, Randy W. Snyder, Michael Starsinic, Michael M. Coleman, Deborah W. Kuehn, and Alan Davis

4. Applications of Fourier Transform IR Spectroscopy in Fuel Science .. 77
   P. R. Solomon, D. G. Hamblen, and R. M. Carangelo

5. Chemistry and Structure of Coals: Diffuse Reflectance IR Fourier Transform (DRIFT) Spectroscopy of Air Oxidation 133
   N. R. Smyrl and E. L. Fuller, Jr.

6. Comprehensive Elemental Analysis of Coal and Fly Ash 147
   R. A. Nadkarni

7. Application of Inductively Coupled Plasma Atomic Emission Spectrometry (ICP–AES) to Metal Quantitation and Speciation in Synfuels 163
   R. S. Brown, D. W. Hausler, J. W. Hellgeth, and L. T. Taylor

8. Determination of Chlorine in Organic Combination in the Coal Substance 185
   J. N. Chakrabarti

9. Electron Probe Microanalysis: A Means of Direct Determination of Organic Sulfur in Coal 191
   Robert Raymond, Jr.

10. Chemical Fractionation and Analysis of Organic Compounds in Process Streams of Low Btu Gasifier Effluents 205
    R. L. Hanson, R. E. Royer, J. M. Benson, R. L. Carpenter, G. J. Newton, and R. F. Henderson

11. Solvent Analysis of Coal-Derived Products Using Pressure Filtration 225
    Bruce R. Utz, Nand K. Narain, Herbert R. Appell, and Bernard D. Blaustein

12. Scanning Electron Microscope-Based Automated Image Analysis (SEM–AIA) and Mössbauer Spectroscopy: Quantitative Characterization of Coal Minerals 239
    F. E. Huggins, G. P. Huffman, and R. J. Lee

13. **Analytical Instruments in the Coal Preparation Industry: Current Status and Development Needs** 259
    Leon N. Klatt

14. **Coal as Energy in the Steel Industry** 281
    Dan P. Manka

15. **Electron Optical and IR Spectroscopic Investigation of Coal Carbonization** 293
    J. J. Friel, S. Mehta, and D. M. Follweiler

Index 311

PREFACE

THE INDUSTRIAL REVOLUTION was based upon efficient use of mechanical energy for production and processing of materials. The initial source of this energy came from the thermal energy stored in coal, but rapid advancements were made, and petroleum soon replaced coal as the prime source of energy to support an increased standard of living for mankind. The world's population has increased to such a staggering extent that the machines that were such a luxury a century ago have now become necessities just to provide essential life-supporting food, clothing, and shelter. Only a degree of "independence" and self-sufficiency can be achieved, and each of us must work to provide services and products that are needed by others. This increase in efficiency allows us to inhabit the earth at a population density and state of well being far in excess of the hunter/farmer status of the past.

However, this mode of operation strains our natural resource supplies to varying degrees. The recent "energy crisis" serves a good purpose: it points out that comprehensive planning and research are required to ensure that mankind can flourish and that the raw materials are carefully and efficiently utilized. The simple political control of one supply of energy (crude oil in this case) from one rather small geographic region has suddenly impacted the global market to "crisis" proportions. Systematic planning and foresight will be required to avoid crises reminiscent of the "energy crisis" that came about when the hardwood forests were decimated to provide charcoal for the steel industry.

We now have adequate information, processing capabilities, technical knowledge, and trained personnel to ensure that smooth transitions can be implemented without the tragic impact of "crisis" situations. Our sources of energy can be brought into play without significant impact on the environment if proper comprehensive analysis is employed. Coal still reigns among sources of energy in terms of the amount and availability. The generically related materials (oil shales, tar sands, lignite, peat, heavy oils, etc.) further extend the importance of fossil fuels for energy bases.

This volume represents a small step toward the research and interdisciplinary communications required to return efficiently to our large reserves of coal for a primary source of energy. Each presentation chosen for this volume represents a significant contribution to either a specific area of technical importance in coal processing and utilization, or a specific application of a given method of analysis as applicable to various coals and/or derived products.

New and/or alternate approaches to studies of the structure and chemistry of coals and related materials will yield increased productivity and improved environments. In our small way this treatise is our current international effort to advance the technology we all need.

E. L. FULLER, JR.
Union Carbide Nuclear Division
Oak Ridge, TN 37830

August 1982

Theoretical and Experimental Approaches to the Carbonization of Coal and Coal Blends

MAGGI FORREST and HARRY MARSH

University of Newcastle upon Tyne, Northern Carbon Research Labs, School of Chemistry, Newcastle upon Tyne, NE1 7SU, England

Mechanisms of carbonization of fluid systems from pitches and coals of different origin and rank to form anisotropic cokes are discussed. The concept of nematic liquid crystals and mesophase is introduced. The origins of optical texture in cokes and the chemical and physical factors which control the size and shape of optical texture are explained. The significance of optical texture in metallurgical cokes is analysed in terms of coke strength and chemical reactivity. Laboratory experimental approaches include control over carbonization procedures, the examination of polished surfaces of resultant cokes by optical microscopy, the use of scanning electron microscopy to monitor changes induced by thermal treatment and gasification of cokes, as well as point-counting of optical texture and the use of microstrength testing procedures. Modern technological approaches to the successful use of coals of several ranks to make metallurgical coke include blending of coals sometimes with pitch additions. The resultant enhancement of coke strength is explained in terms of the development of suitable optical texture from solutions of coal in coal or of pitch in coal. Hydrogen transfer reactions are important here. The use of breeze additions in coal blends is commented upon.

Metallurgical coke is used in the blast furnace as an energy source, a chemical reducing agent and to provide permeability and support for the furnace load. To fulfil these functions and maintain blast furnace performance the coke must maintain its size within an optimum range. It must therefore be able to maintain its mechanical strength and to withstand degradation due to gasification in carbon dioxide, abrasion, compressive forces and thermal shock in the furnace. Prime coking coals of volatile content 19-33% are becoming scarce in Western Europe and Japan and economic necessity dictates the need for the development of blending

0097-6156/82/0205-0001$07.50/0
© 1982 American Chemical Society

procedures which make use of coals of volatile content above and below this critical range. The characteristics of cokes from such blends must fall within stringent specifications to maintain blast furnace performance ([1]). It is therefore necessary to combine the results of fundamental research with experience gained from the past empirical approach to coal blending ([2]). The purpose of this paper is to discuss recent theoretical considerations and experimental studies of coal and coal/pitch blending procedures, gasification and thermal treatment of metallurgical cokes and the effects of pitch coke breeze additives upon coke strength.

The Carbonization Process

The Formation of Anisotropic Carbon. As coal is heated it undergoes depolymerization and decomposition resulting in the evolution of gas and condensible vapours and leaves behind a solid residue of high carbon content ([3]). The ability of some coals to soften and become plastic upon heating, coalescing to form a coherent mass, is the property upon which the formation of coke depends ([4]). Such coals are described as caking coals.

Over the temperature range 623 K to 773 K, caking coals begin to soften, coalesce, swell and then re-solidify into a porous structure which, at temperatures just above the resolidification temperature, is green coke.

Initial softening of the coal is due to increased thermal agitation. At the same time, but independently of the physical process, pyrolysis begins to modify the viscosity of the coal by breakage of the chemical cross-linkages responsible for making coal a polymeric material. Swelling of the mass is due to the evolution of volatile matter which cannot escape from the coal mass and which encounters resistance to its flow through macropores and fissures in the coal ([5]).

Resolidification of the coal is due to chemical cross-linking of molecular constituents so converting the plastic state into a visco-elastic state and finally into a brittle porous solid.

Similarly, coal tar and petroleum pitches, on carbonization, form a fluid melt. The viscosity of this melt initially decreases with increasing temperature and then increases once again as molecular reactivity leads to chemical polymerization. Eventually these materials form the anisotropic graphitizable cokes.

It is during the plastic (fluid) stages of carbonization that the most important features of coke structure are formed, particularly porosity ([6]) and pore wall structure. During the plastic stage, the optically isotropic parent material undergoes a phase transition to an optically anisotropic melt which finally forms an anisotropic coke. The significance of this phase transition to coke properties is profound, as will be shown.

Liquid Crystals and Mesophase

Conventional Liquid Crystals. There exist some (generally organic) systems which do not pass directly from a solid to a liquid on heating but do so via distinct phases, called liquid crystals, having properties intermediate between solids and liquids. The transitional symmetry of a solid crystal is not present in liquids, although dynamic local positional ordering (statistical order) is present, extending over 1 nm and lasting for 10 to 100 molecular interactions. In a liquid crystal phase, however, there is well defined order on a length scale of 10^2 μm although fluid properties are also well defined. Viscosities are measurable and of the order of 10^2 to 10^{11} Nsm^{-2}. Conventional liquid crystals (known since 1880) will dissociate and revert to normal liquids on heating. There are several types of liquid crystal, nematics, smectics and cholesterics, of which only nematics will be discussed here ($\underline{7}$). Conventional nematics are composed of rod-like molecules, polar and without facility of intra-molecular rotation. The long axes of these rods are aligned about a common direction in space, the director vector, $\hat{n}$, which is an axis of rotational symmetry for all macroscopic properties. The positions of the centres of mass of the rods are randomly distributed; there is no long range order and an X-ray diffraction pattern would show no Bragg peaks. A nematic liquid crystal is uniaxial with the optic axis along $\hat{n}$.

Liquid Crystals in Carbonaceous Systems. The formation of anisotropic carbon from the isotropic melt of pitch or coal was first attributed to the presence of a liquid crystal phase by Brooks and Taylor ($\underline{8}$). The isotropic melt undergoes a physical phase transition to a lamellar nematic liquid crystal. There are important differences between the lamellar nematics associated with carbons and conventional rod-like nematics. Their physical properties are essentially identical but chemically they are quite different. Conventional nematics are chemically stable while pitch or coal lamellar nematics in a pyrolysing system are extremely reactive. The important distinction of shape of constituent molecules must also be emphasised; a conventional nematic consists of long rods aligned about $\hat{n}$ while carbonaceous mesophase is composed of plate-like molecules. Here it is the normal to the surface of the molecule (again the axis of symmetry) which is aligned parallel to $\hat{n}$. Further, in conventional liquid crystals the phase change from liquid to liquid crystal is brought about by a reduction in kinetic energy (molecular mobility) by lowering the temperature; in the carbonaceous mesophase molecular mobility is reduced to a point where it is favourable for the phase change to take place because of increasing molecular weight caused by polymerization associated with increasing heat treatment temperature (HTT). Chemical studies of the molecular species present during the fluid phase of carbonization suggest molecular weights of typically 2000 amu (9).

A pyrolysing system about to pass through the liquid crystal phase transition (630-700 K) comprises molecules with a large spread of molecular weights which are continually interacting chemically and polymerising via cross-linkages using dehydrogenative polymerisation reactions. This polymerisation creates planar molecules consisting of sheets of hexagonally linked carbon atoms containing holes, heteroatoms and free electrons. When the molecular size reaches ~1000 amu it becomes energetically favourable for the system to undergo a physical phase transition to a liquid crystal phase in which van der Walls forces maintain stability initially. The transition is seen in the hot stage of a polarized light microscope as the formation of droplets of optically anisotropic material which appear as spheres of yellow/blue colour in a purple isotropic matrix. Continued pyrolysis causes these droplets to grow at the expense of the surrounding isotropic material.

This phase transition is initially reversible, as demonstrated by Lewis (10), but as carbonization progresses chemical cross-linking makes it irreversible. The liquid crystal droplets grow and coalesce until all the isotropic melt has undergone the phase transition. This new phase is then termed the liquid crystal mesophase, *i.e.* the phase intermediate between the isotropic fluid pitch and solid semi-coke.

Optical Texture

The optical anisotropy of cokes gives rise to a characteristic pattern of extinction contours when a polished surface is examined by polarized light microscopy using crossed polars, or reflection interference colours if the polars are parallel and a half wave plate is inserted into the optical system (11, 12). This characteristic pattern is termed the optical texture of the coke. Optical texture increases in size with increasing fluidity (decreasing plasticity) of the mesophase. Mesophase viscosity can be affected by parameters of carbonization such as heating rate, HTT and soak time but the single most important factor is chemical reactivity (13). If reactivities are too high, early polymerisation leads to the formation of isotropic carbon because of randomly aligned interactions. Low reactivities give rise to a low viscosity mesophase which flows and coalesces easily. Hence the size and type of optical texture is predominantly a function of the parent material carbonized and may be used to characterise the coke. It is therefore necessary to define precisely the different types of optical texture seen in cokes. A standard nomenclature has been devised (Table I) which ranges from very fine-grained mosaics (<0.5 μm diameter) to flow domain anisotropy (>60 μm length, >10 μm width).

<u>Significance of Optical Texture</u>. In metallurgical and foundry cokes, subject to stringent specifications of strength, size and reactivity, it is necessary to know whether the development of anisotropy during the mesophase is incidental or essential to coke properties (14, 15, 16).

Strength

Coke strength is intimately related to its porosity and pore wall structure. During the formation of the latter, liquid crystals encounter surfaces (gaseous or solid) within the carbonization system and tend to flow in such a way that the lamellar molecules align with the surfaces (14). Optical microscopy of coke shows layers of anisotropic material forming a 'lining' on internal and external surfaces. Thus a growing pore may develop such a lining which undergoes further polymerization to form a continuous lamellar structure (17). This is equivalent to a large increase in surface tension, severely restricting pore growth and preventing the formation of 'frothy' coke. Similarly, an inert particle may be encased in a strong layer of orientated lamellar molecules. These mechanisms increase coke strength.

Fissure propagation through the pore wall material of coke is also dependent on the size and orientation of anisotropic components present; fissures propagate easily through large optical textures (>60 μm) while mosaic textures (1 to 10 μm) tend to act as fissure stops. Experimental verification of this aspect of fissure generation and propagation will be discussed below.

Reactivity

When carbon dioxide reacts with carbon to produce carbon monoxide at temperatures above 1140 K, in the blast furnace, carbon goes into 'solution' in the gas and solution loss is said to have occurred (18). Solution loss is so termed because carbon monoxide in blast furnace gases represents a loss in reduction capacity. Reduction capacity is maximised with zero CO in the effluent gas. The reaction reduces the impact strength and abrasion resistance of the coke and must be minimized. The reaction does not occur evenly over the coke surface but gives rise to preferential gasification fissuring and pitting (17, 19, 20). This is a direct consequence of the anisotropy of coke whereby reaction rates differ with crystallographic direction. Such fissuring causes premature coke abrasion and breakage, reducing permeability in the blast furnace melting zone and influencing the stability of the race-way. Formation and

Table I

Nomenclature to Describe Optical Texture in Polished Surfaces of Cokes

Isotropic	(I)	No optical activity
Very fine-grained mosaics	(VMF)	<0.5 μm in diameter
Fine-grained mosaics	(Mf)	<1.5 >0.5 μm in diameter
Medium-grained mosaics	(Mm)	<5.0 >1.5 μm in diameter
Coarse-grained mosaics	(Mc)	<10.0 >5.0 μm in diameter
Supra mosaics	(SM)	Mosaics of anisotropic carbon orientated in the same direction to give a mosaic area of isochromatic colour.
Medium-flow anisotropy	(MFA)	<30 μm in length; <5 μm in width
Coarse-flow anisotropy	(CF)	<60 >30 μm in length; <10 >5 μm in width
Acicular flow domain anisotropy	(AFD)	>60 μm in length; <5 μm in width
Flow domain anisotropy	(FD)	>60 μm in length; >10 μm in width
Small domains	(SD)	<60 >10 μm in diameter
Domains, ~isometric	(D)	>60 μm in diameter

D_b is from basic anisotropy of low-volatile coking vitrains and anthracite.
D_m is by growth of mesophase from fluid phase.
Ribbons (R) Strands of mosaics inserted into an isotropic texture.

propagation of fissures formed upon gasification and thermal stressing of the coke are minimized in cokes of small optical texture (fine- and medium-grained mosaics; see Table I), as found in cokes from prime coking coals. Overall, anisotropic carbon is less reactive than isotropic carbon.

The presence of a liquid crystal phase during carbonization is therefore a pre-requisite for the formation of good metallurgical coke.

Blending Procedures

The practice of coal blending is now of central importance to the coke making industry as supplies of prime coking coals become depleted. Blends must consist of coals which have complementary properties to compensate for their individual coking deficiencies The coking properties of the overall blend must be similar to those of a single prime coking coal. Coking potential of a coal blend is determined by:
  (a) Behaviour in the plastic zone (fusing and pore formation).
  (b) Behaviour in the post-plastic zone (fissure formation from differential contraction).
  (c) Compatibility of the blend components.
  (d) Content and size distribution of the mineral matter.

The Plastic Zone. Plastic zone behaviour is quantified by determining dilation in a dilatometer (21); conditions and constraints are not the same as in a coking oven, but results are indicative of the swelling pressure developed and show if it is sufficient to achieve the degree of surface contact of particles necessary to produce adequate coalescence. It has been shown (2) that for charges to equal volatile content, the cohesion improves as total dilation of the blend increases to 40-50% and then remains constant. The target specification for coal blends for metallurgical cokes is therefore a total dilation of not less than 50%. (That of prime coking coals is ∼75%).

The Post-Plastic Zone. It is difficult to quantify blend behaviour (contraction and resultant fissuring) in the post-plastic zone. There is, however, a relationship between the temperature of the first peak of the contraction coefficient of the coke and volatile matter content (2) which suggests target specifications for blends of volatile content in the range ∼20-32% and dilation in the range ∼50-150%.

Compatibility of Blend Components. It is important when blending coals to ensure that the temperature ranges over which the individual coals become plastic overlap. The greater the degree of overlap, the better the fluid mixing and chemical interaction of the components and the more homogeneous and cohesive the resultant coke. Compatibility may be improved by crushing the low volatile blend components to <0.5 mm before mixing or by addition of a bridging coal of intermediate plastic zone.

Mineral Matter Content. Coarse mineral matter content can initiate fissure formation and decrease coke strength and so must be monitored during blending procedures. However, the careful addition of inert material can sometimes be used to improve coke strength. Coke oven breeze or petroleum coke breeze may be added to a coal blend to improve the tensile strength of the coke ([22], [23]). Coke oven breeze, finely ground and added as up to 8% of the charge can improve the tensile strength of some cokes.

Petroleum coke breeze, however, gives progressive increases of tensile strength for breeze additions of up to 50% in some blends. In general, the lower the HTT of the breeze, the better the tensile strength of the coke ([24]). Particle size (affecting the ability of the reactive components to incorporate the particles) also considerably influences coke strength. Increases in coke strength with low percentage additions and small particle sizes are thought to be due to the action of breeze as a filler, increasing the thickness of the pore wall material of the coke and thus increasing the volume of load-bearing material ([24]). This strengthening factor outweighs the weakening due to micro-fissure generation associated with non-uniformity of contraction rates of the breeze and surrounding coal coke.

The Use of Pitch as a Blend Additive. The use of pitch additives in blends is being increasingly investigated as an alternative to using prime coking coal as one blend component. Low viscosity pitch additives are particularly successful when used in place of bridging coals to extend the temperature range of plastic zone overlap of the different blend components.

Pitch materials fall into two main categories; some pitch additives in coal blends have the ability to modify the optical texture of the resultant coke (and hence its bulk properties) significantly. Small percentage additions of such pitches to poor quality coals can significantly improve coking properties and bring them within the specifications for industrial usage. Such pitches are termed 'active' or 'superactive'. Pitches which do not possess this modifying ability are termed 'passive'. Work has been done on the modifying ability of various pitches and makes use of the relationship between optical texture and coke properties ([25]). The ability of a pitch to modify optical texture of a blend to give a texture characteristic of a good coke is easily monitored using techniques of optical microscopy; optical texture is thus an ideal parameter for initial characterization of blends. This work is reported below.

Hydrogen Transfer

As chemical compositions of coals and pitches are extremely complex no exact explanation exists of differences in size and shape of optical texture of the coke in terms of the pyrolysis chemistry of the carbonization process (26). It has been suggested that the presence of naphthenic groups and aromatic ring systems in molecular constituents promote the growth of larger anisotropic areas (optical texture) while the presence of heteroatoms and functional groups promote the growth of smaller optical textures (27). Recently, the concept of 'hydrogen shuttling' has been introduced in discussions on the development of mesophase during co-carbonization of coals with pitches (28, 29). It is suggested that the hydrogen may stabilize certain free radicals produced during pyrolysis (30). This stabilization of the otherwise reactive molecules prevents early formation of solid isotropic carbon and allows the temperature of the carbonization system to be increased such that viscosity of the system is decreased. This in turn facilitates the phase transition to a nematic liquid crystal. The mesophase growth units increase in size and coalesce and an anisotropic coke results (31). Larsen and Sams (32) consider coals to be good hydrogen donors and acceptors and that only heat and the proper medium are necessary to carry out extensive internal hydrogen transfer reactions. These reactions produce more condensed aromatic material, the released hydrogen forming aliphatic structures. If a pitch is hydrogenated and then carbonized then the viscosity of the mesophase is significantly decreased (33). Optical texture is therefore increased in size. This is probably associated with hydrogen transfer reactions (hydrogen shuttling) involving movement of hydrogen from the pitch to the radicals of the carbonization process. Thus, studies of hydrogen transfer reactions involved in coal liquefaction and of free radical concentrations in carbonizing systems may be related (29). These considerations are relevant both to carbonizations of single pitches and, more importantly, to co-carbonization systems of coal and pitch in which the latter is used to up-grade the coking ability of the coal (27, 34, 35).

Experimental Techniques of the Study

Optical Microscopy and Scanning Electron Microscopy. Cokes are prepared for optical microscopy by mounting in resin, grinding a flat surface and polishing with progressively finer grades of alumina powder. These polished surfaces are then examined under a Vickers M41 polarized light microscope using parallel polars with a half-wave retarder plate to give reflection interference colours.

Coke samples to be examined by SEM are mounted on a stub and gold coated to prevent charging in the electron beam.

Point-counting. Optical textures seen under the optical microscope are classified according to the definitions in Table I (25). Metallurgical coke is heterogeneous and often contains several different types of optical texture. In order to obtain a statistically reliable description of optical texture, each coke is point-counted. The frequency of occurrence of each type of texture at points on a 'grid' is recorded on a Swift Automatic point counter.

Strength Testing. The relative strengths of cokes are measured using a microstrength apparatus suitable for laboratory use. This consists of two metal cylinders containing 12 steel ball bearings, ~8 mm in diameter. The tubes are rotated at 25 rev min^{-1} for 400 revolutions. The resultant coke particles are subjected to careful sieve analysis and are classified as three percentages of the total weight in size groups 600 μm - 1.17 mm (R_1), 212 - 600 μm (R_2) and <212 μm (R_3).

Materials Used in the Study

Coals
1. Blaenhirwaun Pumpquart anthracite (NCB CR 101).
2. Tilmanstone vitrain (NCB CR 204).
3. North Celynon Meadow Vein (NCB CR 301a).
4. Roddymoor Ballarat (NCB CR 301b).
5. Six Bells (NCB CR 301a).
6. Cortonwood Silkstone (NCB CR 401).
7. Maltby Swallow Wood (NCB CR 502).
8. Manton Parkgate (NCB CR 602).
9. Nailstone Yard (NCB CR 902).

Comparison of the UK and USA coal ranking systems is in Table II.

Pitches
Ashland A200 petroleum pitch.
Ashland A170 pitch coke (green).

Industrially Prepared Metallurgical Cokes
Spencer Works Wharf coke)
Clyde Ironworks coke   ) carbonized industrially to 1350 K.

Experimental Procedures of the Study

Gasification and Heat Treatment. Two polished surfaces of each of the industrially prepared metallurgical cokes listed above were prepared and examined under the optical microscope.

A representative area of each was chosen, photographed and its position recorded. The same areas were located and photographed in the scanning electron microscope (SEM). The technique of examining the same area before and after experimentation eliminates much of the statistical variation inherent in such a heterogeneous material as metallurgical coke.

Gasification in Carbon Dioxide. One sample of each coke was gasified in CO_2 in a horizontal electrical tube furnace for 16 h at 1173 K (900°C).

Heat Treatment. One sample of each coke was heat treated under argon in a graphite resistance furnace for 0.5 h at 2073 K (1800°C).

The selected areas of all samples were then relocated and photographed by SEM.

Breeze Additives. A170 pitch coke breeze was further carbonized to HTTs of 900 K, 1000 K, 1100 K and 1200 K at 3 K min^{-1} under N_2 with a soak time of 0.5 h.

The five resultant pitch cokes and the coals listed above (numbers 5, 6 and 7) were ground to sieve sizes 600 μm - 1.17 mm. Blends of coal and pitch coke (9:1 by weight) and the single coals were carbonized in a horizontal electrical tube furnace under N_2 to 1250 K (977°C) at 3 K min^{-1} with 0.5 h soak time.

Each coke was mounted in resin and a polished surface prepared which was examined and photographed under the optical microscope. Fracture surfaces of each coke were examined by SEM.

Analyses of these coals are in Table III.

Strength Measurements. Each coke was ground to particle sizes of 600 μm - 1.17 mm. 2 g of each coke was tested in the microstrength apparatus and then sieved.

Blending of Pitch with Coals. The coals listed above (with the exceptions of those numbered 5, 6 and 7) were ground to a particle size of <250 μm and were then mixed with 25 wt % A200 pitch. These blends were carbonized in a horizontal electrical tube furnace to heat treatment temperatures (HTTs) of 1273 K (1000°C) at a heating rate of 5 K min^{-1} under N_2 with a 0.5 h soak. The resulting cokes were mounted in resin and polished surfaces prepared. The surfaces were examined optically and point-counted. The cokes from coals were also point-counted after being carbonized singly under identical conditions for comparison.

Analyses of these coals and of the A200 pitch are in Tables IV and V.

Table II

Comparison of U.K. and U.S.A. coak ranking systems

U.K.	U.S.A.
N.C.B. coal rank	
100-102 *V.M. 2-9%	Meta anthracite Anthracite
200-206 V.M. 9-19.5%	Semi anthracite Low volatile bituminous
†300-306 V.M. 19.5-32%	Medium volatile bituminous
400-700 V.M. 32-44%	High volatile bituminous
800-900 V.M. 32-47%	Sub bituminous

* V.M. = volatile matter content.
† Coals 304-306 heat affected.

Table III

Analyses of Coals

Material	Volatile Content (wt % d.a.f.)	C (wt % d.a.f.)	H (wt % d.a.f.)
Six Bells coal (CR 301a)	24.1	90.0	4.7
Cortonwood Silkstone coal (CR 401)	35.4	86.8	-
Maltby Swallow Wood coal (CR 502)	37.4	86.8	-

Table IV

Analyses of Vitrains

Vitrain	Air-dried basis (wt %) Moisture	Ash	Dry ash-free basis (wt %) C	H	N	S	O^a	VM	B.S. Swelling No.	NCB Coal Rank Code No.
Blaenhirwaun Pumpquart	1.6	0.5	93.6	3.2	1.2	0.7	1.3	4.6	N.A.	101
Tilmanstone	1.0	1.3	89.7	4.6	1.4	1.0	3.3	19.3	8½	204
North Celynon Meadow Vein	1.0	2.5	88.6	4.9	1.5	1.1	3.9	24.4	9	301a
Roddymoor Ballarat	0.9	0.4	88.4	5.3	1.7	0.7	3.5	29.8	9	301b
Manton Main Parkgate	8.1	1.2	82.8	5.4	1.8	1.2	8.8	35.3	4	602
Nailstone Yard	13.2	3.6	78.2	5.2	1.5	1.1	14.0	42.5	1	902

N.A. Non-agglomerating

Table V

Analysis of Ashland A200 Pitch

Softening RB (°C)	199	Benzene-insolubles (wt %)	35.1
Volatile matter (wt %)	35.5	Heptane-insolubles (wt %)	93.5
Sulphur (wt %)	2.9	α-resin, Quinoline insolubles (wt %)	0.45
CERCHAR Conradson carbon (wt %)	71.6	α + β resins (wt %)	20.8
Asphaltenes (wt %)	58.4		

Elemental analysis (dry ash-free basis):

Carbon	Hydrogen	Sulphur	Nitrogen	Oxygen^a	Ash
91.7	5.0	2.9	0.1	0.3	0.1

^{a}By difference

Results and Discussion

Gasification and Heat Treatment. Examination under the optical microscope showed the Spencer works and Clyde Ironworks cokes to have optical textures mainly consisting of fine- and medium-grained mosaics with some coarse flow anisotropy and isotropic inert material. Of particular interest are the fissures which develop in different types of optical texture and those occurring at the anisotropic-inert interface. SEM examination of these polished surfaces before experimentation shows all of them to be flat and featureless.

Gasification. Optical microscopy of the gasified cokes was not possible as their surfaces were too roughened. SEM examination of the same areas of Spencer and Clyde cokes originally studied by optical microscopy revealed extensive gasification fissuring over their surfaces. Figure 1 is an optical micrograph of Spencer coke before gasification; Figure 2 is a SEM micrograph of the same area of coke surface after gasification in CO_2 for 16 h at 1173 K. Comparison of the two Figures indicates that, in general, size, shape and orientation of the gasification fissures occurring in optically anisotropic areas are strongly dependent upon the character of the optical texture in which they occur (Position A). In regions of medium-grained mosaics, fissures develop which are typically of the same size as the optical texture; these fissures are orientated parallel to the layer planes of the anisotropic carbon and tend not to cross isochromatic boundaries.

The bonding between the inerts and surrounding anisotropic material appears to be a region of weakness which gasified preferentially. Large fissures (~5 μm wide) are found at these interfaces (Position B). Inert material often gasifies completely leaving a void (Position C); this may be due to the presence of a catalyst in some inorganic inerts. Other inerts gasify uniformly or not at all; there is no evidence of the preferential gasification seen in the anisotropic areas.

Heat Treatment. Optical microscopy showed two changes to have taken place in these cokes upon heat treatment to 2073 K. Firstly, both showed a strengthening of the interference colours. This effect may be caused by an increase in the perfection of ordering of the graphitic layer planes as the heat treatment approaches graphitizing temperatures. Secondly, a network of fissures could be seen on the coke surface, particularly in the anisotropic areas. These fissures were more clearly visible using SEM. Figure 3 is an optical micrograph of the Clyde coke; Figure 4 shows the same area examined by SEM after heat treatment. Once again, comparison shows fissures of the same size and orientation as the optical texture (predominantly medium-grained mosaics) to have developed (Position D). These fissures appear

1. FORREST AND MARSH *Carbonization of Coal and Coal Blends* 15

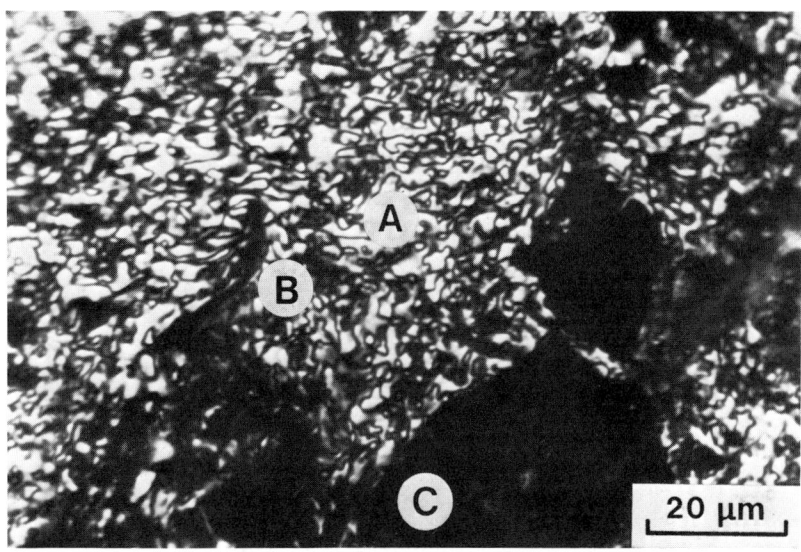

Figure 1. Optical micrograph of Spencer Wharf coke. Positions A, B, and C discussed in text.

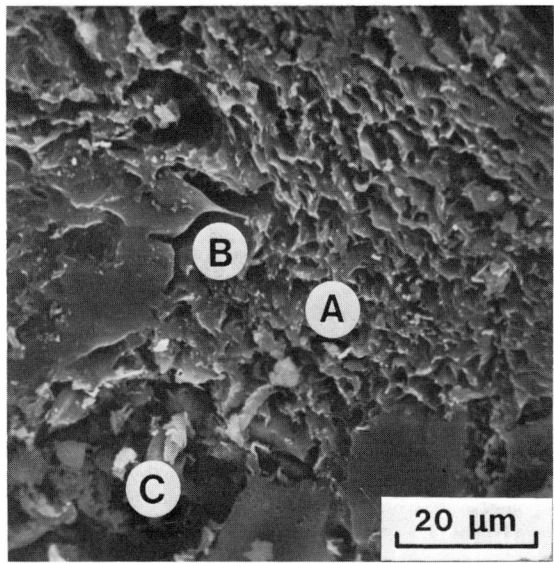

Figure 2. SEM micrograph of Spencer Wharf coke after 16 h gasification in CO_2 at 1173 K. Positions A, B, and C discussed in text.

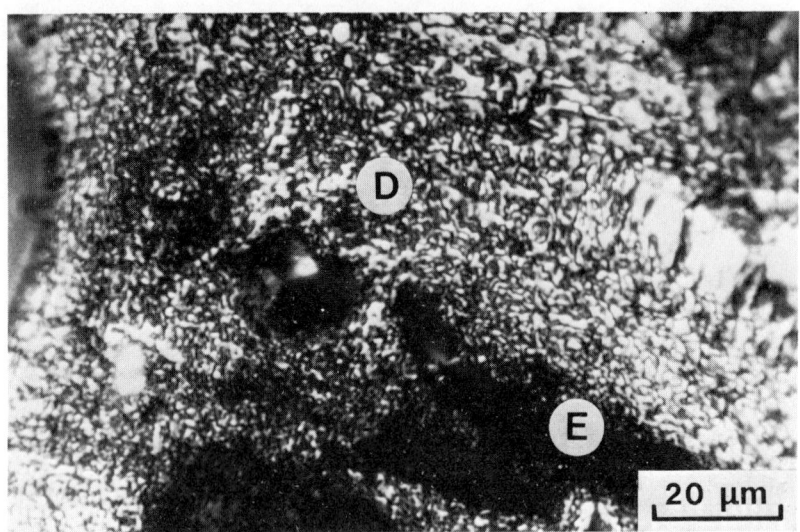

Figure 3. Optical micrograph of Clyde Ironworks coke. Positions D and E discussed in text.

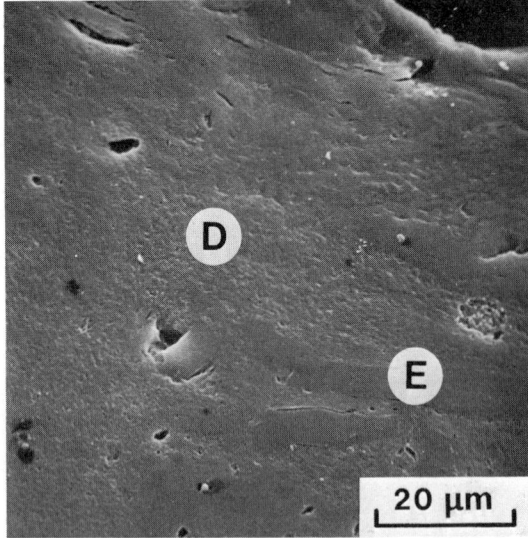

Figure 4. SEM micrograph of Clyde Ironworks coke after heat treatment to 2073 K under argon. Positions D and E discussed in text.

to be bounded by the isochromatic boundaries across which few of them propagate. Hence a rapid change in orientation of, or discontinuity in, the layer planes of the coke appears to act as a fissure stop. The few larger fissures (~30 μm x 5 μm) which were found in these cokes occurred mainly in the infrequent areas of coarse flow anisotropy. In the absence of rapid changes in structural orientation in this type of optical texture the propagation of fissures is possible to a much larger extent than in the regions of smaller mosaics.

Again, the inert-anisotropic boundary is a region where fissuring occurred (Position E); the inerts themselves show no surface fissuring.

Differences in coke behaviour in the blast furnace (not detectable by cold testing prior to the charging of the blast furnace) may be attributable to differences in the mode of gasification of the coke as a result of combined effects of thermal and gasification fissuring. These results indicate that mosaic optical textures are preferable to flow anisotropy in terms of fissure containment and also show that inert particles can act as centres of fissure generation.

Breeze Additives. Optical microscopy of the Six Bells (CR 301a), Cortonwood (CR 401) and Maltby (CR 502) cokes with the A170 pitch coke breeze additives shows that the breeze additives with predominantly flow domain anisotropy optical texture are easily distinguishable from the surrounding coal coke of predominantly fine-grained mosaics. The effect of carbonization to 1200 K upon the petroleum coke breeze is to increase progressively the number and size of fissures within the breeze particles.

Figures 5 and 6 are of coke from Six Bells coal co-carbonized with green A170 pitch coke breeze and 1200 K A170 pitch coke breeze respectively. Position C shows the optical texture of the coal-coke, while Position P shows that of the pitch coke breeze.

Figure 5 shows that the green pitch coke is well bonded to the surrounding coal coke (Positions E, F); there are no fissures at the interface. Fissures within the pitch coke do not propagate across the interface into the coal coke. As the HTT of the pitch coke breeze is increased, the interface becomes progressively more fissured, until for the 1200 K pitch coke breeze (Figure 6, Positions L, M) the breeze particle is almost entirely surrounded by voids. Where breeze particles of HTT 1200 K are bonded to the surrounding coal coke, fissures originating in the breeze propagate across the interface into the surrounding material. These results are confirmed by SEM examination of fracture surfaces of all the cokes produced and are true for all three coals used.

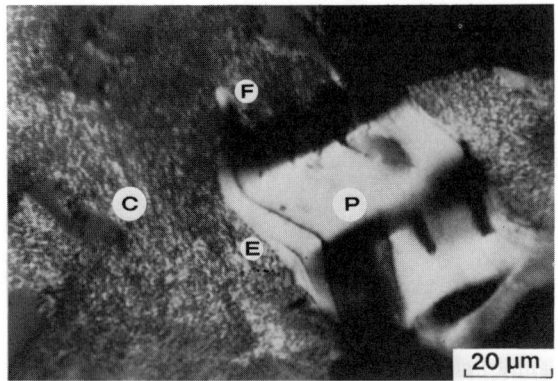

Figure 5. Optical micrograph showing green pitch-coke breeze in coke from Six Bells coal. Positions C, E, F, and P discussed in text.

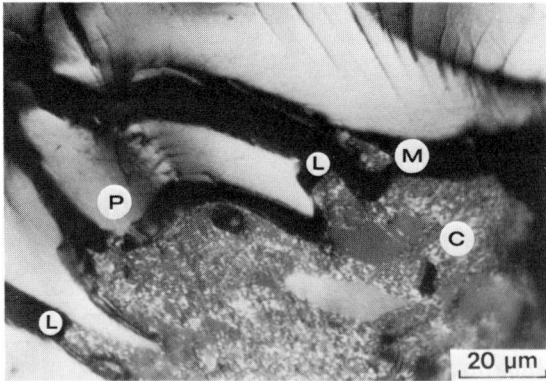

Figure 6. Optical micrograph showing pitch-coke breeze (1200 K) in coke from Six Bells coal. Positions C, L, M, and P discussed in text.

Data from the microstrength measurements are given in Tables VI, VII and VIII. The higher the values of the size fractions R_1 and R_2, the stronger is the pore wall material of the coke. These results indicate an interesting feature of the effects of breeze additions upon coke strength. For all three coals used it is true that the lower the HTT of the pitch coke breeze added, the stronger the resultant coke. Strength decreases progressively with increasing HTT of pitch coke breeze. However, comparison of the microstrength results for the prime coking Six Bells coal when carbonized singly, with those for the same coal carbonized with any of the breeze additives, shows the single substance coke to be significantly stronger than those cokes with breeze additives. Strength of coke from the prime coking coal is decreased by addition of pitch coke breeze.

For the Cortonwood and Maltby coals the single coal cokes are significantly weaker than those cokes with breeze additives, even 1200 K breeze additives.

It therefore appears that pitch coke breeze additives can enhance the coke strength of poorer quality caking coals and that the pitch coke breezes of lower HTT (higher volatile content) produce the best results. It should be emphasized that only one variable (HTT of the breeze) has been investigated here; other variables include size and shape of the breeze particles and percentage by weight added to the carbonization system.

Blending. The coal-pitch and coal-coal blends were analysed in terms of their optical texture by point-counting. Care is necessary to describe accurately the type of anisotropy seen and for this purpose a standard nomenclature has been devised (Table I). Results of the point-counting of cokes from the six coal-pitch blends are shown in Figures 7 to 10.

From these results it is evident that 25 wt % addition of A200 to coal carbonizations can significantly modify the optical texture of the coke from the parent coal. Results can be placed in 5 categories.

Firstly, co-carbonization of anthracites (CR 101) with A200 gives a coke showing two distinct types of anisotropy or phases. One phase is coke from the anthracite not modified by the A200 pitch and which shows basic anisotropy. The other phase is coke from the A200 pitch (flow domain anisotropy) which appears to act as a binder.

Secondly, for high rank low volatile caking coals (CR 203-204), cokes produced with 25% A200 pitch also show two phases or types of anisotropy but with no sharp boundary between the two. Optical texture progressively decreases in size from the flow

Table VI
Microstrength results from Six Bells coal plus 10% A170 pitch coke

Coke constituents	R_1	R_2 (%)	R_3
Six Bells Coal	3.4	68	26
Six Bells + A170 (green)	2.4	57	39
Six Bells + A170 (900 K)	1.9	60	37
Six Bells + A170 (1000 K)	1.2	55	41
Six Bells + A170 (1100 K)	0.85	53	44
Six Bells + A170 (1200 K)	1.0	50	49

Table VII
Microstrength results from Cortonwood coal plus 10% A170 pitch coke

Coke constituents	R_1	R_2 (%)	R_3
Cortonwood Silkstone coal	0.1	28	69
Cortonwood + A170 (green)	0.8	48	49
Cortonwood + A170 (900 K)	0.9	50	46
Cortonwood + A170 (1000 K)	0.5	47	51
Cortonwood + A170 (1100 K)	0.2	32	66
Cortonwood + A170 (1200 K)	0.2	42	57

Table VIII
Microstrength results from Maltby coal plus 10% A170 pitch coke

Coke constituents	R_1	R_2 (%)	R_3
Maltby Swallow Wood coal	0	21	77
Maltby + A170 (green)	0.6	39	59
Maltby + A170 (900 K)	0.5	41	58
Maltby + A170 (1000 K)	0.3	38	60
Maltby + A170 (1100 K)	0.2	31	57
Maltby + A170 (1200 K)	0.1	32	66

1. FORREST AND MARSH *Carbonization of Coal and Coal Blends* 21

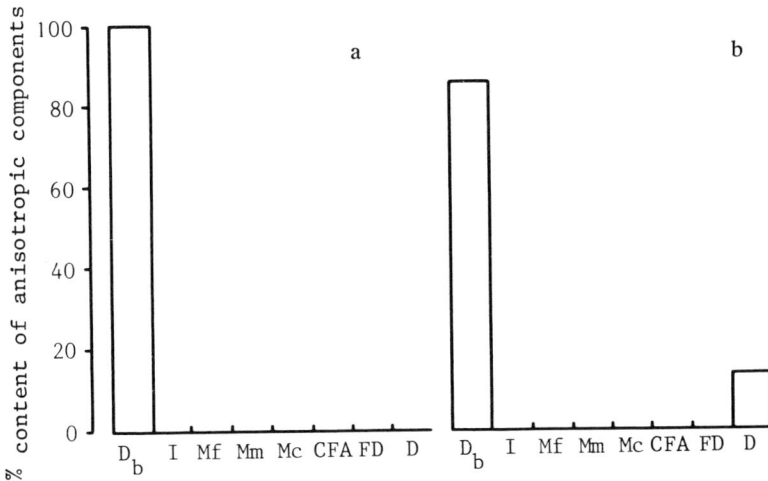

Figure 7. Point-counting analyses of optical textures of cokes. Key: a, Blaenhirwaun Pumpquart vitrain (N.C.B. rank 101); and b, Blaenhirwaun Pumpquart vitrain + 25% A200 pitch.

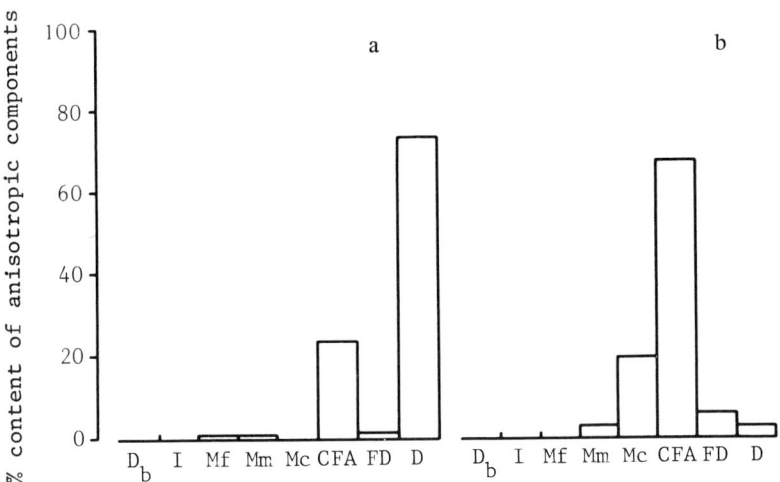

Figure 8. Point-counting analyses of optical textures of cokes. Key: a, Tilmanstone vitrain (N.C.B. rank 204); and b, Tilmanstone vitrain + 25% A200 pitch.

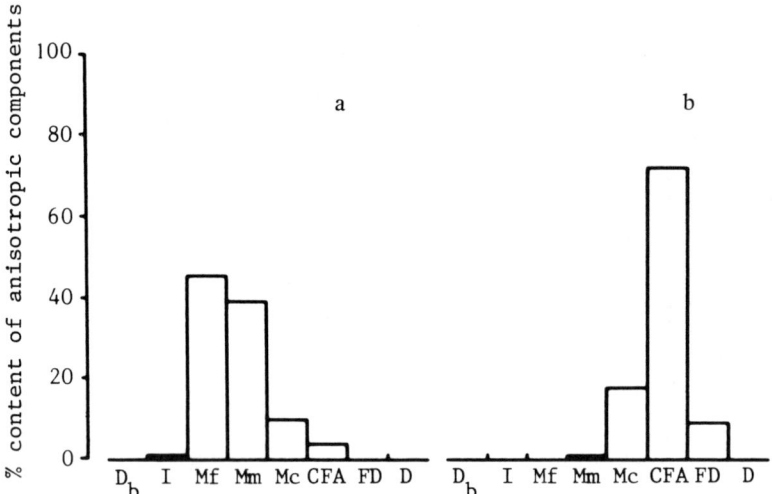

Figure 9. Point-counting analyses of optical textures in cokes. Key: a, North Celynon, Meadow Vein vitrain (N.C.B. rank 301a); and b, North Celynon, Meadow Vein vitrain + 25% A200 pitch.

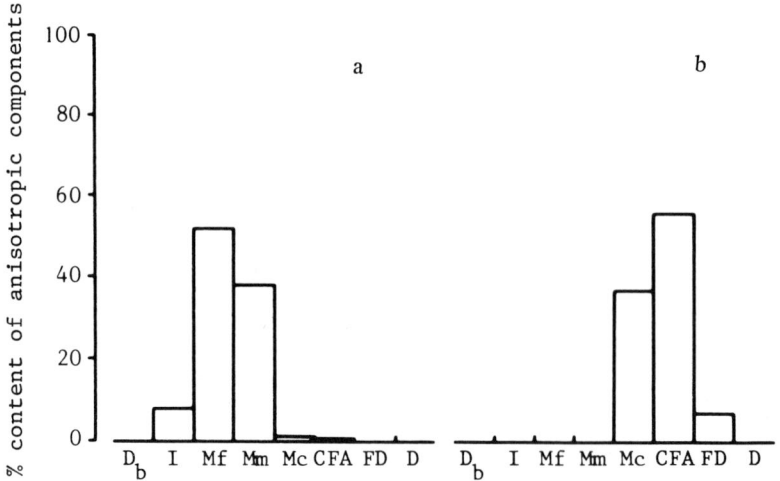

Figure 10. Point-counting analyses of optical textures of cokes. Key: a, Roddymoor Ballarat vitrain (N.C.B. rank 301b); and b, Roddymoor Ballarat vitrain + 25% A200 pitch.

1. FORREST AND MARSH   *Carbonization of Coal and Coal Blends*

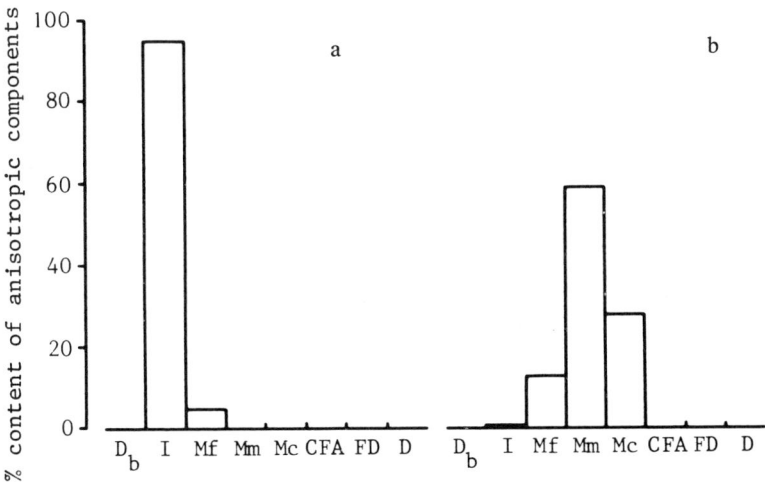

Figure 11. Point-counting analyses of optical textures of cokes. Key: a, Manton Parkgate vitrain (N.C.B. rank 602); and b, Manton Parkgate vitrain + 25% A200 pitch.

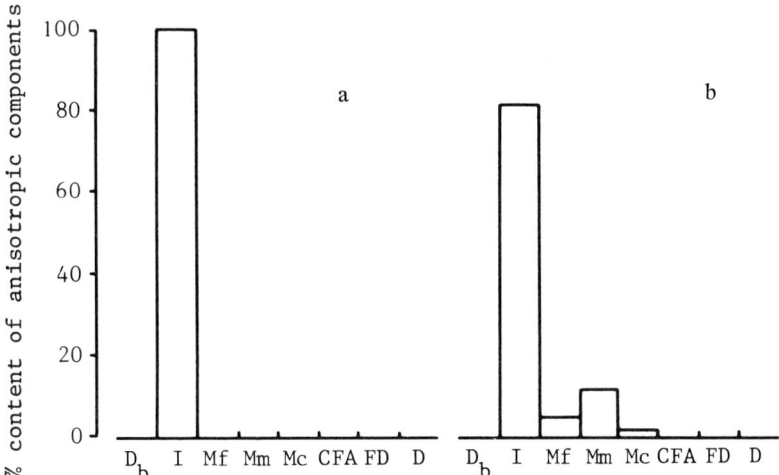

Figure 12. Point-counting analyses of optical textures of cokes. Key: a, Nailstone Yard vitrain (N.C.B. rank 902); and b, Nailstone Yard vitrain + 25% A200 pitch.

domain of coke from the A200 pitch through coarse- and fine-grained mosaics which extend about 10 μm into the unaltered domain anisotropy of the coke from the coal.

Thirdly, when A200 is blended with prime coking and medium volatile caking coals (CR 301-501) the resultant coke shows an increase in size of optical texture compared to that of the coal coke when carbonized singly. This new optical texture is intermediate in size between that of the coal coke and pitch coke.

Fourthly, A200 pitch when co-carbonized with high volatile caking coals (CR 601-701) almost completely modifies the isotropic texture of the coal coke to give an optical texture predominantly of medium-grained mosaics. Again, the optical texture is uniform rather than having two phases.

Fifthly, the effect of A200 addition upon high volatile, very weakly caking and non-caking vitrains (CR 801-901) is to reduce the isotropic content of the coke and to increase the content of fine- and medium-grained mosaics.

Thus, for most coal ranks some modification in the optical texture is produced by addition of A200 to the carbonization system; A200 is an active pitch. It is suggested that the mechanism of modification is the suppression of the recombination reactions of free radicals created during carbonization. This suppression can be facilitated by addition of hydrogen donors to the carbonization system. It therefore appears that for pitches such as A200 to be good modifiers they should have the ability to donate hydrogen to the products of thermolysis of coal.

Acknowledgements

This study was carried out with financial support from the European Coal and Steel Community, Grant No. ECSC 7220-EB-807. The authors also acknowledge support from BCURA Ltd (UK) and the ACS which made possible the presentation and publication of this paper. The support of Mrs. Marion Poad and Mrs. Patricia M. Wooster is appreciated.

Literature Cited

1. Grainger L. "COMA Yearbook", Mexborough, U.K., 1975, p. 282.
2. Gibson J., Gregory D.H. "COMA Yearbook", Mexborough, U.K., 1978, p.159.
3. Gibson J., Gregory D.H. "Carbonization of Coal", M & B Monograph CE/4., Mills and Boon, Ltd., 1971.
4. Rantell T.D., Clarke J.W. Fuel 1978, 57, 147.
5. Hays D., Patrick J.W., Walker A. Fuel 1976, 55, 297.

6. Patrick J.W., Wilkinson H.C. "COMA Yearbook", Mexborough, U.K., 1977, p. 245.
7. Priestley E.G. (Ed.) "Introduction to Liquid Crystals", Plenum, New York, 1975.
8. Brooks J.D., Taylor G.H. "Chemistry and Physics of Carbon", Ed. Walker P.L., Jr., Marcel Dekker Inc., N.Y., 1968, Vol. 4 243.
9. Brooks J.D., Taylor G.H. Carbon 1965, 3, 185.
10. Lewis R.T., Ext. Abs. 12th Biennial Conf. on Carbon, American Carbon Society, Pittsburgh, 1975, p. 215.
11. Forrest R.A. M.Sc. Thesis, University of Newcastle upon Tyne, U.K., 1977.
12. Forrest R.A., Marsh H. Proc. 5th London Int. Carbon and Graphite Conf., S.C.I., London, 1978, Vol. 1, p. 328.
13. Marsh H., Smith J. "Analytical Methods for Coal and Coal Products", Ed. Clarence Karr, Jr., Academic Press, N.Y. 1978, Vol. II p. 371-414.
14. Marsh H. Fuel 1973, 52, 205.
15. Marsh H. Proc. 4th London Int. Carbon and Graphite Conf., S.C.I., London 1976, p. 2.
16. Pacheco L., French M., Marsh H., Ragan S. Proc. 5th London Int. Carbon and Graphite Conf., S.C.I., London 1978, p. 219.
17. French M.A. Ph.D. Thesis, University of Newcastle upon Tyne, U.K. 1979.
18. Strassburger J.H. (Ed.) "Blast Furnace Theory and Practice", Gordon and Breach Science Publishers Vol. 1, 1969.
19. Adair R.R., Boult E.H., Marsh H. Fuel 1972, 51, 57.
20. Marsh H., Mochida I. Fuel 1981, 60, 231.
21. van Krevelen D.W. Fuel 1959, 38, 165.
22. Patrick J.W., Stacey A.E. Fuel 1975, 54, 256.
23. Patrick J.W., Stacey A.E. Fuel 1978, 57, 258.
24. Triska A.A., Schubert C.D., Proc. Int. Cong. Charleroi, 1966, p. 402.
25. Grint A., Swietlik U., Marsh H. Fuel 1979, 58, 642.
26. Yokono T., Miyazawa K., Sanada Y., Marsh H. Fuel 1979, 58, 691.
27. Mochida I., Marsh H., Grint A. Fuel 1979, 58, 803.
28. Marsh H., Yokono M., Yokono T., Carbon '80, Deutsche Keramische Gesellschaft 1980, p. 13.
29. Marsh H., Neavel R.C. Fuel 1980, 59, 511.
30. Petrakis L., Grandy P.W. Fuel 1981, 60, 115.
31. Yokono T., Miyazawa K., Obara T., Sanada Y., Marsh H. Ext. Abs. 15th Conf. on Carbon, Philadelphia, American Carbon Society 1981, p. 134.
32. Larsen J.W., Sams T.L. Fuel 1981, 60, 272.
33. Marsh H., Mochida I., Scott E., Sherlock J. Fuel 1980, 59, 517.
34. Yokono T., Marsh H. Fuel 1981, 60, 607.
35. Lopez H., Marsh H. Ext. Abs. 15th Conf. on Carbon, Philadelphia, American Carbon Society 1981, p. 124.

RECEIVED April 30, 1982

Characterization of Alkanes in Extracts of Coals, Lignites, and Related Fuels

K. D. BARTLE
University of Leeds, Department of Physical Chemistry, Leeds LS2, 9JT, England

D. W. JONES and H. PAKDEL
University of Bradford, School of Chemistry, Bradford BD7, 1DP, England

>The application of ^{1}H and ^{13}C N.M.R. spectroscopy, gas chromatography (G.C.) and mass spectrometry (M.S.) in the separation and identification of alkanes extracted from fossil fuels is illustrated with three Turkish lignites (including one extracted by supercritical gas), coal tar and petroleum crude. Elution of hydrocarbons from a silica-gel column may be monitored by ^{1}H N.M.R. and molecular-sieve sub-fractionation into normals and branched/cyclics by G.C., together with ^{13}C N.M.R. T_1 measurements. G.C. (e.g. with a eutectic packed column) can enable individual normal, iso-prenoid and cyclic alkanes, valuable as geochemical indicators, to be identified. ^{13}C N.M.R. chemical shifts are consistent with G.C.-M.S. identifications of acyclic isoprenoids in several fuels.

Alkanes in Fossil Fuels. Sedimentary rocks, which range in age from Recent ($\sim 10^4$ years) to the pre-Cambrian period ($\sim 3 \times 10^9$ years), contain the major reservoir of organic (and reactive) carbon in the crust of the earth, deposited as fossil fuels: these include oil shale, coal, petroleum, tar sands, natural asphalts and natural gas. The organic matter in fossil fuels contains, in addition to carbon, several percent each of hydrogen and oxygen; coals and related fuels typically also include 1-5% nitrogen and 1-10% sulphur (and small amounts of many other elements). These complicated coal materials are thought to have been generated by anaerobic degradation of plant and animal materials by micro-organisms in a reducing environment. From asphaltites to high-rank coals, the carbon content increases and hydrogen content decreases (Table I); crude oils, to which some reference will also be made here, have higher hydrogen contents (and much lower mineral contents) than the solid fuels.

Table I

Alkane Content of Hydrocarbon Minerals and Other Sediments

Source	Atomic H/C Ratio of Organic Matter	% Alkanes in Organic Matter
Recent sediments	1.7-1.9	0.8-36.9
Crude oils	1.5-2.0	30-45
Oil shales	1.3-1.6	6
Turkish asphaltites	1.3	4-10
Bituminous coals	0.5-0.7	0.4-0.7
Lignites	0.7-1.0	0.01-1.0
Turkish Montan Wax	0.80	0.05

Saturated hydrocarbons amount to 30 or 40% of petroleum crudes and are thus of direct economic importance. Although the much smaller proportions of alkanes in coals represent very little direct commercial value, a knowledge of the alkanes extracted from coal liquids can be of considerable help in determining the behaviour during processing of commercial fuel products, as well as in organic geochemical investigations of the nature and origin of fuels. The alkanes are generally present as a multi-component mixture of (a) acyclics-normals (predominant), singly-branched (*iso* and *anteiso*), and multiply branched (mainly acyclic isoprenoids)— and (b) cyclics (naphthenics)— mono-, di-, tri-, tetra-, and pentacyclics, including isoprenoids (Table II).

Analytical Spectroscopy and Spectrometry of Fossil Fuels. Investigation of the composition and structure of fuels has for long provided one of the major fields of industrial application of molecular spectroscopy (1). More recently, the demand for detailed structural analyses of fossil-fuel extracts has been stimulated both by increased interest in energy sources in general, especially their organic geochemistry and the environmental consequences of combustion, and, also, for coal in particular, by the need for information on the chemical structure of the starting material and of subsequent stages of new coal-conversion processes. A wide range of spectroscopic methods (2), including nuclear magnetic resonance (N.M.R.) (in solution and in solid), electron spin resonance (E.S.R.), infrared (I.R.), ultraviolet (U.V.), X-ray and luminescence, has been applied to the structural analysis of fossil fuels. However, for materials as complex as coals, preliminary subdivision of extracts is an essential pre-requisite for successful exploitation of these spectroscopic techniques (3). Thus, when alkanes have been isolated, their further sub-fractionation into n and branched-cyclic alkanes may be achieved by 5Å molecular-sieve adsorption, a method which we generally prefer to thiourea- or urea-adduction.

TABLE II Saturated Hydrocarbons

Type		Typical Carbon Structure	Formula	Group
(a) acyclic				
(i)	Normal		C_nH_{2n+2}	-
(ii)	Iso		C_nH_{2n+2}	Singly-branched
	Ante-iso		C_nH_{2n+2}	Singly-branched
(iii)	Acyclic isoprenoid		C_nH_{2n-2}	Poly-branched
(b) cyclic				
(iv)	Monoterpane C_{10}		C_nH_{2n}	Monocyclic
(v)	(Sesquiterpane C_{15}		C_nH_{2n-2}	Dicyclic
	(Diterpane C_{20}		C_nH_{2n-2}	Dicyclic
(vi)	Diterpane C_{20}		C_nH_{2n-4}	Tricyclic
(vii)	Sterane		C_nH_{2n-6}	Tetracyclic
(viii)	Triterpane		C_nH_{2n-8}	Pentacyclic

Gas chromatography (G.C.) by packed and capillary column then enables individual normal (homologous series) and acyclic isoprenoid alkanes to be identified from, in the present work, coal tars and Turkish lignite, wax and asphaltites (and also petroleum crudes).

Over the last two decades, developments in G.C., mass spectrometry (M.S.), N.M.R. spectroscopy and other physical techniques have appreciably extended the ability of chemists to undertake detailed analyses of fractions extracted from coal and other fossil fuels. In this paper, we survey some of the characterization techniques for alkanes, emphasising particularly N.M.R. and M.S. Illustrations of their application are mainly taken from materials less fully covered in other parts of the Symposium, including commercial coal products, crude petroleum, and Turkish asphaltites (which have some affinities with North American tar sands), lignites and Montan wax.

Experimental

    Materials and Methods of Extraction. The solid Montan wax (58.8% carbon in dry ash-free material) was Soxhlet-extracted from Turkish (Demircikoy) lignite (4) with benzene/isopropyl alcohol, and the resins and asphaltenes were removed by Dr. E. Ekinci (Technical University of Istanbul). Asphaltite from the Avgamasya vein (Sirnak in Siirt Province, Turkey) was Soxhlet-extracted with chloroform and benzene/ethanol and gave elemental analysis C 56.8, H 5.7, N 0.9, S 8.0% (5). The Elbistan (Turkish) lignite sample, included for comparison of 100 MHz ^{1}H N.M.R. spectra of total organic extracts of fuels, was extracted by supercritical toluene at 350°C, a mild method which has been shown (4,6) to possess advantages over Soxhlet extraction in minimising degradation of alkanes and securing appreciable and representative quantities of alkanes, valuable as potential geochemical markers. The Rexco coal tar is a low-temperature (750°C) carbonization product (81.4% C d.a.f.) obtained by the Rexco (smokeless fuel) process from U.K. (Thoresby, CRC 901) coal (7). The above viscous starting materials were partition-solvent-extracted with distilled chloroform (or tetrahydrofuran for the solid Montan wax) and successively separated and extracted with n-hexane until aromatic insoluble material no longer separated.

    The Kuwait crude-petroleum sample, obtained from Dr. D.F. Duckworth (B.P. Ltd.), was distilled up to 192° under ordinary pressure to separate off straight-run gasoline and naphtha (b.p. 32-190°C). The naphtha-free degassed oil distillate was then vacuum-distilled on a molecular still, successively at 90° and 1.8 Torr, 130° and 0.9 Torr, and 130° and 0.08 Torr. Unsaturateds were largely removed from the solution in n-hexane by alternating separations and treatments with concentrated sulphuric acid.

Neutral oils from solid fuels and petroleum crude were extracted with 10% sulphuric acid and 10% sodium hydroxide solutions.

Separation and Sub-fractionation of Alkanes. Saturated hydrocarbons were separated from the neutral oil by silica-gel (60-120 mesh, dehydrated at 150°C for 5 h) chromatography in a 1 m x 30 mm i.d. column eluted with distilled n-hexane. n-Alkanes were separated from iso-octane solutions of total alkanes by adsorption for one week on 5 Å molecular sieve (freshly dehydrated for 24 h at 400°C). Washing with iso-octane, followed by Soxhlet extraction, freed the molecular sieve from unwanted non-adsorbed compounds; n-alkanes were recovered by desorption after refluxing the molecular sieve for several hours with n-hexane. For the Kuwait crude and fluidized-bed tar, the molecular-sieve treatment was preceded by urea-adduction of n-alkanes and thiourea-adduction of branched-chain alkanes.

Gas Chromatography, Spectroscopy and Spectrometry. Gas chromatography on a Pye 105 instrument with flame-ionization detector was used for monitoring the separation of normals from total alkanes and for identification of individual n-alkanes. With a 4 m x 5 mm i.d. glass packed column, a 30% eutectic mixture of inorganic salts (54.5 wt% KNO_3, 27.3% $LiNO_3$ and 18.2 wt% $NaNO_3$ (Analable)) supported on Chromosorb W ([4],[8]) proved quicker and more effective as stationary phase than organic phases. G.C.'s were also run with a 50 m x 0.25 mm i.d. glass WCOT capillary column coated with OV-101.

Gas-chromatography coupled to mass spectrometry (G.C.-M.S.) was carried out on Rexco and Kuwait branched/cyclic fractions with a Pye 104 chromatograph (WCOT glass-capillary column coated with OV-1) on a Varian MAT 44 spectrometer (with SS 144 computer) at 70 eV ionization voltage. For Turkish Montan wax and asphaltite, G.C.-M.S. were run with the eutectic packed column on an MS 50 spectrometer. Field-desorption (F.D.) M.S. for Kuwait and Rexco samples were recorded on a Varian CH5D spectrometer (calibrated with perfluorkerosene) linked to a Varian Spectrosystem 100 data output.

^{1}H N.M.R. spectra of 20% w/v solutions in deuteriated chloroform were recorded at 100 MHz on a JEOL MH-100 spectrometer and 220 MHz on a Varian HR-220 with T.M.S. as reference. ^{13}C N.M.R. spectra were recorded at 15 MHz on a JEOL FX 60Q spectrometer for asphaltite branched/cyclic (IRFT spectra), at 22.6 MHz on a Brüker HX-90E spectrometer for branched/cyclics of Kuwait and Rexco and asphaltite extract, and at 25.2 MHz on a Varian XL-100 for Rexco branched/cyclics (IRFT spectra).

Infrared (IR) spectra of the tar extract and crude-oil distillate were run as KBr disks on a Perkin-Elmer 257 grating spectrometer over the range 600-3500 cm^{-1}.

Application of Instrumental Techniques

Ultraviolet (U.V.) and Infrared (I.R.) Spectroscopy.
Observation of U.V. absorption (30,000-45,000 cm^{-1}) was largely confined to confirming the absence of unsaturateds during the earlier elution with n-hexane of saturated hydrocarbons through the silica gel. For monitoring the progress of this separation, it had some advantage over ^{1}H N.M.R. in speed and sensitivity to alkenes.

Although overlap prevents much use of conventional I.R. spectroscopy (i.e. other than in combination with G.C. (9) in identification of individual alkanes, I.R. can aid differentiation between saturated and unsaturated hydrocarbons and, to a small extent, between classes of alkanes. Slight differences include narrower 1470 cm^{-1} absorption in n-alkanes than branched/cyclics, loss of n-alkane 735 cm^{-1} absorption following molecular-sieve treatment, and emergence of a weak 2880 cm^{-1} C-H band in branched/cyclic alkanes. The I.R. spectrum of Montan wax (which has the surprisingly high sulphur content, mostly elemental, of 27%) shows absorptions in the region characteristic of -S-S-, C-S, S=O, S=C, O-H (presumably primary, secondary and tertiary alcohols), C=O, N-H (amide I and II), as well as weaker absorptions in the saturated C-H region.

^{1}H N.M.R. Spectroscopy.
For comparing mixtures of saturated hydrocarbons from different fuels and for monitoring the effectiveness of silica-gel column chromatography in the separation of normal alkanes from non-normals by 100 and, especially, 220 MHz ^{1}H N.M.R., the emphasis is on the chemical-shift profile (10) and peak areas rather than on spin-spin coupling constants (Table III). Although the 100 MHz spectra indicate that the n-hexane-soluble part of Montan wax in CCl$_4$ has rather similar hydrogen distributions to the chloroform-soluble part, about 54% of Montan wax was soluble in n-hexane; presumably the n-hexane-insoluble fraction contains all the alkanes, as well as polycyclic aromatics. The spectra of n-hexane- and chloroform-soluble fractions of Turkish asphaltite indicate hydrogen distributions of about 7.8 and 12.2 H$_A$, 21.2 and 22.0 H$_\alpha$, 46.0 and 43.3 H$_\beta$, and 25 and 22.5% H$_\gamma$.

Availability of ^{1}H N.M.R. spectra at 220 MHz (Figure 1) increases the chemical-shift separation and so enhances the differences between distinct kinds of hydrogen. For all the fuel extracts examined, removal of n-alkanes by molecular sieve from the total alkanes causes small changes in the methyl absorptions as a result of changing proportions of spin-spin triplets (from CH_3-CH$_2$-), doublets (from CH_3-CH-), and singlets (from CH$_3$-C$\lessapprox$). At 220 MHz, the CH_3 region of the branched/cyclic Kuwait crude (Figure 1(b)) is closely similar to the CH_3 region of the branched/cyclic fraction from a U.K. low-temperature-carbonization classical coal tar.

Table III

Examples of Hydrogen Distribution in Fossil-Fuel Alkane Extracts Measured from 100 MHz ^{1}H N.M.R. Spectra:
(i) total-alkane extract (before n-alkane removal);
(ii) branched-chain/cyclic extract (after n-alkane removal by molecular sieve.

Sample	H_{β_1} %[a] (i)	(ii)	H_{β_2} %[b] (i)	(ii)	H_γ %[c] (i)	(ii)
Kuwait crude	50	45	16	18	34	37
Rexco coal tar	69	53	9	15	21	32
Turkish asphaltite (Avgamasya)		38		27		35
Turkish lignite (Elbistan)		48		21		31
Turkish Montan-wax	47		19		34	

[a] Branched-chain CH_2 (plus small amounts of CH_3 α to a ring (10)).
[b] Branched-chain CH and cyclic CH_2 hydrogens.
[c] CH_3 in linear and branched-chains and in cyclics β or further from ring.

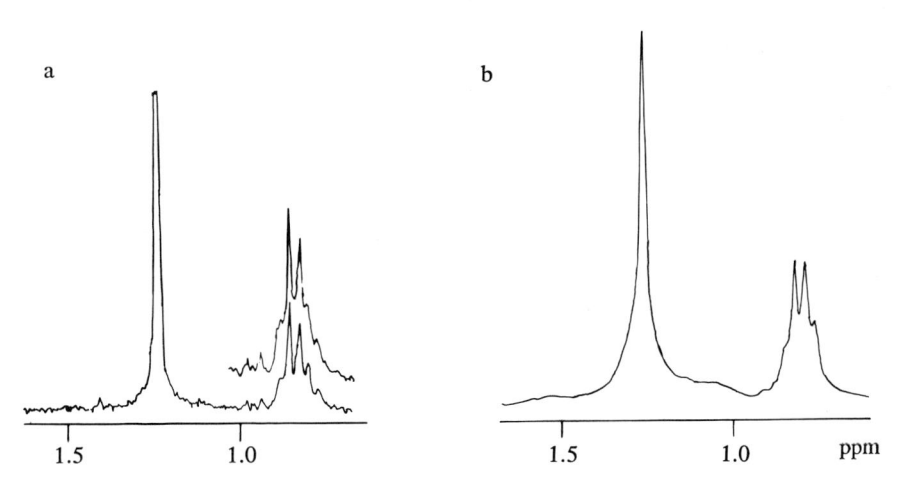

Figure 1. Alkane regions of 220-MHz ^{1}H NMR spectra of a, Montan wax total alkanes; and b, Kuwait crude branched cyclics in CCl_4 solution (shifts in ppm downfield from TMS).

During elution of the n-hexane-soluble fractions of fossil fuels, saturated alkanes, with strong resonances between 0.5-2.0 p.p.m., may be distinguished readily by areas on 100 MHz spectra from unsaturated hydrocarbons, with peaks in the regions of 4-6 p.p.m. and 6-8 p.p.m. Thus, the change in profile of alkane resonances during elution of a Kuwait crude-oil distillate fraction (free of unsaturateds) provides a very effective monitor of the separation process. In the spectrum of the branched/cyclic alkane fraction (i.e. after removal of n-alkanes) chain methylenes absorb at slightly higher field than ring methylenes and are also sharper (because of the free rotation of the magnetically equivalent hydrogens) than cyclic methylenes in fixed conformations (in which the hydrogens are not magnetically equivalent). Thus the ^{1}H N.M.R. spectrum of a complex mixture which contains both ring and chain hydrocarbons shows a broad band for cyclic methylene absorption and a narrower one for chain methylenes (which sharpens as the number of methylenes increases). The later the eluate, the higher is the area ratio $CH_3/(CH_2 + CH)$. Successive 10 cm^3 eluates show a steady decrease in the proportion of CH_2 (from the combination of normals, which are removed first, and branched) and then a steady increase in $(CH_2 + CH)$ ($\delta \sim 1.3$-2.0 p.p.m.), which comes from branched (CH) and cyclic (CH_2,CH) peak areas. Since they are mostly eluted in the last few fractions, cyclics can have their concentration reduced by silica-gel chromatography, a procedure which would favour identification of acyclic isoprenoids by ^{13}C N.M.R. For Montan wax, about one-third of the alkane-peak area in the 100 MHz ^{1}H spectrum is H_γ (CH_3), a much higher proportion than in coal tars; the appearance of the methyl region at 220 MHz also points to appreciable branched chains in this material.

^{13}C N.M.R. Spectroscopy. As an N.M.R. nucleus, ^{13}C suffers from low "receptivity" (11) and relatively long (and different) spin-lattice relaxation times, T_1, but these disadvantages are overcome by pulse excitation and fast Fourier transformation of the free-induction decay. Advantage may thus be taken of the narrow well-resolved ^{13}C N.M.R. lines (with a much wider spread of chemical shifts than for ^{1}H and with ^{13}C-^{1}H spin coupling eliminated by noise-modulated ^{1}H decoupling) from, in principle, all the chemically distinct carbon atoms in a complex natural-abundance fossil-fuel extract. Consequently, ^{13}C N.M.R. has been much used for petroleums and is now increasingly applied to sediments and coals. For alkanes, whether single compounds or mixtures, isomers may often be distinguished by means of ^{13}C shifts and intensities, taken with substituent-shift calculations based on additivity rules. Thus, in the acyclic isoprenoid pristane, whereas CH and CH_2 resonances are not resolved in the ^{1}H spectrum, nine ^{13}C resonances can be resolved (12). Broadband ^{1}H-decoupled ^{13}C spectra have provided convincing support for the identification of six acyclic isoprenoids in alkane mixtures derived from fluidized-bed low-temperature-carbonization (Rexco) coal tar and from Kuwait crude (13); these assignments are supported by recent T_1 measurements (see later). Both for n-alkanes (or straight-side-chain sections of branched or substituted cyclics) and for acyclic isoprenoids, the close agreement between observed ^{13}C shifts and those calculated shows that the additivity relation, well-tested for shorter-chain single compounds (12), is also applicable to mixtures of long-chain alkanes. For acyclic isoprenoids, the majority of the absorptions for particular carbon environments occur in several different compounds; indeed, about seventeen shifts accounted for all the resonances in all isoprenoids, as well as for those in n-alkanes. For cyclic alkanes, on the other hand, which figure significantly in the branched-chain/cyclic fractions of Kuwait and asphaltite, for example, ^{13}C absorptions are distributed over the approximate range 10-55 p.p.m. and there is less coincidence between chemical shifts of different homologues. Thus identification of individual cyclic alkanes, if achieved, will be unequivocal, but the resonances tend to be broad (from overlap of close shifts) and weak (7).

For Rexco branched/cyclics, gating of the ^{1}H-noise decoupling (so that it is applied only during signal acquisition) in order that nuclear Overhauser enhancement (n.O.e.) is suppressed causes the intensities of the methylene (29.85 p.p.m.) to decrease relative to that of the methyl (14.15 p.p.m.). Presumably different relaxation mechanisms predominate in the two cases (14); for the methylene group, relaxation is mainly dipolar, with positive Overhauser enhancement, whereas the methyl group may relax by spin-rotation with negative Overhauser enhancement.

In quantitative measurements, the problems posed by variable n.O.e. and by rather long and different spin-lattice relaxation times, T_1, can be largely overcome, as with oil products (15), if T_1s are reduced by the addition of paramagnetic relaxation agent, $Cr(acac)_3$, to the sample, together with the use of gated coupling. For the gated-decoupled ^{13}C spectrum of branched/cyclic alkanes from a U.S. fluidized-bed coal tar (FMC COED), addition of 150 mg $Cr(acac)_3$ had greater effect on the isoprenoid resonances than on those attributed to straight-chain sections. Problems due to low concentration of isoprenoids and to broadening and overlap of lines, may still remain, of course.

As an example of off-resonance ^{1}H-decoupling of ^{13}C spectra, whereby the multiple structures due to spin-spin coupling to directly-bonded hydrogens are retained but longer-range couplings are eliminated, the Rexco branched-chain/cyclic fraction spectrum (Figure 2) contains ten multiplets (with individual line widths up to 7 Hz), identified (from shifts in the fully decoupled spectrum and from splittings) as alkane (mainly acyclic-isoprenoid) resonances. Quartet 1 (14.0 p.p.m.) arises from methyl, and triplets 2, 3 and 4 (22.8, 28.5 and 29.8 p.p.m.) from methylene absorptions of straight-chain sections. The fairly intense

Figure 2. Off-resonance ¹H-decoupled ¹³C NMR spectrum of Rexco coal–tar branched chain alkanes in CDCl₃ solution. Quartet 1 (14.0 ppm) arises from methyl, and triplets 2,3,4 (22.8, 28.5, and 29.8 ppm) arise from methylene absorptions of straight-chain sections.

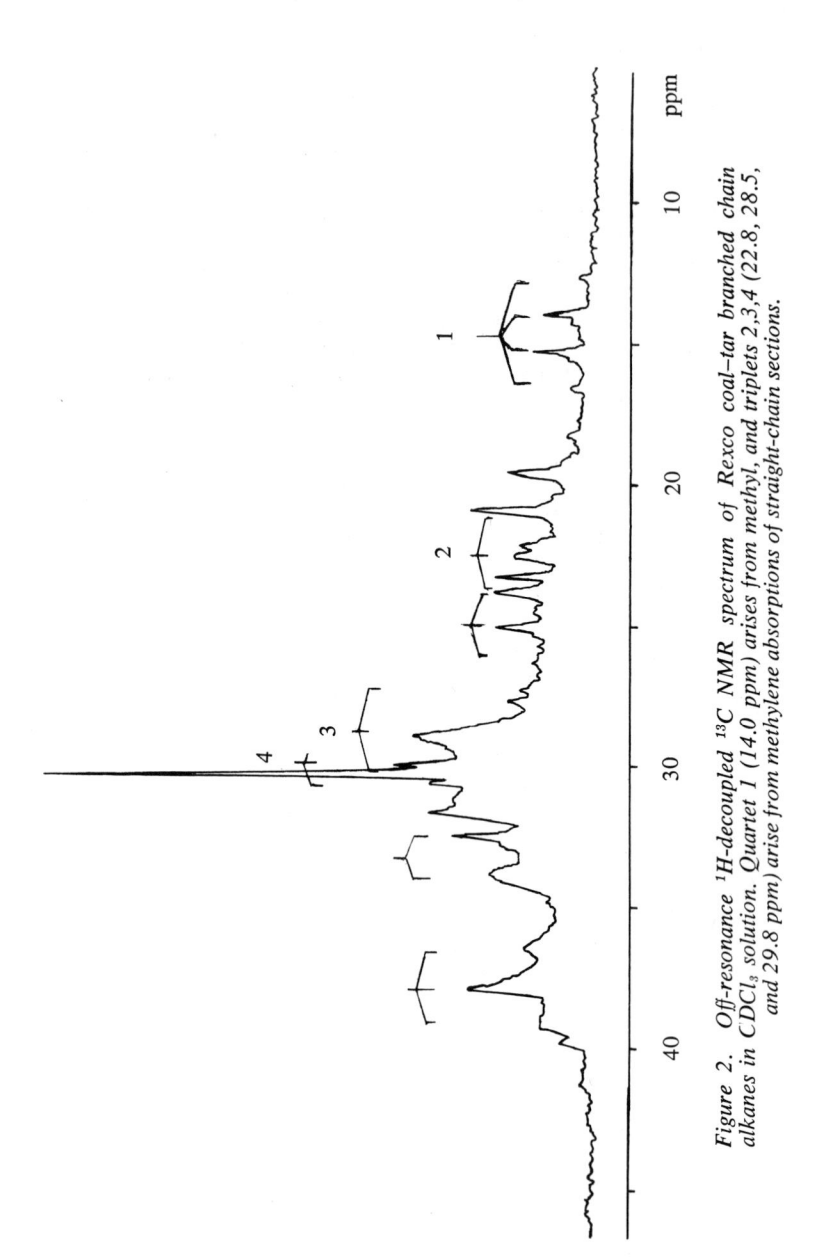

branched-chain methyl, methylene and methine carbon resonances at 19.7 ($-CH_3$), 24.6 ($-CH_2-$), 28.1 ($-CH_2$), 32.9 ($-CH$), 37.4 ($-CH_2$) and 39.5 p.p.m. ($-CH_2$) were also identified with the aid of their multiplet patterns in the off-resonance spectrum.

^{13}C Spin-lattice Relaxation-Time Measurements. Determination of ^{13}C spin-lattice relaxation times (T_1) for individual carbons of alkanes provides a direct probe of backbone segmental motion and of localised motion of side-chain and terminal main-chain carbons (16,17). In a given alkane, enhanced segmental motion of the chain causes T_1 of CH_2 groups to decrease as a function of the distance from the end of the chain, and T_1 for a given carbon to decrease progressively as the chain length is increased. Thus T_1 is likely to depend on the molecular weight of the sample, as well as on its viscosity. By changing the reorientational characteristics of the nearest carbons, branching alters their relaxation times, an effect which decreases with increasing numbers of bonds away from the branching point.

T_1 measurements at normal probe temperature were made by the inversion-recovery Fourier-transform (IRFT) technique on deuteriated chloroform solutions of pristane (2,6,10,14-tetramethylpentadecane) and of branched-chain/cyclic alkanes from Rexco coal tar and a Turkish asphaltite (Avgamasya). Inversion-recovery traces were recorded by using 200 cycles of the 180°-90°-PD pulse sequence with a pulse delay time (PD) of about 20s at a series of intervals, τ (e.g. 1,3,5s, Figure 3). T_1 values (±0.5s) (Table IV) were measured by plotting $\ln(A_o-A)$ vs. τ, where A_o is the equilibrium amplitude in a normal FT spectrum and A is the amplitude in an IRFT spectrum (Figure 4).

T_1s for the branched-chain carbons from Rexco coal-tar alkanes are generally rather lower (0.7-2.0s) than those of the n-alkanes (1.0-3.5s) in the mixture. For singly branched alkanes (e.g. 10-methylnonodecane) (17) and multiply branched alkanes (18), the decrease in the T_1s at (or near) the branch point of the chain has been associated with a decrease in internal motion. For pristane, our T_1s (Table IV) measured in deuteriated chloroform solution are rather larger than, but generally in the same sequence as, those reported by Long et al. (18) for neat samples at about the same temperature; the shortest T_1s correspond to the midchain methylenes C_5, C_7 and C_8 and the longest to the (branched) methine C_2. T_1s attributed to resonance in straight-chain sections are rather shorter in Rexco and longer in the asphaltite than those reported by Birdsall et al. (19) and Lyerla et al. (16) in the neat n-alkanes hexadecane, octadecane and eicosane. Now it is to be expected that, in general, T_1s will be shorter for systems of higher molecular weight and of lower viscosity. Since neither the molecular weights nor the viscosities of the two samples are thought to differ appreciably, one may infer that some of the chain carbons in Rexco alkanes are fairly close to methine carbons (branch points), so that internal

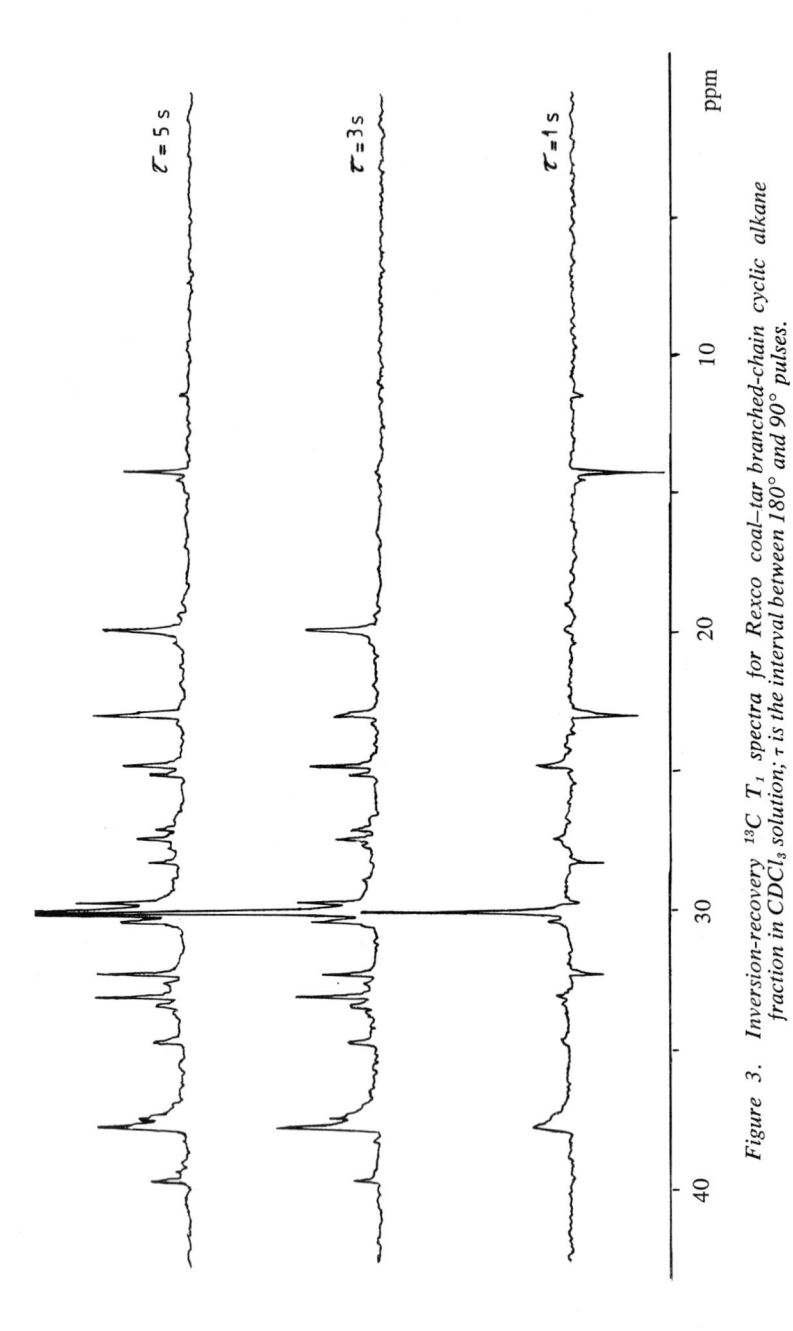

Figure 3. Inversion-recovery ^{13}C T_1 spectra for Rexco coal–tar branched-chain cyclic alkane fraction in $CDCl_3$ solution; τ is the interval between 180° and 90° pulses.

Table IV

^{13}C spin-lattice relaxation times, T_1/s, of branched-chain/cyclic fraction of Rexco coal tar, Turkish (Avgamasya) asphaltite and pristane (all measured in deuteriated chloroform solution).

Chemical shift /ppm from TMS	T_1 (coal tar) /s	T_1 (asphaltite) /s	Example of carbon type+	Corresponding T_1 in pristane /s
11.4	4.0	---	C_{16}(phytane)	---
14.1	3.5	5.3	$\underline{CH_3}$	---
19.8	1.5	2.1	C_{17},C_{18}(phytane)	1.9 (C_9)
22.6	1.7	NM	C_1(phytane)	2.4 (C_1)
22.7	2.0	3.9	$CH_3-\underline{C_2}-$	---
24.5	0.7	NM	C_4(phytane)	1.9 (C_4)
24.9	1.3	NM	$C_{8,12}$(phytane)	1.5 (C_8)
26.9	1.2	2.1	(cycloalkane)	---
27.2	1.2	NM	C_{12}(norpristane)	---
28.1	2.5	7.8	C_2(phytane)	3.2 (C_2)
29.4	1.5	2.4	$CH_3-(CH_2)_2-\underline{CH_2}-$	---
29.8	1.0	1.9	$CH_3-(CH_2)_3-(\underline{CH_2})_n-$	---
30.2	1.0	1.2	(cycloalkane)	---
32.0	1.7	3.5	$CH_3-CH_2-\underline{CH_2}$	---
32.9	1.0	2.5	$C_{6,10}$(phytane)	2.5 (C_6)
33.3	1.3	NM	C_{13}(norpristane)	---
34.5	1.3	NM	C_{14}(phytane)	---
37.1	0.8	2.0	C_{13}(phytane)	---
37.4	0.7	2.0	$C_{5,9,11,7}$(phytane)	1.7 (C_5,C_7)
39.5	1.5	NM	C_3(phytane)	2.7 (C_3)

+ Subscripts indicate carbon-atom numbers.   NM = not measured.

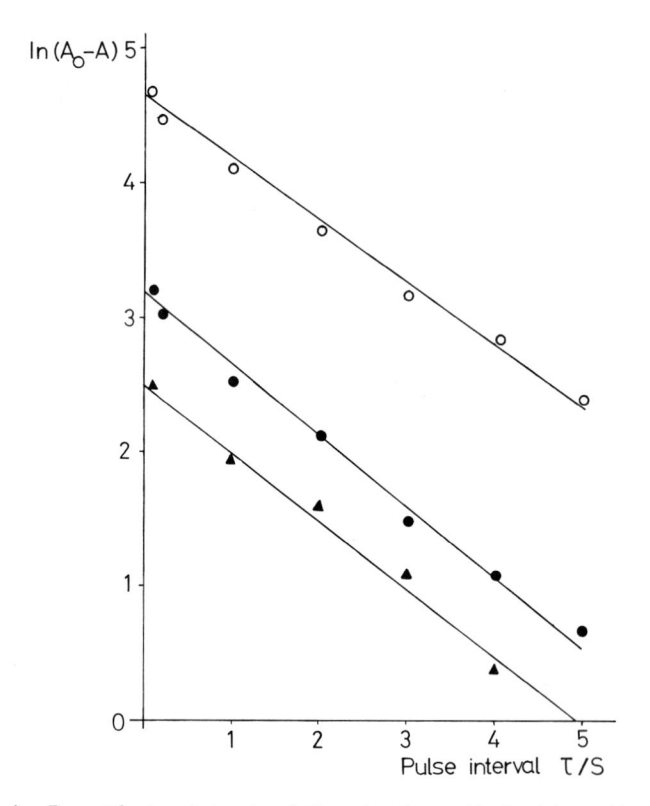

Figure 4. Logarithmic plots of relative signal amplitude, $(A_0 - A)$ vs. pulse interval, τ, for determination of ^{13}C spin-lattice relaxation times (T_1) of three species of carbons in Avgamasya-asphaltite alkanes. Key to T_1: ○, 2.2 s; ●, 1.9 s; and ▲, 2.0 s.

motions are reduced and consequently T_1s are lower than in a free n-alkane chain; for asphaltite alkanes, on the other hand, chain-carbon peaks may be superimposed on those of carbon atoms with longer relaxation times.

Gas Chromatography. Whereas N.M.R. effectively senses the functional groups present in a mixture, differences between retention times enable G.C. to identify individual members (provided they are moderately volatile) of a homologous series such as the n-alkanes. Positive identification requires calibration by co-injection with authentic samples (here n-alkanes) or coupling of M.S. with, preferably, a capillary G.C. column. Although G.C. stick diagrams of n-alkane distributions have been used for comparing extracts from fossil fuels, we advocate acyclic-isoprenoid-hydrocarbon distributions, and particularly the pristane/phytane ratio, as more reliable indicators of fossil fuel maturation (6).

G.C. was carried out on all the fossil-fuel extracts but the analysis by eutectic packed column and by capillary-column (OV-1 WCOT) of the total alkane fraction from Turkish Montax wax illustrates (4) the unambiguous assignment of series of n-alkanes (C_{12}-C_{34}, with odd carbon numbers predominating over evens from n-C_{22} upwards) and of acyclic isoprenoids (C_{18}-C_{20}: norpristane, pristane and phytane); cycloalkanes were also identified by G.C.-M.S. For Avgamasya asphaltite, the total-alkane G.C. (eutectic column) shows an approximately Gaussian distribution of n-alkanes in the C_{12}-C_{32} range (centred on n-C_{17}) with a large background peak due to the relatively high cycloalkane content. The G.C. of the branched/cyclic fraction assisted retention-time confirmation of the presence of five acyclic isoprenoids (C_{15}, C_{16}, C_{18}, C_{19} and C_{20}) indicated in the G.C.-M.S. From the G.C. of the total-alkane fraction of another Turkish asphaltite (Harbulite) (5), it was possible to recognise three series, namely n-alkanes, acyclic isoprenoids (C_{15}, C_{16}, C_{18}, C_{12}, C_{20}), and cyclic isoprenoids (C_{27}, C_{29}, C_{30}, C_{31} and C_{32}), all on the same chromatogram. For the Rexco coal tar, G.C.s of total-alkane (Figure 5(a)) and branched/cyclic alkanes (Figure 5(b)) demonstrate the effectiveness of the molecular sieve in removing n-alkanes; the remaining trace of n-alkanes would be eliminated by refluxing in the molecular sieve for a few more hours.

Gas Chromatography - Mass Spectrometry. Coupling to a mass spectrometer can greatly reduce the ambiguities in extracting structural information from the small quantities of materials employed in gas chromatography. With G.C.-M.S., a complete mass-spectrometric analysis can be obtained in essentially the time required for an ordinary G.C. run; with computer data-processing, each component can be analysed and the homogeneity of a single G.C. peak can be checked by obtaining successive mass

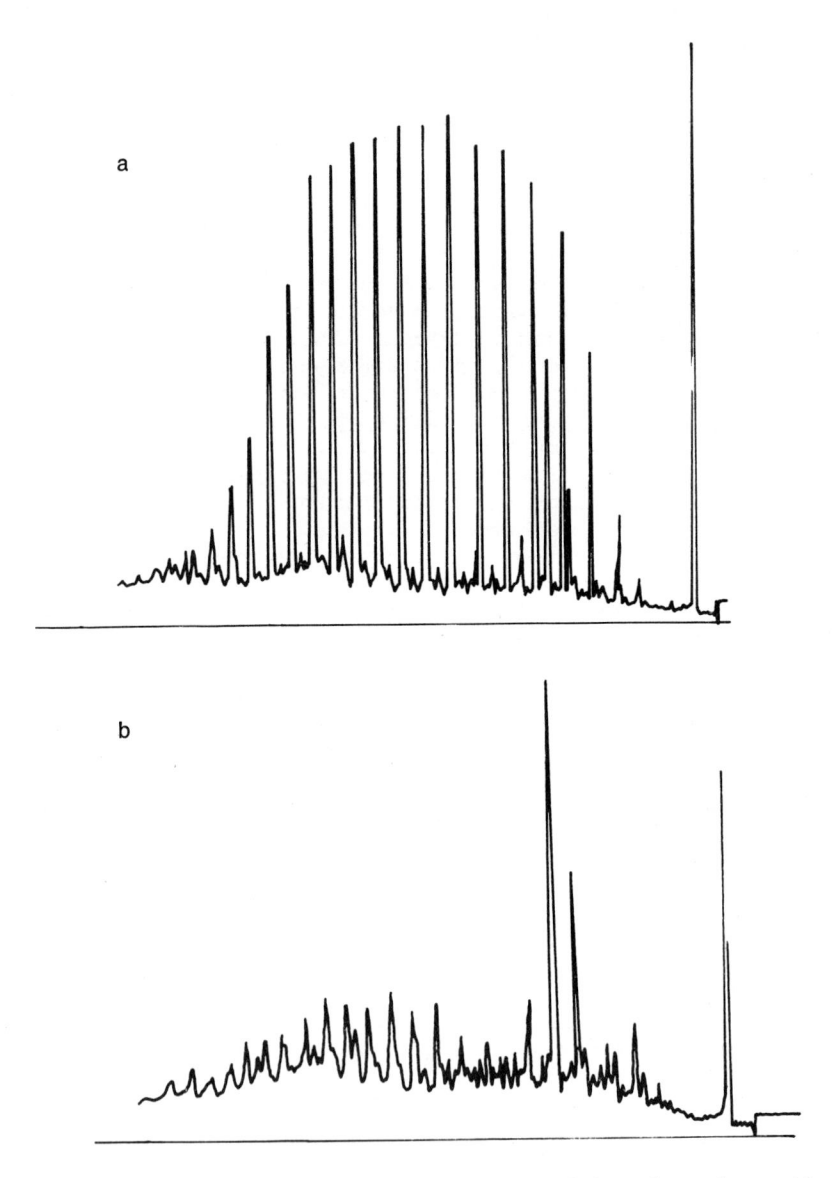

Figure 5. Gas chromatograms of Rexco coal–tar recorded on glass column with eutectic salt on Chromosorb. Key: a, total alkanes; and b, branched cyclic alkanes following molecular-sieve removal of most of the n-*alkanes.*

spectral scans. In this way, it has been possible to establish the presence, in alkane fractions from a range of coal tars and lignites, of up to six acyclic isoprenoids (13) and a number of cyclic isoprenoids (7). Thus for the branched/cyclic alkane fraction from Turkish Montan wax, G.C.-M.S. with a eutectic packed column allowed identification of 21 n-alkanes (C_{13}-C_{33}), three acyclic isoprenoids (C_{18}-C_{20}), and three cyclic alkanes (two C_{31} triterpanes and one dicyclic alkane) (4). Figure 6 shows the total-ion-current trace computer-reconstructed from the G.C.-M.S. analysis. The n-alkane distribution is distinctive, with a marked preference of odd over even carbon number (carbon preference index, 1.9, as in low-rank coals) in the range above C_{25} but with a smooth, nearly Gaussian, distribution for the lower range (peaked at C_{16} or C_{17}); also, the pristane/phytane ratio is high (4.5), as in sediments of marine origin.

Field-Desorption-Mass-Spectrometry. Analysis of G.C.-M.S. spectra from complex mixtures such as coal-derived extracts recorded by electron-impact (E.I.) ionization spectrometers is often rendered difficult by the profusion of fragment-ion peaks, weak molecular-ion peaks and background G.C. peaks; moreover less volatile compounds may not be sensed. Field-desorption (F.D.) M.S., though not without experimental and interpretative problems (fragment-ion peaks are not available to aid isomer elucidation), permits ionization (by an electric field) of higher-boiling compounds involatile under E.I. M.S. conditions. For the branched/cyclic fraction of Kuwait crude, F.D. M.S. enabled the molecular-weight ranges for acyclic and mono-, di-, tri-, tetra- and pentacyclic alkanes (including C_{28}, unusual in fossil fuels) to be established. F.D. M.S. has also been valuable in indicating the presence of a range of mono-, di- and tricyclic alkanes (some tetracyclics (steranes) and pentacyclics (triterpanes) had already been established by G.C. M.S. (7)) in the branched/cyclic extract from Rexco coal tar. F.D. M.S. spectra of CS_2 extracts of coal macerals (20) also provide evidence for the presence in coal of n-alkanes with n as high as 50 or 60.

Acknowledgements

We express thanks to Drs. E. Ekinci and D.F. Duckworth for samples, to Dr. D. Games for the field-desorption mass spectra, and to Drs. J.E. Beynon and F.W. Wehrli and to the PCMU Harwell for ^{13}C N.M.R. spectra.

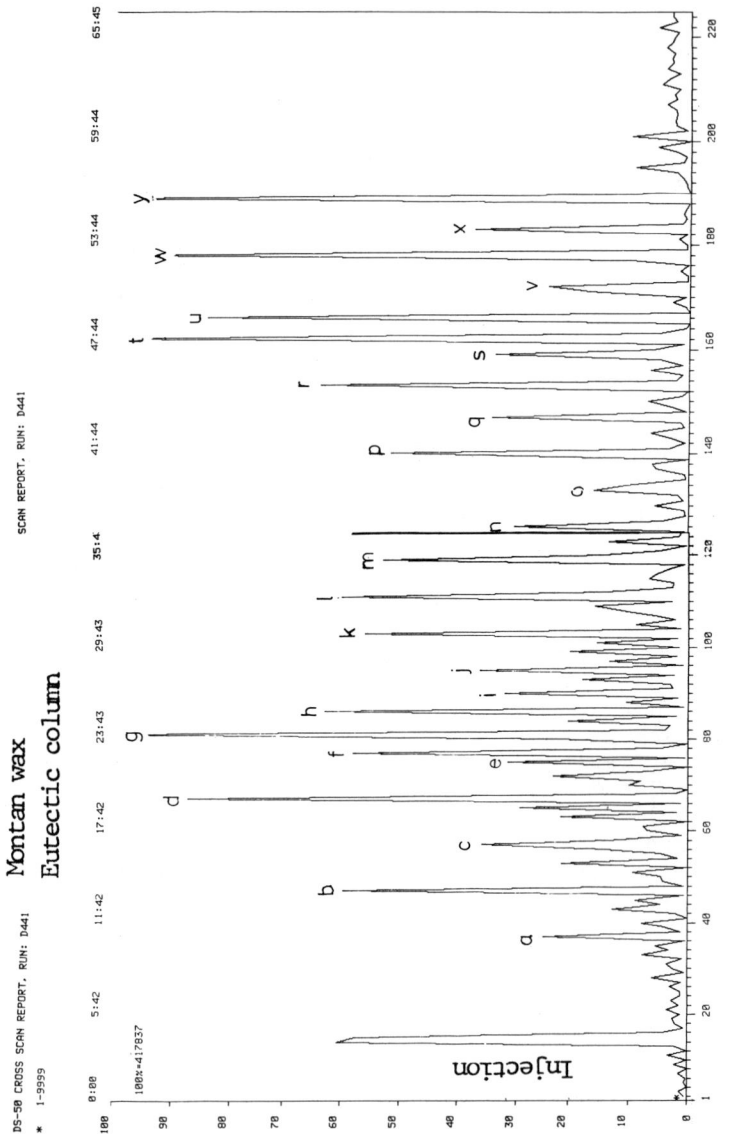

Figure 6. Total ion current constructed from GC MS spectra of Montan wax total-alkane fraction.

Literature Cited

1. Friedel, R.A. "Spectrometry of Fuels"; Plenum, New York, 1970.
2. Speight, J.G. "Analytical Methods for Coal and Coal Products"; Karr, C. Jr., Ed., Academic, New York, 1978; Vol. 2; p.75.
3. Bartle, K.D.; Jones, D.W.; Pakdel, H. "Analytical Methods for Coal and Coal Products"; Karr, C. Jr., Ed., Academic, New York, 1978; Vol. 2; p.210.
4. Bartle, K.D.; Jones, D.W.; Pakdel, H. Separation Science and Technology 1982, in press.
5. Bartle, K.D.; Ekinci, E.; Frere, B.; Mulligan, M.; Sarac, S.; Snape, C.E. Chem. Geol., in press.
6. Bartle, K.D.; Jones, D.W.; Pakdel, H.; Snape, C.E.; Calimli, A.; Olcay, A.; Tugrul, T. Nature 1979, 277, 284.
7. Jones, D.W.; Pakdel, H.; Bartle, K.D. Fuel, in press.
8. Snowden, L.R.; Peake, E. Anal. Chem. 1978, 50, 379.
9. Abelson, H.I.; Gordon, J.S.; Lowenbach, W.A. "Analytical Methods for Coal and Coal Products"; Karr, C. Jr., Ed., Academic, New York, 1978; Vol. 2; p.563.
10. Bartle, K.D. Rev. Pure Appl. Chem. 1972, 22, 79.
11. Harris, R.K. "N.M.R. and the Periodic Table"; Harris, R.K.; Mann, B.E. Eds., Academic, London, 1979.
12. Lindemann, L.P.; Adams, J.Q. Anal. Chem. 1971, 43, 1245.
13. Bartle, K.D.; Jones, D.W.; Pakdel, H. "Molecular Spectroscopy"; West, A.R., Ed., Heyden, London, 1977; p.127.
14. Abraham, R.J.; Loftus, P.L. "Proton and Carbon-13 N.M.R. Spectroscopy"; Heyden, London, 1979.
15. Gillet, S.; Rubini, P.; Delpuech, J.I.J.; Escalier, J.-C.; Valentin, P. Fuel, 1981, 60, 221.
16. Lyerla, J.R.; McIntyre, H.M.; Torchia, D.A. Macromolecules 1974, 7, 11.
17. Lyerla, J.R. Jr.; Horikawa, T.T. J. Phys. Chem. 1976, 80, 1106.
18. Long, R.V.; Goldstein, J.H.; Carman, C.J. Macromolecules 1978, 11, 574.
19. Birdsall, N.J.M.; Lee, A.G.; Levine, Y.K.; Metcalf, J.C.; Partington, P.; Roberts, C.K. J.C.S. Chem. Comm. 1973, 757.
20. Drake, J.A.G.; Games, D.; Jones, D.W. unpublished measurements.

RECEIVED April 30, 1982

Fourier Transform IR Spectroscopy

Application to the Quantitative Determination of Functional Groups in Coal

PAUL C. PAINTER, RANDY W. SNYDER, MICHAEL STARSINIC, MICHAEL M. COLEMAN, DEBORAH W. KUEHN, and ALAN DAVIS

Pennsylvania State University, College of Earth and Mineral Sciences, University Park, PA 16802

Fourier transform infrared (FTIR) spectroscopy is potentially a powerful tool for the characterization of coal. Although the optical advantages of these instrument compared to traditional dispersive devices are important, we believe that the most significant results can be obtained by applying sophisticated data analysis programs. However, if these programs are applied uncritically there is the real possibility of serious error, particularly in quantitative work. Consequently, in this paper we will attempt an assessment of the application of FTIR to the quantitative determination of functional groups, with particular emphasis on O-H and C-H groups. We will consider the use of programs for spectral subtraction, derivative spectra, curve resolving and factor analysis.

In the late 1950's and early 1960's infrared spectroscopy was a widely used analytical tool for investigating the structure of coal. Unfortunately, this technique was limited in part by the problems associated with investigating highly absorbing systems but predominantly by the overlap and superposition of the vibrational modes characteristic of such complex multicomponent systems. In fact, by 1973 there was a general decline in the use of infrared spectroscopy for chemical studies as other techinques (eg NMR) came to the fore. This state of affairs prompted H.A. Laitinen, in an editorial in Analytical Chemistry (_1_), to draw an analogy between the seven ages of an analytical technique and Shakespeare's seven ages of man, using infrared spectroscopy as a prime example. Senescence and an early demise seemed assured. However, for analytical

0097-6156/82/0205-0047$08.25/0
© 1982 American Chemical Society

techniques there is always the possibility of reincarnation and at the same time as Laitinen's depressing prognosis appeared commercial Fourier transform infrared instruments were being delivered to a number of laboratories. The subsequent resurgence of infrared spectroscopy has been remarkable and has led to a renewed interest in applying this technique to the characterization of coal structure.

A description of Fourier transform infrared (FTIR) instruments together with an account of the underlying theory and optical advantages compared to conventional dispersive spectrometers has been given by a number of authors (2-4). Although the improvements in the quality of the spectra that can be obtained from materials such as coal are useful, in our opinion the most significant advances have been made through the use of the dedicated on-line minicomputer that is an integral part of the system. Programs capable of a range of manipulations, from the simple such as spectral subtraction to the complex like factor analysis can now be routinely applied. Naturally, this type of analysis is not unique to FTIR and many sophisticated techniques were developed in the 1960's for use with digitized data obtained from dispersive instruments. Nevertheless, computer methods were not routinely applied and the type of analysis that depends on data manipulation has subsequently become associated with FTIR.

Havelock Ellis remarked that "What we call progress is the exchange of one nuisance for another nuisance". The advent of computerized instruments has certainly made routine a number of difficult spectroscopic measurements. Unfortunately, the sophistication of the procedures now available can result in erroneous results or interpretations if they are not used with caution. A classic example is the use of curve-resolving techniques. There is always a suspicion that a good fit between an observed spectral profile and a number of bands can be obtained providing that a sufficient number of the latter are included in the analysis. Clearly, in such cases a prior knowledge of the number of bands and their frequency would significantly increase our confidence in the results.

In this review we will attempt to critically assess the application of FTIR procedures to the characterization of the structure of coal. To an extent much of what we have to say is not original. Difficulties, such as those mentioned above for curve resolving, were encountered and addressed a number of years ago. However, with the advent of any new instrumentation there is a tendency to ignore segments of previous work that were obtained on inferior machines and in effect spend a considerable amount of time and effort "reinventing the wheel". Accordingly, we will not simply review the results of the FTIR studies of coal published to date, but first consider the use of certain computer routines. We will then consider how such routines can be applied to the quantitative determination of specific functional groups in coal. Published work concerning the analysis of mineral matter in coal or chemical changes such as oxidation will only be considered in passing, to the extent that they illustrate the usefulness of a given procedure.

Sample Preparation

When commercial FTIR instruments were first introduced the widely touted optical advantages of the instrument sometimes gave the impression that one could place a brick in the sample chamber and still obtain a spectrum. The preparation of samples was not emphasized, but if anything the data manipulations that have since became routine have made careful and consistent sample preparation even more important, particularly for quantitative work.

Coal samples are most conveniently prepared by dispersion in an alkali halide matrix, usually KBr or CsI. Estep et al (5) examined the problems involved in this method of preparation with specific reference to the analysis of mineral matter in coal. These authors preground the mineral or ash under investigation and then simply mixed a measured weight of the material with dried CsI. For optimum absorption and the minimization of problems associated with particle scattering it is necessary that the size of the sample particles are less than the wavelength of the infrared radiation and that they are evenly dispersed within the matrix. Estep et al (5) noted that this method of sample preparation did not give maximum absorption, grinding of the CsI was also necessary. Consequently, in work in this laboratory we also grind the alkali halide (usually KBr) together with the coal using a Perkin Elmer Wig-L-Bug. [For hard minerals and shale it is necessary to also pregrind the samples (6,7)]. The grinding time necessary to achieve optimum absorption will vary from laboratory to laboratory according to the equipment available. An initial grinding study should be conducted to determine optimum conditions. For example, Figure 1 is a plot of the integrated absorption of the aromatic C-H out-of-plane bending modes of a vitrinite concentrate (1.3 mg of maceral in 300 mg KBr) plotted as a function of grinding time. In recently acquired equipment optimum absorption was reached after grinding for 30 seconds, although on older equipment approximately 20 minutes grinding was required.

Grinding into KBr is a convenient and relatively straightforward method of sample preparation, but it is not without problems. In terms of coal analysis the most severe difficulty is the ubiquitous presence of water, since the strong water absorption near 3400 cm^{-1} overlaps the coal bands due to OH and NH groups. The effect of water can often be minimized by heating the pellets under vacuum, but in this laboratory we have never succeeded in completely eliminating water absorption by such procedures. This problem was addressed some years ago by Friedel (8) who observed that heating pellets to 175°C is

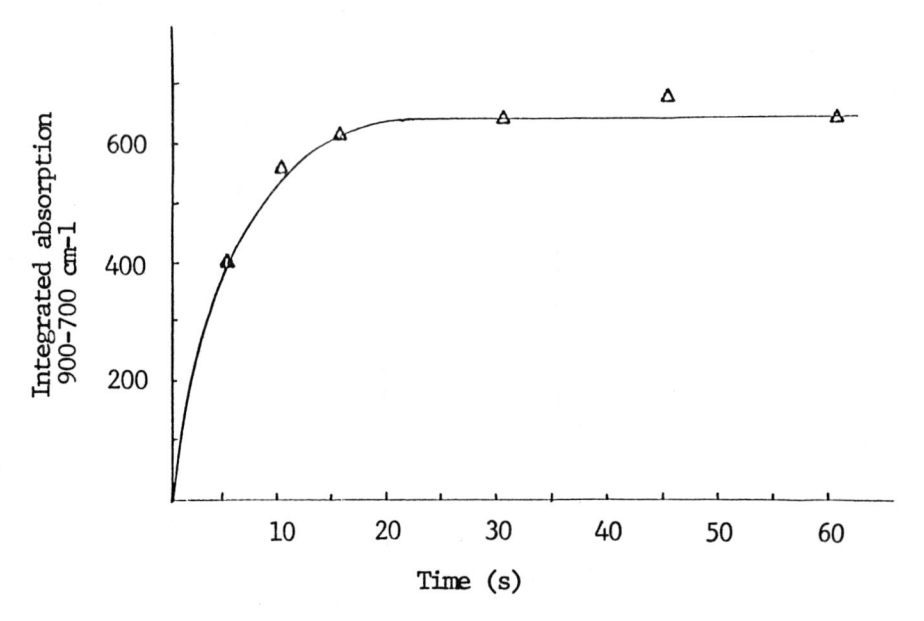

Figure 1. Plot of the integrated absorption in the region 900–700 cm^{-1} in the spectrum of a coal sample vs. grinding time in KBr pellet preparation.

required to completely remove water bands, which nevertheless reappear upon cooling. Breger and Chandler (9) also reviewed this problem and reported that they could not eliminate the presistent 3400 cm^{-1} water absorption despite numerous attempts at dehydration. It was suggested that this band may not only be due to water, but also hydroxyl groups substituted in the alkali halide matrix, an interpretation also proposed by Durie and Dzewczyk (10).

The problems associated with water in alkali halide preparations suggest that other preparative techniques are worth pursuing. Mulling techniques were originally used in coal studies but require long grinding times. Furthermore, quantitative studies are difficult if not impossible on the resulting "smears" that are usually used for analysis.

Recently, Fuller and Griffiths (11) have demonstrated that diffuse reflectance could be extremely useful. Photoacoustic spectroscopy also has potential (12-15). However, at this time the application of these techniques to the quantitative determination of functional groups in coal remains to be demonstrated.

Corrections Or Adjustments To The Spectra

In order to obtain quantitative measurements of the functional groups present in coal by FTIR it is necessary to first account for the minerals that may be present. Solomon (16,17) reported that coal spectra were corrected by subtracting the contributions of kaolinite and illite and scaling the spectra to give the absorbance for 1 mg of coal dmmf. However, it is often the case that the spectra of mineral components are dominated by clays such as kaolinite and illite (18,19), but these may constitute only 20 to 30% by weight of the mineral matter present. Furthermore, pyrite does not have absorption bands in the mid-infrared region used in coal studies, so that the contribution of this mineral would not be measured by simple subtraction procedures. The most accurate method for determining mineral content and adjusting the spectra to account for all the materials that may be present is to first obtain the low temperature ash and then adjust the spectrum to the equivalent of 1 mg of organic material using the fractional amount of ash in the coal so determined. The spectrum of this ash can then be directly subtracted from that of the coal so as to eliminate the bands of the mineral constituents (19). Occasionally, organic sulfur and nitrogen can be fixed as inorganic sulfate and nitrate in the ashing process, but these minerals can readily be detected and quantitatively measured by FTIR methods (19,20).

The second adjustment to coal spectra that concerns us, since we believe it may lead to spectral artifacts, is the correction of the baseline for scattering. In his seminal paper on the infrared spectra of coal, Brown (21) discussed the origin of the sloping background and concluded that scattering was not the only source. Microscopic examination of various KBr pellets showed no significant variation in average coal particle size, but there was a general trend to higher background absorption with increasing rank. Reviewing the work of Brown and other similar studies, Dryden (22) concluded that at least in higher rank coals part of the background can be attributed to the 'wings' of electronic absorption bands extending into the infrared. The background absorption may therefore vary with frequency in a non-linear fashion. In fact, if we examine the spectrum of a vitrinite sample shown in Figure 2, a substantial sloping baseline is apparent between 3800 and about 1850 cm^{-1}, but the spectrum flattens out and apparently slopes in the opposite direction between 1800 and 500 cm^{-1}. This type of variation in background is apparently not unique to coal. Maddam's (23) has pointed out that various workers have employed parabolic functions to fit baselines. Clearly, if we straighten the baseline according to the slope in one region (e.g. between 3800 and 1850 cm^{-1}, as shown in Figure 3, there will be a distortion in the remaining part of the spectrum. Solomon (16,17) used this approach and obtained spectra similar to those reproduced here, with the absorbance minima between 1000 and 500 cm^{-1} clearly raised above the new (straight) baseline. In subsequent curve resolving an extremely broad band (half width approximately 200 cm^{-1}) centered near 700 cm^{-1} was determined. This band was tentatively assigned to a carbonyl group and was stronger than the characteristic C=O stretching mode near 1700 cm^{-1}. However, the characteristics of the 700 cm^{-1} band do not correspond to any known carbonyl group frequency. We suggest that it may be an artifact of the method used to straighten the baseline. The bottom spectrum shown in Figure 3 displays the result of using two separate sloping straight baselines, illustrated in the top spectrum, to adjust the spectra between 3800 and 1850 cm^{-1} and between 1850 and 500 cm^{-1}, respectively. In this adjusted spectrum there is clearly no broad (200 cm^{-1} half width) underlying absorption centered near 700 cm^{-1}.

If coal spectra are to be curve resolved or if spectra of materials having different backgrounds are to be accurately compared, it may prove necessary to adjust the baseline. However, the spectra reproduced in Figure 3 indicate that such procedures are more dependable if applied separately to specific local regions of the spectrum.

Computer Applications

The primary function of the computer in FTIR instruments is to perform the Fourier transformation that converts an interferogram to a recognizable spectrum. However, the availability of an on-line mini-computer has opened the door to routine data manipulations. In this section we will review procedures that have been or promise to be useful in coal characterization. Certain data analysis operations, such as numerical integration

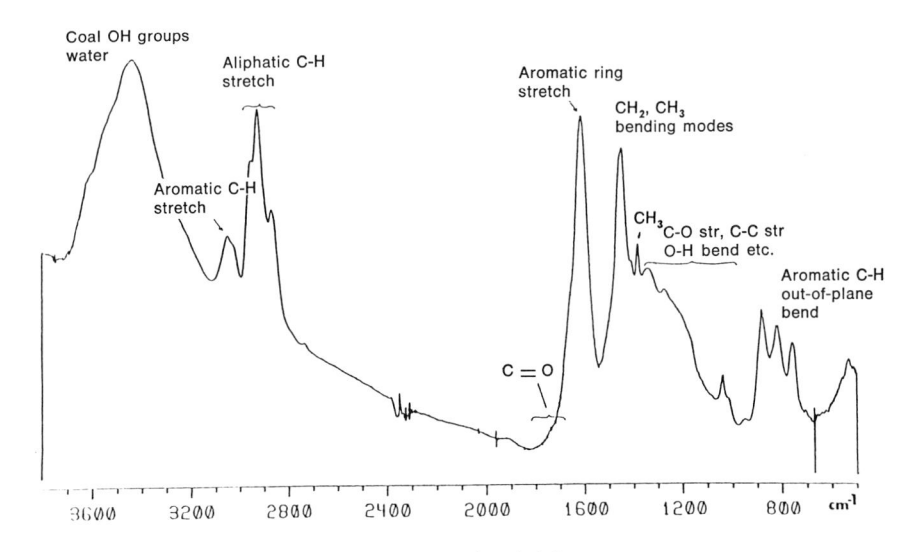

3. PAINTER ET AL. *Functional Group Determination Using FTIR* 53

Figure 2. FTIR spectrum of a vitrinite concentrate.

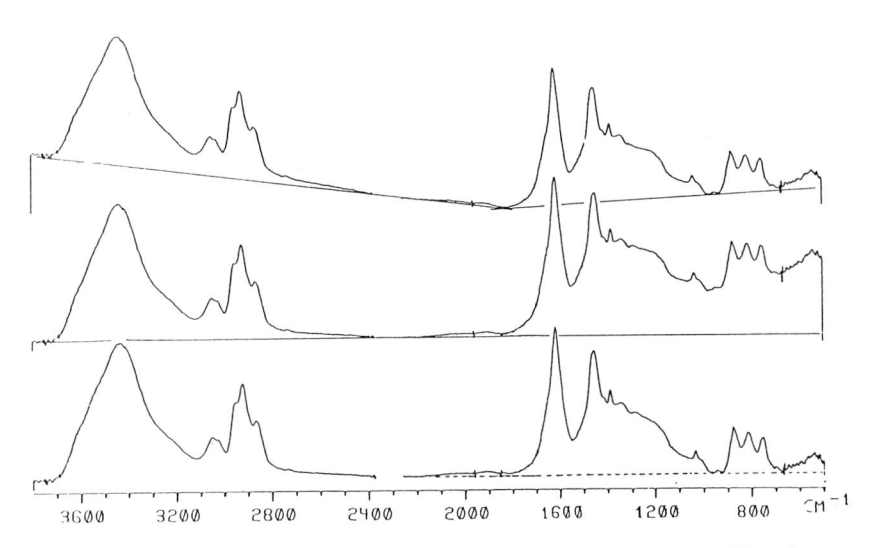

Figure 3. FTIR spectra of vitrinite concentrate. Key: top, spectrum with estimation of baseline positions; middle, spectrum corrected for single, sloping straight-line spectrum; and bottom, regions of spectrum corrected separately for background.

between defined limits, are to some degree trivial and will not be discussed.

Difference Methods

The ability to subtract one spectrum from another to obtain a difference spectrum has proved very useful in coal studies, particularly the detection of oxidation products (24,25) and the analysis of mineral matter in coal (18-20, 26). Spectra are stored in digital form in computerized systems and it is a relatively straight forward task to multiply these spectra by appropriate scaling factors and subtract one from another. For example, Figure 4 compares the infrared spectrum of a fresh coal sample to that of the same sample oxidized for two hours at 150°C. The only apparent difference is the appearance of a shoulder near 1695 cm^{-1} in the spectrum of the oxidized sample. However, more information can be obtained from the difference spectrum, shown in the same figure. The difference spectrum was obtained by multiplying the spectrum of the unoxidized coal by a scaling factor (usually a fractional number) and subtracting from the spectrum of the oxidized sample. This procedure was repeated on a trial-and-error basis until the bands of the mineral matter (near 1000 cm^{-1}) were eliminated. The 1695 cm^{-1} is resolved as a separate band in the difference spectrum and a prominent new band near 1575 cm^{-1} is now revealed. In addition, bands due to aliphatic C-H vibrational modes near 2900 cm^{-1} appear negative or below the baseline, indicating that the functional groups responsible for these absorptions are disappearing upon oxidation.

There are two major problems associated with this procedure. First is the choice of bands to be eliminated or used as an internal standard. In the example used above bands due to mineral matter were chosen. This could lead to errors in that only small quantities of sample (about 1mg) are used to obtain a spectrum. Consequently, any variation in mineral matter content among the coal particles will lead to an improper choice of scaling factor. This problem can in principle be solved by obtaining the spectrum of the low temperature ash (LTA) of the coal. This spectrum is then subtracted from each of the spectra of the coal samples. The resulting difference spectra can then be adjusted to the equivalent of 1 mg of organic material by multiplying by scaling factors calculated from the weight of coal in the original KBr pellets and the scaling factor determined from elimination of the mineral bands. All subtractions between the spectra oxidized and fresh samples can then be performed on a 1/1 weight basis to determine true differences.

The second problem is more general to all difference methods and involves the correct determination of scaling parameters. Trial-and-error subtractions are limited by the essentially subjective judgement of when bands have been eliminated. This procedure is tedious but reasonably accurate if strong fairly isolated bands can be used as an internal standard. If there are only slight differences between the spectra of interest, potential errors are naturally much larger. Koenig and co-workers have addressed this problem with respect to the analysis of polymers and developed a ratio method (27) and least-squares spectral fitting procedures (28). These programs have yet to be applied to the analysis of organic functional groups in coal, although the least squares method offers some potential in the analysis of mineral matter (25). We will consider additional examples of the use of difference methods below in our discussion of methods for the determination of hydroxyl groups in coal.

Factor Analysis

The analysis of the spectra of coals would be considerably assisted by a knowledge of the number of principal components that contribute to the spectrum. Theoretically, this knowledge can be obtained by the method of factor analysis, described in detail by Malinowski and Howery (29) and adapted to FTIR by Koenig and co-workers (30). A requirement of the method is that the spectra of a number of samples are available in which the concentration of the components varies from sample to sample. Naturally, the number of samples (or spectra) must be greater than the number of components. If the spectra are written in matrix form so that the absorbance values at designated frequency intervals are the elements of the matrix, then the rank of this matrix should be equal to the number of components. The rank or number of components is usually determined by forming an appropriate square matrix by multiplying the matrix constructed from the spectra by the transpose of itself and determining the number of non-zero eigenvalues.

One useful application of this method is the determination of the number of components in a set of mineral mixtures (for example, a set of low temperature ash samples from a given suite of coals). To determine the usefulness of this program in coal studies we made up a set of ten mixtures, each consisting of six minerals. The composition of each mixture was different. The results are illustrated in Figure 5, where the value of the log of each eigenvalue is plotted as a function of component number as recommended by Antoon et al.(30). Clearly, there are six non-zero or significant eigenvalues indicating the presence of six components.

Although a knowledge of the number of major components contributing to a set of spectra would be important in performing an accurate analysis, there is a more intriguing aspect to factor analysis. It is theoretically possible to reconstitute the spectra of the components and determine the concentration of each in all mixtures by the determination of the eigenvalues and eigenvectors of appropriate matrices. This would require a 'targeting' of the eigenvectors to the spectra of mineral standards (17).

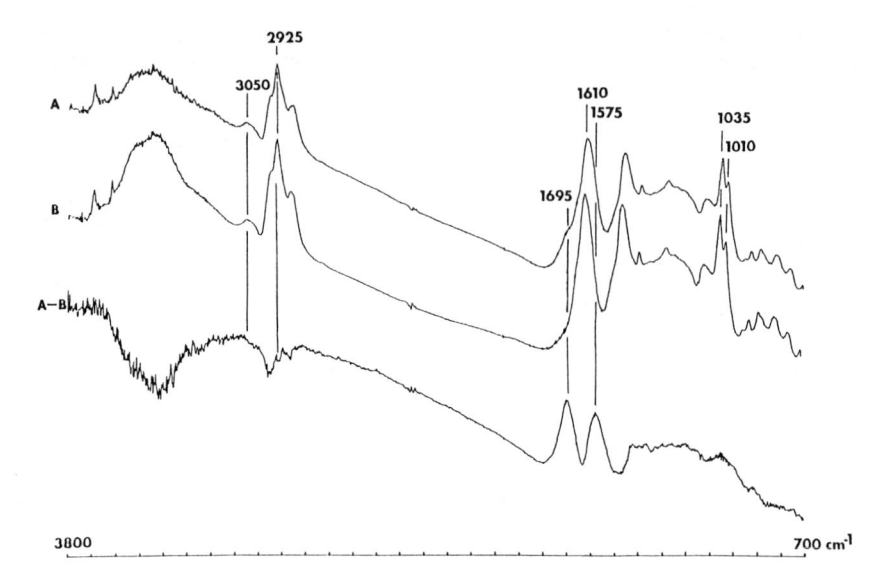

Figure 4. FTIR study of coal oxidation. Key: A. FTIR spectrum of oxidized PSOC 377 coal (HVA) (1 h, 150°C in air); B, FTIR spectrum of fresh coal; A − B, difference spectrum.

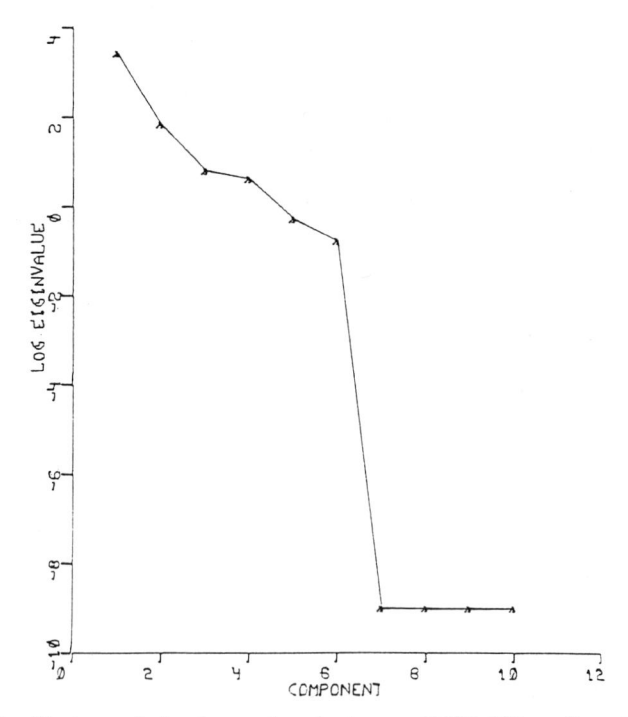

Figure 5. Factor analysis of ten mineral mixtures (1700–500 cm^{-1}) each containing six minerals. Plot of log eigenvalue vs. component number.

This approach has been attempted some years ago in the study of reaction kinetics, but at the time on-line mini-computers were not available. The procedure involved obtaining a laborious point-by-point description of each spectrum which was then used as input to a large computer. Consequently, the method was not actively pursued. The use of FTIR with its on-line computer should theoretically allow us to use this approach to analyze mineral mixture, but at this time the procedure has yet to be adapted. We are presently attempting to develop appropriate programs.

The application of factor analysis to the determination of mineral matter in coal is relatively straightforward. In principle, this technique should be of value in the analysis of the organic structure of coal samples obtained from different positions in a seam or liquefaction products from a coal obtained as a function of different liquefaction conditions. However, mineral mixtures are simple in that the spectra are accurately reproduced by a sum of the spectra of the constituents. There appears to be no vibrational coupling or significant interactions between the components. This is probably not the situation in studies of the organic structure of coal. Nevertheless, even if the complete infrared spectrum is too complex for such an analysis, specific regions consisting of well-known, isolated modes might prove more amenable to factor analysis. In preliminary work we have examined the out-of-plane bending modes between 950 and 750 cm^{-1} of 13 vitrinite concentrates. The spectrum of one of these was shown in Figure 2, above. All of these spectra have three main bands in this region (near 850, 810 and 750 cm^{-1}) with the suggestion of weak additional shoulders in certain samples. The results of the factor analysis are illustrated in Figure 6. The first eigenvalue has a significant non-zero value, but there is a considerable drop in magnitude to the second and third eigenvalues, followed by a further drop to the other components. This type of result is not easy to interpret. It suggests that all the spectra can be reproduced by the spectrum of one principle component. This component would have the three bands shown in Figure 1. Minor variations are then accounted for by the presence of two other components represented by the second and third eigenvalues. This in turn implies that the relative intensities of the out-of-plane bending modes and hence the relative concentration of aromatic hydrogen groups (lone C-H, two adjacent C-H etc) are fairly constant over the range of samples examined, although the absolute concentration of aromatic hydrogen varies. Unfortunately, at this point in our studies we are not certain whether we are guilty of the sin of over-interpreting the data. More work on the application of this method is required, but it clearly has interesting potential.

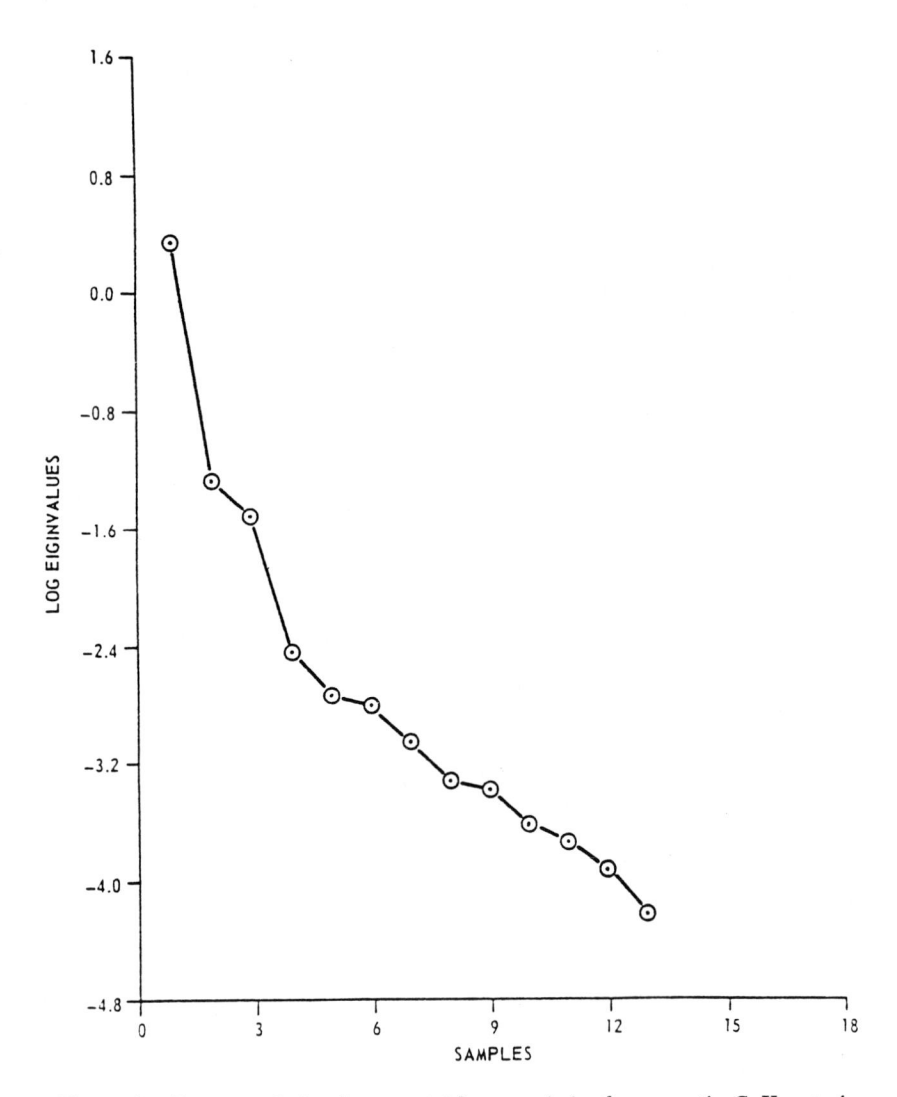

Figure 6. Factor analysis of a set of 13 macerals in the aromatic C–H out-of-plane bending mode region (950–700 cm^{-1}). Plot of log eigenvalue vs. component number.

Derivative Spectroscopy

Derivative spectroscopy has been used for many years as a means of enhancing the appearance of minute shoulders in spectra (31). The derivative of digital data can be computed by a Newton-Gregory interpolating polynominal (32). Applications such as the detection of weak bands in the initial stages of coal oxidation are potentially of great value. However, in our laboratory the major application of this technique to coal studies has been the determination of the number of bands contributing to a given spectral profile. It was mentioned in the introduction to this review that in order to have confidence in the results of curve-resolving procedures it is necessary to have a prior knowledge of the number of bands in the spectral region of interest and, if possible, an estimate of the position of each band and its width at half height. Within certain limitations this information can be obtained from the second derivative of the absorbance spectrum.

As an example of the use of this method Figure 7 displays the result of adding two overlapping bands of equal halfwidth and intensity. Three spectra are shown in which the two peak positions are separated by an amount equal to the halfwidth, half the halfwidth and one third the halfwidth of the bands, respectively. Superimposed upon each of these spectra is the profile of the second derivative. The minima in the second derivative profile corresponds to the position of bands and shoulders (23). It can be seen that this method is not capable of differentiating between bands separated by less than half their halfwidth. The fourth derivative of the spectrum is capable of a higher degree of resolution, as illustrated in Figure 8. (The maxima in the fourth derivative profile corresponds to the peak position). However, the signal-to-noise ratio degrades considerably with every successive differentiation, so that the fourth derivative of a typical coal spectrum is often so noisy as to be uninterpretable. Nevertheless, as long as these limitations are kept in mind derivative spectroscopy can still be extremely useful. We will consider some specific examples below when we discuss the analysis of functional groups.

Curve Resolving

Coal spectra consist of broad overlapping peaks, so that if we are to gain some insight into structure it would seem essential that attempts are made to resolve individual bands associated with specific functional groups. It is usually a smiple task to obtain a good fit between a synthesized spectrum consisting of a sum of a sufficient number of bands and an observed spectral profile. However, the validity of the results are open to question unless a number of key problems are addressed.

The first problem is the choice of mathematical function most suitable for characterizing the observed band shapes. In a

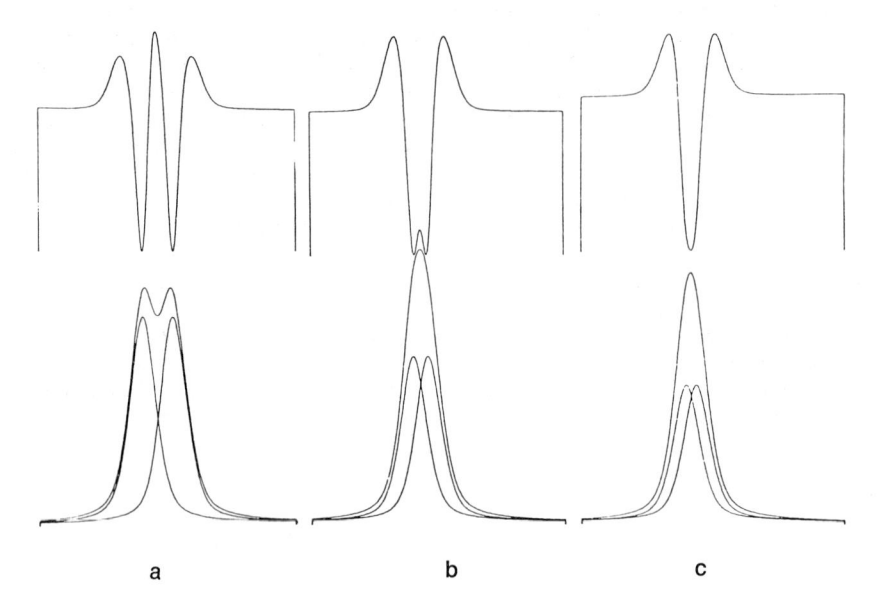

Figure 7. The second derivative of spectra consisting of two component bands separated by a, $D = \Delta X_{1/2}$; b, $D = \frac{1}{2} \Delta X_{1/2}$; and c, $D = \frac{1}{3} \Delta X_{1/2}$.

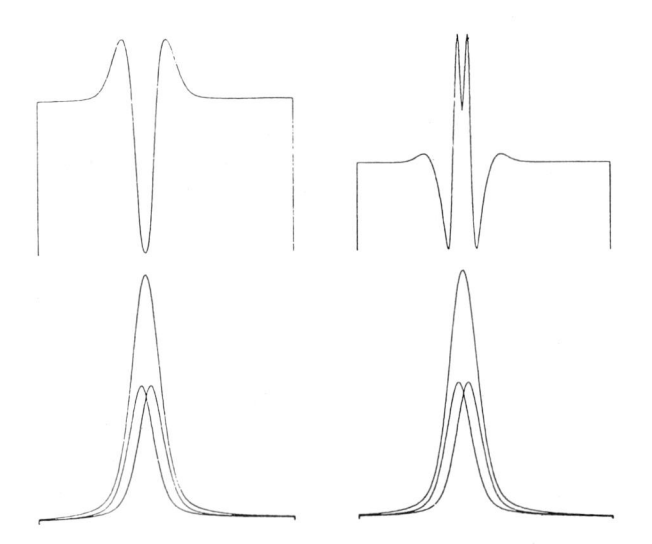

Figure 8. The second (F'', left) and fourth (F'''', right) derivatives of a spectrum consisting of two bands separated by $D = \frac{1}{3} \Delta X_{1/2}$.

review of the literature Maddams (23) has pointed out that the evidence in favor of Lorentzian band shapes in the spectra of homogeneous materials is very strong. Gauss-Lorentz sum and product functions have often been used, as with dispersive instruments a shift from Lorentzian to Gaussian band shapes was found with increasing slit width. Although FTIR instruments do not have a set of slits, bandshapes can be affected by instrument factors such as the apodization function used in the Fourier transformation and by factors associated with the material under consideration; for example, interactions between the constituents of a complex multicomponent system. Consequently, an empirically determined bandshape would seem to offer the best choice in coal studies. Jones and co-workers (33-35) examined a number of different bandshapes in their infrared studies and expressed a preference for a function that is a sum of Gaussian and Lorentzian bandshapes. On this basis we decided to utilize a similar function in which the Gaussian and Lorentzian constituents that make up a given band have equal half widths and are present in the proportion f to (1-f);

$$A = fA_o \exp\left[-\ln 2\left[\frac{2(X-X_o)}{\Delta X_{1/2}}\right]^2\right] + \frac{(1-f)A_o}{1 + [2(X-X_o)/\Delta X_{1/2}]^2} \quad (1)$$

where A is the peak height, X is the wavenumber coordinate of the peak and $X_{1/2}$ is the bandwidth at half height. Initial estimates of these parameters are input to a program that fits the parameters for a set of bands to the experimental spectral profile by a least squares optimization procedure, as described by Fraser and Suzuki (36).

Other problems with curve resolving are associated with the initial choice of parameters as input to the type of program outlined above. It is important to have a prior knowledge of the number of bands in the region of the spectrum that is to be resolved and good initial estimates of the peak position of each band and its width at half height. The problems that arise when such information is not initially obtained are best illustrated by an example. This will be presented in the following section where we consider the application of FTIR to the determination of functional groups in coal.

The Application Of FTIR To The Quantitative Determination Of Functional Groups In Coal

Because of the complexity of the infrared spectra of coals most quantitative studies have been limited to the determination of hydroxyl groups and aliphatic and aromatic C-H. Brown (21) attempted to obtain a measure of the relative proportions of these latter groups by measuring the ratio of the peak heights of the aromatic and aliphatic C-H stretching modes at 3030 and

2925 cm^{-1}, respectively. This approach requires a knowledge of the extinction coefficients, e_{ar} and e_{al}, which Brown estimated from measurements on a number of low molecular weight model compounds. An average value of 0.5 for e_{ar}/e_{al} was used, although measured values ranged from 0.3 to 1.0. In a later study Brown and Ladner (37) applied both infrared and proton magnetic resonance to the study of soluble coal products and proposed the Brown-Ladner equation to determine aromaticities from infrared measurements. Other authors (38-40) have attempted to improve the infrared technique by using integrated absorption measurements as opposed to simple peak heights and by considering other regions of the spectrum, principally the aromatic C-H out-of-plane bending modes between 900 and 700 cm^{-1}. Although a comparison with the results of NMR studies indicated fairly consistent results for certain coals, there are discrepancies. For example, Retcofsky (38) calibrated the C-H aromatic out-of-plane vibrations of pyridine and carbon disulfide coal extracts using the results of proton magnetic resonance studies, but obtained different curves for the two sets of extracts. These differences could be due to variations in the values of extinction coefficients with structure. However, there is a remarkable similarity in the spectra of coals and their extracts or liquefaction products. In addition, the main features of the spectra of coals do not show significant changes as a function of rank, varying only in the relative intensities of some features and the appearance of prominent bands due to carbonyl groups and carboxylic acids in low rank, (or oxidized) material. (Anthracites could be considered an exception to this generalization because the presence of an intense background absorption which has to be regarded as a key feature.) On the basis of this similarity we would intuitively expect only minor variations in the values of the extinction coefficients of the characteristic C-H group frequencies. We suggest that the observed variation could be due to the procedures used to make most measurements. For example, Retcofsky (38) and Durie, et al. (39) considered the integrated absorption between 680 and 920 cm^{-1} where there are typically three major aromatic C-H out-of-plane bending modes, near 850, 810 and 750 cm^{-1}. This implicity assumes that the extinction coefficients of these three bands are the same or that the relative intensities of the three bands remains unaltered from coal to coal. The latter possibility is clearly not true when a range of coals from different sources is examined. There are significant differences in the relative intensities of the out-of-plane C-H modes that reflect different degrees and types of aromatic substitution. Consequently, it would be more accurate to define the absorbance in the range 920 to 680 cm^{-1} as the sum of (usually) three components, each with a distinct extinction coefficient. Variations in the degree of aromatic substitution among distinct sample sets (e.g. pyridine soluble or carbon disulfide **soluble material**) would then result in different calibration curves. This problem also extends to the aliphatic C-H stretching region of the spectrum, where the bands between 3000 and 2800 cm^{-1} can be considered as a superposition of CH, CH_2 and CH_3 modes.

Recently, Solomon (16,17) has used FTIR to examine the relationship between coal structure and thermal decomposition products. In this work the spectra of a number of coals were "corrected" in order to account for particle scattering and mineral content. Each spectrum was then fitted to a selected set of 24 Gaussian bands whose individual widths and positions were held constant so that only their relative intensities were allowed to vary. A quantitative measure of aliphatic and aromatic C-H content was determined using the area under the peaks between 3000 and 2800 cm^{-1} and between 900 and 700 cm^{-1}.

The papers of Solomon are the first to report the use of the data handling capabilities of FTIR in an attempt to obtain quantitative results. In several earlier studies with dispersive instruments, for example the work of Durie, et al. (39) and Von Tschamler and de Ruiter (41), similar quantitative measurements of aliphatic and aromatic C-H contents have been made using the integrated intensity of the same peaks as Solomon (16,17). However, there are significant differences in the value of the factor, k, relating to the ration of the integrated areas to the corresponding ratio of aliphatic to aromatic hydrogen, H_{al}/H_{ar}. Solomon determined a value of 1.06 whereas Durie, et al. determined values ranging from 0.52 to 0.86. Although it was certainly more difficult for this latter group to obtain data with the same degree of accuracy as can be determined with modern instrumentation, their measurements cannot be easily dismissed. Cailbration of peak areas to hydrogen content was obtained using proton magnetic resonance studies on extracts representing approximately 40% of the original coal. On the basis of the range of values of k obtained, Durie, et al. suggested that any use of infrared spectroscopy to obtain H_{al}/H_{ar} values be treated with caution.

Solomon (16,17) has used a different method to obtain extinction coefficients. Essentially, total hydrogen content from elemental analysis and hydroxyl content from measurements of the area of the O-H stretching band near 3450 cm^{-1} were used in conjunction with the peak areas of aliphatic and aromatic bands to obtain a plot from which extinction coefficients can be determined. In principle, this approach appears to be sound, but there are a number of problems. One difficulty, discussed above, is general to all infrared methods that have been employed so far; what errors are introduced by summing peak areas over a number of bands, each of which has an individual extinction coefficient, and essentially averaging such coefficients for the total area? Other problems involve the correct use of curve resolving techniques and the measurement of hydroxyl groups, which we will now consider in more detail.

The Determination Of Hydroxyl Groups In Coal By FTIR

There have been a number of attempts to measure the -OH content of coal by infrared spectroscopy, either by direct measurement of the intensity of the O-H band near 3450 cm^{-1} (16,17,42) or by measuring the intensities of characteristic bands introduced by chemical reaction, e.g. acetylation (43). Osawa and Shih(42) determined a relationship between the specific extinction coefficient of the 3450 cm^{-1} absorption and the hydroxyl content, which was subsequently applied in FTIR studies of coal by Solomon (17,18). There are a number of problems with this approach. The extinction coefficient reported by Osawa and Shih relates to the peak height of the 3450 cm^{-1} band. Solomon (16) apparently applied this coefficient to the area under this absorption. Assuming the coefficient was adjusted to reflect areas rather than peak heights, it is still difficult to see how accurate results can be obtained. Not only were different methods used to estimate the baseline, but Solomon applied the same extinction coefficient to the summed areas of four separately resolved bands. Furthermore, in plots of optical density at 3450 cm^{-1} vs weight of coal sample (42) there is an intercept at positive values of optical density. Osawa and Shih proposed that this 'residual' absorption was due to traces of water remaining in the KBr disk, despite the care taken in the preparation and subsequent drying of the sample. We have discussed this problem in the section concerning sample preparation. However, water absorbed in KBr is not the only problem in using the intensity of the 3450 cm^{-1} band to quantitatively determine the number of OH groups. In addition to water absorbed by the KBr disks, water is bound to the coal itself in a complex manner that is apparently a function of rank. The broadening of the hydroxyl band due to hydrogen bonding (the degree of which could vary from coal to coal) and the possible presence of NH groups are also factors that conspire to make the interpretation of quantitative measurements suspect. Finally, it is not always desirable to heat coal samples prior to analysis. Coking coals, for example, are easily oxidized and we have detected the formation of carbonyl groups and loss of aliphatic C-H gorups at low levels of oxidation (1 hour at 150°C) (25). Consequently, this band may be useful in the qualitative comparison of the hydroxyl group content of different coals, providing that the KBr pellets are prepared in an identical manner, but for quantiattive measurements we should consider other procedures.

In view of these factors we suggest that it will ultimately prove more useful to measure OH groups in coal by a combination of FTIR and chemical procedures. Durie and Sternhell (43) reported an infrared study of acetylated coal twenty years ago. Although some useful linear plots were obtained, the method was complicated by the overlap of the acetyl bands with those of the original coal. This made the determination of baselines and the measurement of peak intensities subject to possible error. The problem is illustrated in Figure 9, which compares the infrared spectrum of an Arizona HVC coal (PSOC 312) to that of the same sample subsequent to acetylation. FTIR is capable of solving many problems of this type (band overlap) by simple spectral subtraction. Figure 9 also shows the difference spectrum obtained by subtracting the spectrum of the original coal from that of the acetylated product. The characteristic acetyl bands are now relatively well-resolved and it is a straight-forward task to draw an appropriate baseline and measure peak heights, or even make integrated absorption measurements of, for example the 1370 CH_3 mode. We will discuss the details of the use of acetylation reactions in conjunction with FTIR in a separate publication. Here we wish to point out the potential of FTIR for discriminating between different types of OH groups in the original coal, since this brings us back to the sensitive subject of curve resolving.

In an initial analysis of the carbonyl region of the difference spectrum shown in Figure 9, three bands were defined for curve fitting at the initial estimated positions of 1770, 1725 and 1605 cm^{-1}. Although a 'good' fit in least squares terms was obtained the results are probably meaningless. Figure 10 shows the difference spectrum between 1900 and 1550 cm^{-1} thgether with the three component bands determined by the program to give the best fit to the data. A strong, very broad band near 1725 cm^{-1} is the major component and appears stronger than the band assigned to acetylated phenolic-OH groups near 1765 cm^{-1}. This result does not make sense in terms of what is known of the structure of coal. It is thought that phenolic OH groups are by far the major hydroxyl component. Although a large difference in the extinction coefficients of various types of ester linkages could account for this discrepancy, it is far more likely that the curve resolving procedure is in substantial error. There are two probable sources of such error. First, that we have not included a sufficient number of bands in the analysis; second, that the 1765 cm^{-1} is significantly asymmetric, as also illustrated schematically in Figure 10. These initial errors and misconceptions on our part allows us to emphasize a key point in curve resolving. It is usually possible to obtain a reasonable fit between a synthesized spectrum consisting of a sum of a sufficient number of bands an observed spectral profile. However, if we are to have confidence in the results it is desirable or even essential to have an initial knowledge of the number of bands. We reviewed one method for obtaining a knowledge of the number of bands in a profile above, namely derivative techniques. The application of this method to the carbonyl region of the difference spectrum of an acetylated coal (PSOC 272, Kentucky No. 9, HVB coal) is shown in Figure 11. The spectrum and its second derivative are both presented in this figure. The minima in the second derivative profile correspond to the positions of

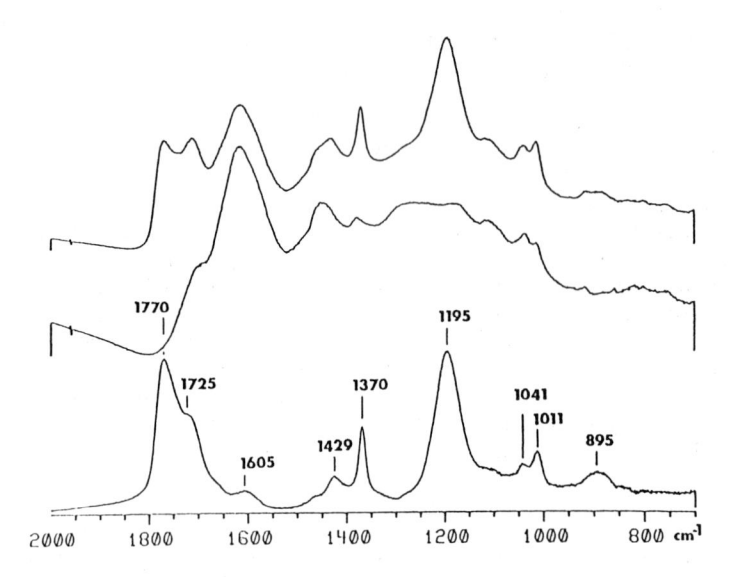

Figure 9. FTIR spectra of an Arizona HVC coal. Key: top, acetylated coal; middle, original coal; and bottom, difference spectrum.

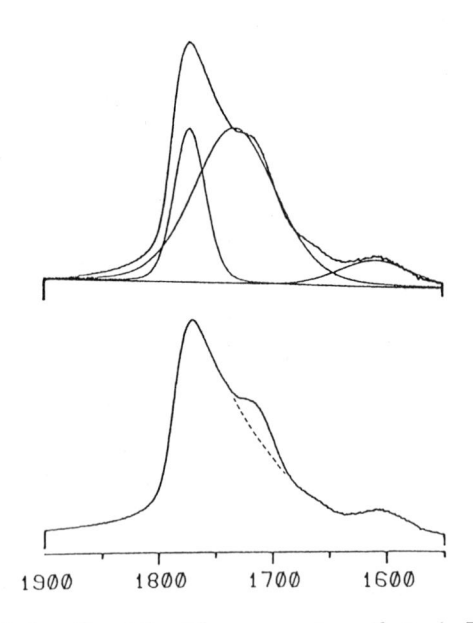

Figure 10. Spectral profile of the difference spectrum shown in Figure 9 between 1900 and 1550 cm^{-1} fitted to three curves (top); and spectral profile considered with an asymmetric major peak (bottom).

bands and shoulders (23). The presence of most of the bands can be confirmed by inspection of the spectral profile. The only questionable assignment is of a band near 1747 cm^{-1}, where the second derivative minima is of the order of magnitude of the noise level. However if the difference spectrum is carefully examined an inflection near 1747 cm^{-1} can be observed. Maddams (23) has pointed out that the value of the eye and brain should not be underestimated in such judgements. Less subjective evidence, however, is provided by examination of other coals where the components have different relative intensities. For example, the difference spectrum obtained from an acetylated lignite is shown in Figure 12. The presence of a band near 1740 is clearly indicated from the second derivative curve shown in the same figure. It is satisfying to observe that bands in approximately the same position are obtained from spectra of coals of different rank. The small differences in frequency are to be expected in materials that vary in chemical structure.

In addition to providing initial values of the peak positions for the least squares refinement, the second derivative curves can be used to obtain an initial estimate of the width at half height from a measure of the distance between inflection points. It is essential to have good initial estimate of these values if convergence to a good (and meaningful) least squares fit is to be obtained. Figure 13 and 14 show the curves resolved in the carbonyl region of the difference spectra of the bituminous coal and lignite considered above. Figure 13 also shows the effect of describing the 1769 cm^{-1} band as asymmetric according to:

$$A = A_o \exp\left(-\ln 2 \; \frac{\ln(1+2B) \; (X-X_o)/\Delta X_{1/2}}{B}\right)^2 \qquad (2)$$

Where B is a factor which determines the degree of asymmetry, as described by Fraser and Suzuki (36). It can be seen that the effect of asymmetry is to reduce the intensity of the 1745 cm^{-1} band. However, it has been emphasized that in studies of this type bandshapes should not be skewed unless there are 'a priori' grounds for doing so (23). Consequently, a fit that has symmetric bands is to be preferred. This aspect of the problem is presently being checked by a study of model compounds and the results, together with evidence supporting a complete assignment of the resolved bands, will be presented in a future publication. It is pertinent to point out here, however, that the band near 1720 can be assigned to residual acetic acid, so that previous methods of measuring OH groups, for example using ^{14}C labelled acetic anhydride, could be in error. The bands near 1773, 1740 and 1660 cm^{-1} can be assigned to acetylated phenolic OH, alkyl OH, and $>$ N-H groups, respectively. Accordingly, this preliminary work indicates that FTIR should not only be useful in determining total OH, through measurements of the intensity of the CH$_3$ and

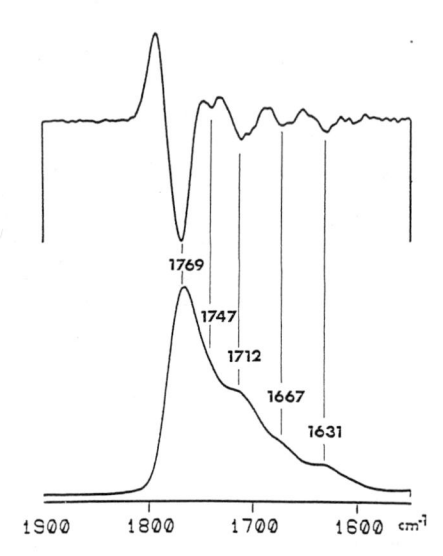

Figure 11. Difference spectrum between 1900 and 1550 cm⁻¹ obtained from acetylated PSOC 272 coal (bottom) and second derivation of spectrum (top).

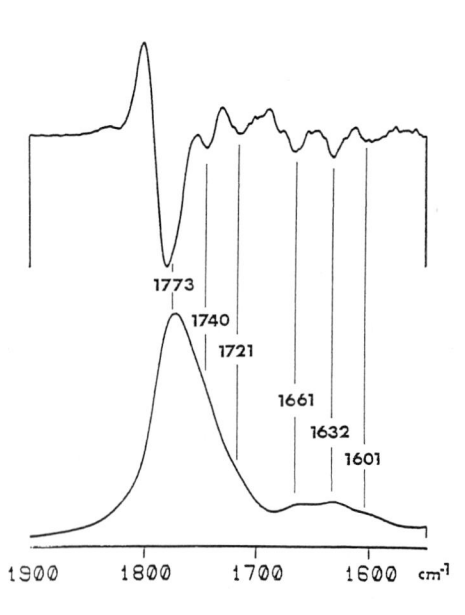

Figure 12. Difference spectrum obtained from an acetylated lignite (bottom) and second derivative of spectrum (top).

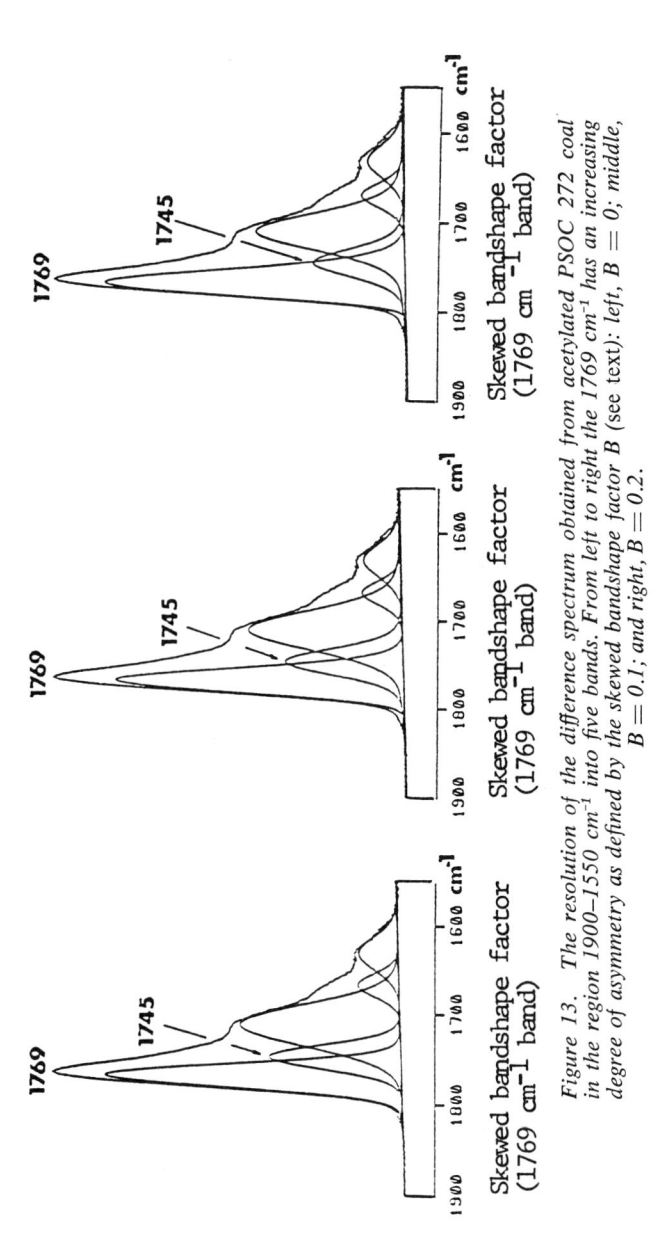

Figure 13. *The resolution of the difference spectrum obtained from acetylated PSOC 272 coal in the region 1900–1550 cm⁻¹ into five bands. From left to right the 1769 cm⁻¹ has an increasing degree of asymmetry as defined by the skewed bandshape factor B (see text): left, B = 0; middle, B = 0.1; and right, B = 0.2.*

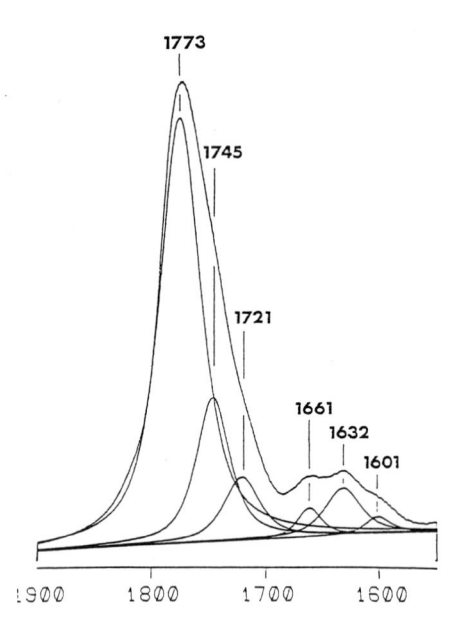

Figure 14. The resolution of the difference spectrum obtained from the acetylated lignite into six bands.

C-O bands, but also has the clear potential for discriminating between types of OH groups and also measuring N-H groups through an analysis of the C=O stretching region, providing that curve resolving methods are applied with circumspection.

The Determination Of C-H Groups

In a number of studies attempts have been made to obtain structural information by measuring the peak heights or integrated intensities of bands assigned to aromatic and aliphatic C-H modes. As we mentioned above, there is some variation in values of extinction coefficients determined by different groups (16, 21, 37-41). We have suggested that at least in part this variability is due to the methods used to obtain data, namely measuring integrating intensities over an entire spectral region thus assuming that the extinction coefficients of each band is the same (or that the bands maintain a constant relative intensity from coal to coal). At the beginning of this section we used the aromatic C-H out-of-plane bending modes to illustrate this point. The same argument applies to measurements in the aliphatic C-H stretching region where it has become conventional to measure the integrated absorption between 3000 and 2800 cm^{-1} as a measure of total aliphatic C-H concentration. This region of the spectrum has to be considered as the sum of contributions from three components, CH, CH_2 and CH_3. If the spectral contribution of these groups could be separated, then not only would we have obtained a greater insight into the structure of coal and its variation with rank, but more consistent aliphatic to aromatic C-H ratios might be obtained by taking into account differences in the extinction coefficients of these groups.

In order to obtain a meaningful resolution of the bands in the C-H stretching region we followed the procedures outlined above for analysis of the carbonyl region of the difference spectra of acetylated coals. Figure 15 shows a scale expanded plot of the aliphatic C-H stretching region of the spectrum of the vitrinite concentrate considered in Figure 2. Also displayed in Figure 15 is the second derivative plot, clearly indicating the presence of five bands. With an appropriate initial choice of band position and width at half height obtained from this plot, the C-H stretching region can be resolved into the five components shown in Figure 16. At this point it is tempting to make the obvious assignment of the 2956 and 2864 cm^{-1} modes to asymmetric and symmetric CH_3 stretch, the 2923 and 2849 cm^{-1} bands to the asymmetric and symmetric CH_2 stretch and the 2891 cm^{-1} band to lone C-H groups. However, we have to consider one more limitation to curve resolving methods. The identification of peaks in a profile by second derivative techniques will obviously depend upon the relative intensities of the peaks involved and their separation relative to their half widths, as discussed previously. If we carefully consider established group

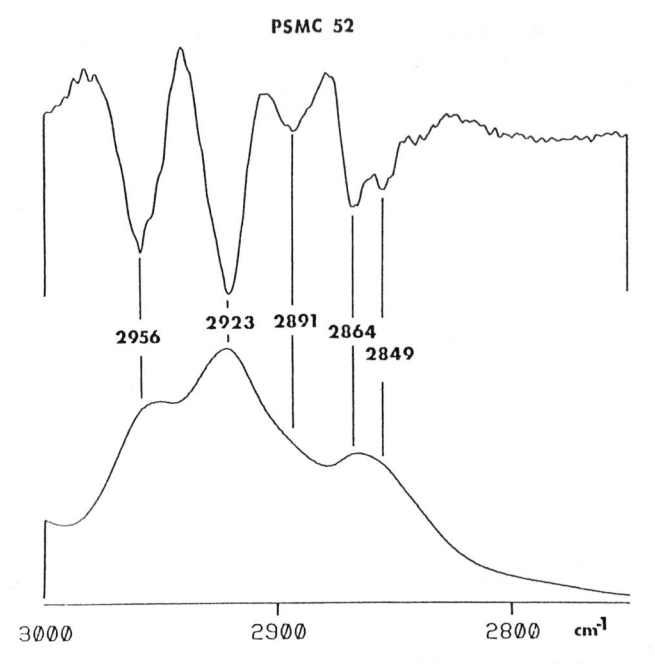

Figure 15. Scale expanded aliphatic C—H stretching region of the spectrum of a vitrinite concentrate (bottom); and second derivative of the spectrum (top).

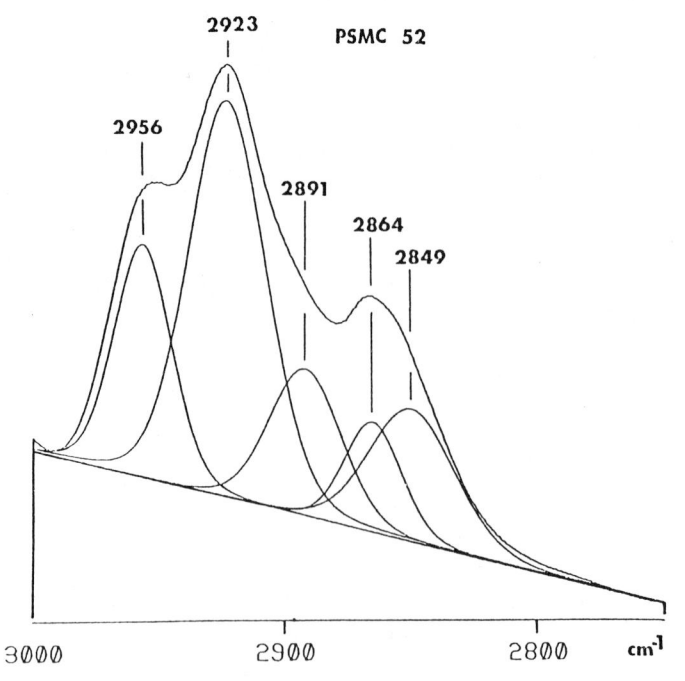

Figure 16. The resolution of the aliphatic C—H stretching region into five bands.

frequencies in conjunction with what is known of the structure of coal we have to conclude that the 2923 cm^{-1} band is a composite of two contributions, the asymmetric CH_2 stretching mode and the asymmetric CH_3 stretching mode of methyl groups attached directly to aromatic rings. However, the position of the symmetric CH_3 stretching mode appears to be less sensitive to local environment, appearing near 2865-2875 cm^{-1} for methyl groups attached to alkyl chains or aromatic rings. Consequently, if appropriate extinction coefficients can be determined it should prove possible to use the symmetric CH_2 and CH_3 modes to determine these groups and there is then the further possibility of using the 2956 cm^{-1} and 2923 cm^{-1} bands to determine the distribution of methyl groups. (The intensity of 2923 cm^{-1} band would have to be considered as the sum of the contributions from CH_2 and CH_3 groups). Finally, the 2891 cm^{-1} could be applied to the determination of tertiary hydrogen.

These initial results indicate that good curve resolving methods are capable of allowing a greater insight into coal structure through a more detailed analysis of the infrared spectrum. However, there remains the difficult task of obtaining extinction coefficients. Methods used in previous studies, mainly the use of model compounds and calibration by means of soluble extracts characterized by proton magnetic resonance, should prove useful, but we wish to suggest a third method. Essentially, this builds on the approach used by Von Tschamler and de Ruiter (41) and more recently by Solomon (16,17). We propose to equate the experimentally determined elemental hydrogen to the sum of contributions from hydrogen containing functional gorups, as measured by the intensities of appropriate infrared bands multiplied by a conversion factor (equivalent to an extinction coefficient converted to weight units);

$$H = \sum_n I_n e_n$$

where H is the weight percent hydrogen content, I_n is the intensity of the nth band (e.g. a band assigned to CH_2 groups) and e_n is a conversion factor relating band intensity to the weight percent hydrogen percent at this specific functional group. If we consider the curve resolving results discussed above, it is possible to identify three bands associated with aromatic hydrogen, three bands associated with various aliphatic groups (CH, CH_2 and CH_3) and bands in the spectra of acetylated samples that can be assigned to NH, alkyl OH and phenolic OH. In low rank coals bands near 1700 cm^{-1} can be readily assigned to COOH groups. Clearly, if we have sufficient coal (or maceral) samples chosen so as to cover a range of rank and hence hydrogen content, then theoretically the problem could be solved as a set of simultaneous equations. In fact, we should be able to "over-determine" the problem by analyzing a sufficient number of samples.

This approach has a number of advantages. Model compounds will not be required to obtain extinction coefficients. If a sufficiently large data base is used, any systematic variations in extinction coefficients with rank and hence chemical structure should be apparent providing that a method for determining variations in measured and calculated (by FTIR) hydrogen is included in the procedure.

Summary And Conclusions

One of the major advantages of FTIR in the analysis of complex systems is the ready application of sophisticated programs made possible by the necessity of an on-line mini-computer in these instruments. We have considered the application of selected programs and pointed out that they have to be applied with considerable caution. In particular, we have attempted to critically assess the problems involved in determining functional groups by curve resolving methods. There is a large body of work in the literature concerning this problem and by applying the results of these studies we can conclude that;
 a) Empirically determined bandshapes should be calculated using least squares curve fitting programs.
 b) In order to have confidence in the results it is necessary to have a good initial estimate of the number of bands in a profile, their frequency and width at half height.
 c) These methods should be applied separately to specific regions of the spectrum.

We have considered the application of these criteria to the determination of OH and CH groups in coal. By combining FTIR measurements with acetylation procedures it appears feasible not only to measure total (reactive) OH and NH content, but to distinguish between phenolic and alkyl OH groups by curve resolving methods. Finally, we have argued that the determination of consistent extinction coefficients for aliphatic and aromatic C-H bands for various coal samples would be facilitated by the use of curve resolving procedures and a determination of the relationship between the extinction coefficients of individual components. In previous studies it has been general practice to sum the integrated intensity of a number of bands and hence implicitly assume that each mode has the same extinction coefficient.

Acknowledgements

The authors gratefully acknowledge the financial support of the Department of Energy under contract No. De-AC22-OP30013 and the Penn State Cooperative Program in Coal Research.

Literature Cited

1. Laitinen, H.A. Analytical Chemistry, 1973, 45(14), 2305.
2. Griffiths, P.R. Chemical Infrared Fourier Transform Spectroscopy. John Wiley and Sons, New York (1975).

3. Koenig, J.L. Applied Spectroscopy, 1975, 29, 293.
4. Chenery, D.H. and Sheppard, N. Applied Spectroscopy, 1978, 32, 79.
5. Estep, P.A., Kovach, J.J. and Karr C., Jr., Anal. Chem, 1968, 40 (2), 358.
6. Elliott, J.J., Brown, J.M., and Baltrous, J.P. Private communication, to be published (discussed in reference 7).
7. Painter, P.C. and Coleman, M.M. Digilab FTS/IR Notes, No. 34, Aug. 1980.
8. Friedel, R.A. in Applied Infrared Spectroscopy, (p. 312), Edited by D.N. Kendall. Reinhold, New York (1966).
9. Breger, I.A. and Chandler, J.C. Anal. Chem, 1969, 41 (3), 506.
10. Durie, R.A. and Szewczyk, J. Spectrochim Acta, 1959, 13, 593.
11. Fuller, M.P. and Griffiths, P. Anal. Chem, 1978, 50, 1906.
12. Rockley, M.G., Davis, D.M., Richardson, H.H. Applied Spectroscopy, 1981, 35 (2), 185.
13. Rockley M.G. and Devlin, J. Applied Spectroscopy, 1980, 34, 407.
14. Rockley, M.G. Applied Spectroscopy, 1980, 34, 405.
15. Vidrine, D.W. Applied Spectroscopy, 1980, 34, 314.
16. Solomon, P.R. ACS Division of Fuel Chemistry Preprints 24 #3, 184 (1979) and Advances in Chemistry series (to be published).
17. Solomon, P.R. paper presented at meeting, Chemistry and Physics of Coal Utilization, Morgantown, WV, June 2-4 (1980) (in press).
18. Painter, P.C., Coleman, M.M., Jenkins, R.G., Whang, P.W. and Walker, P.L., Jr., Fuel, 1978, 57, 337.
19. Painter, P.C., Coleman, M.M., Jenkins, R.G. and Walker, P.L., Jr. Fuel, 1978, 57, 125.
20. Painter, P.C., Youtcheff, J. and Given, P.H. Fuel, 1980, 59, 523.
21. Brown, J.K. J. Chem. Soc. March 1955 (5562) 744.
22. Dryden, I.G.C. in 'The chemistry of coal utilization' (H.H. Lowry, Ed.) Suppl. Vol. John Wiley and Sons, New York, p. 232 (1963).
23. Maddams, W.F. Applied Spectroscopy, 1980, 34 (3), 245.
24. Painter, P.C., Snyder, R.W., Pearson, D.E. and Kwong, J. Fuel, 1980, 59, 282.
25. Painter, P.C., Coleman, M.M., Snyder, R.W., Mahajar, O., Komatsu, M. and Walker, P.L. Jr., Applied Spectroscopy 1981, 35, (1), 106.
26. Painter, P.C., Rimmer, S.M., Snyder, R.W. and Davis, A. Applied Spectroscopy, 1981, 35 (1), 292.
27. Koenig, J.L., D'Esposito, L. and Antoon, M.K. Applied Spectroscopy, 1977, 31, 292.
28. Antoon, M.H., Koenig, J.H. and Koenig, J.L. Applied Spectroscopy, 1977, 31, 518.

29. Malinowski, E.R. and Howery, D.G. "Factor Analysis in Chemistry". Wiley Interscience, New York (1980).
30. Antoon, M.K., D'Esposito, L. and Koenig, J.L. Applied Spectroscopy, 1979, 33, 351.
31. Martin, A.E. Spectrochim Acta, 1959, 14, 97.
32. Gillette, P.C., Kormos, D., Antoon, M.K. and Koenig, J.L. "Selected Computer Programs With Application to Infrared Spectroscopy", Digilab Users Group Conference, 1979.
33. Jones, R.N., Appl. Optics, 1969, 8, 597.
34. Jones, R.N., Pure Appl. Chem., 1969, 18, 303.
35. Jones, R.N., Seshadri, K.S., Jonathan, N.B.W., Hopkins, J. W. Can. J. Chem., 1963, 41, 750.
36. Fraser, R.D.B. and Suzuki, E. in Physical Principles and Techniques of Protein Chemistry. Part C, 301 (1973). S.J. Leach, Editor, Academic Press, New York.
37. Brown, J.K. and Ladner, W.R. Fuel, 1960, 39, 87.
38. Retcofsky, H.L. Applied Spectroscopy, 1977, 31, 116.
39. Durie, R.A., Shewshyk, Y. and Sternhell, S. Fuel, 1966, 45, 99.
40. Retcofsky, H.L. and Friedel, R.A. Fuel, 1968, 47, 487.
41. Tschamler, Von H. and Ruiter, E. de., Brennstoff-Chemie, 9, 280 (1963).
42. Osawa, Y. and Shih, J.W. Fuel, 1971, 50, 53.
43. Durie, R.A. and Sternhell, S. Australian J. Chem., 1959 12, 205.

RECEIVED April 30, 1982

Applications of Fourier Transform IR Spectroscopy in Fuel Science

P. R. SOLOMON, D. G. HAMBLEN, and R. M. CARANGELO

Advanced Fuel Research, Inc., East Hartford, CT 06108

Petroleum reserves are limited and supplies are becoming expensive and unstable. Extensive efforts are underway in the United States and abroad to find alternative sources for fuel and chemicals by employing coal, oil shale, tar sands, and biomass. But complicated processing of these raw materials is required to deliver environmentally acceptable products at a competitive price. It is, therefore, increasingly important to obtain better characterization of these raw materials and better understanding of their transformation in a process.

Fourier Transform Infrared (FT-IR) Spectroscopy is one of the most versatile techniques available for providing analytical data on the raw materials, the process chemistry and the products. Dispersive infrared spectroscopy has traditionally been an important tool in fuel characterization since most organic and mineral components absorb in the IR. Discussions of applications to coal may be found in Lowry ([1]), van Krevlen ([2]), Friedel ([3]), Brown ([4]), Brooks, Durie and Sternhell ([5]) Friedel and Retcofsky ([6]) and references cited therein. But FT-IR with its advantages in speed, sensitivity and data processing has added new dimensions.

The FT-IR permits rapid routine quantitative characterizations of solids, liquids and gases. The FT-IR's speed (a complete spectrum can be obtained in 80 msec) provides the possibility of following chemical transformations (such as coal pyrolysis) as it occurs or permits on-line analysis of products subject to separation techniques such as LC, GC or solvent separation. The FT-IR provides high sensitivity because of its high signal throughput and by co-adding spectra to produce good signal to noise. This feature permits measurement of highly absorbing materials such as coal or the use of difficult techniques such as photoacoustic or diffuse reflectance spectroscopy. The latter techniques allow measurement of solids with minimal sample preparation.

Other advantages of the FT-IR are digital storage of spectra and the availability of many data analysis routines which were developed to take advantage of the computer which an FT-IR requires. These routines permit such operations as base line corrections, smoothing, spectral comparisons, spectral synthesis, factor analysis, correlation techniques, solvent subtraction, mineral subtraction, display and plotting flexibility and programmed control of experiments. These techniques have proved so useful that dispersive instruments are now being offered with add-on computers.

As a result of the above advantages of FT-IR, investigators have started to reexamine applications in fuel science and technology. Painter et al. (7, 8) have proposed techniques for quantitatively analyzing mineral components in coals and low temperature ash by using subtraction routines and a quantitative mineral library. Painter also has used FT-IR for studying coal oxidation (9) and liquefaction products (10, 11). Solomon has considered the analysis of organic constituents. Calibration factors were determined for computing the aliphatic and aromatic hydrogen concentrations from the integrated areas under the peaks near 2900 cm^{-1} and 800 cm^{-1} respectively (12) and for determining the hydroxyl concentrations from the absorbance at 3200 cm^{-1} (13). From the aliphatic hydrogen concentration, a reasonable determination can also be made of the aliphatic and aromatic carbon concentration using the Brown-Ladner relation (14) and an assumed aliphatic stoichiometry. Applications of quantitative analysis of organic components have been applied to coal proximate analysis (15), oil shale yields (16), pyrolysis yields (17) and coal structure (18). Applications of FT-IR for analysis of pyrolysis gases has been discussed by Erickson et al (19) and by Solomon and co-workers (20-23). The FT-IR's rapid data acquisitions (80 m sec/scan) also allows pyrolysis kinetics to be followed (20 - 23).

Finally, the FT-IR system operates by coding the infrared source with an amplitude modulation which is unique to each infrared frequency. The detector is sensitive to the modulated radiation so that unmodulated stray radiation is eliminated from the experiment, permitting the use of the FT-IR as an in-situ detector in many experiments. For example, an FT-IR has been used to monitor the evolution of coal pyrolysis products within a drop tube furnace (24) and within an entrained flow reactor (25). The latter has been operated up to 1200°C.

These applications of FT-IR in fuel science are reviewed in the following pages. The paper describes techniques of sample preparation, quantitative analysis of spectra for solids, liquids and gases and applications to hydrocarbon analysis and conversion.

Apparatus

Most of the measurements described in this paper were made with a Nicolet model 7199 Fourier Transform Infrared Spectrometer using a Globar source and a liquid nitrogen cooled mercury-cadmium telluride detector. For solids and liquids, good quality spectra were obtained at 4 wavenumber resolution by co-adding 32 scans with the IR beam transmitted through the sample ratioed to a background of 32 scans co-added in the absence of the sample. Spectra are converted to absorbance since under the conditions of the measurement, Beer's Law is expected to hold and the absorbance is proportional to sample quantity. For gases, high resolution spectra were obtained at 0.5 wavenumbers and rapid scan data were obtained at 8 wavenumbers.

Analysis of Solid Samples

Preparation of KBr Pellets

Quantitative FT-IR transmission spectra of solids were obtained using finely ground samples pressed in KBr pellets. Ten to fifty mg of sample taken from a ground (100 mesh or finer), well mixed, representative sample were placed in a stainless steel grinding capsule, dried in vacuum for several hours, backfilled with dry nitrogen, sealed in the capsule and ground. The length of grinding time necessary to obtain particles which are sufficiently fine (for the IR radiation to penetrate) depends on the shaker, amount and characteristics of the sample and the grinding capsule. A good rule to follow is to grind so that further grinding does not change the intensity of the absorption. For most samples, 20 minutes using a "Wig -L-Bug" shaker is sufficient. Materials like tar sands which have large mineral grains and soft organic matter are among the most difficult to handle and must be ground cold (to freeze the organic matter).

A small sample (typically 1.0 mg but as low as 0.25 mg for high carbon content coals or chars) of this finely ground dry sample is weighed (to $\pm.01$ mg) in a dry box and added to a weighed amount (about 300 mg determined to $\pm$ 0.1 mg) of KBr. The KBr and coal are then mixed by grinding for 30 seconds and pressed into a pellet in an evacuated die under 20,000 lbs pressure. The pellet is then weighed and the sample weight per cm^2 of pellet area is determined.

The spectra for a coal (approximately 1 mg sample in 300 mg KBr) prepared in this manner are shown in Fig 1. The spectra for undried pellets show the presence of absorbed water (peaks at $3400 cm^{-1}$, $1640 cm^{-1}$ and $600 cm^{-1}$). The more KBr used and the longer the mixing duration, the larger the water peaks.

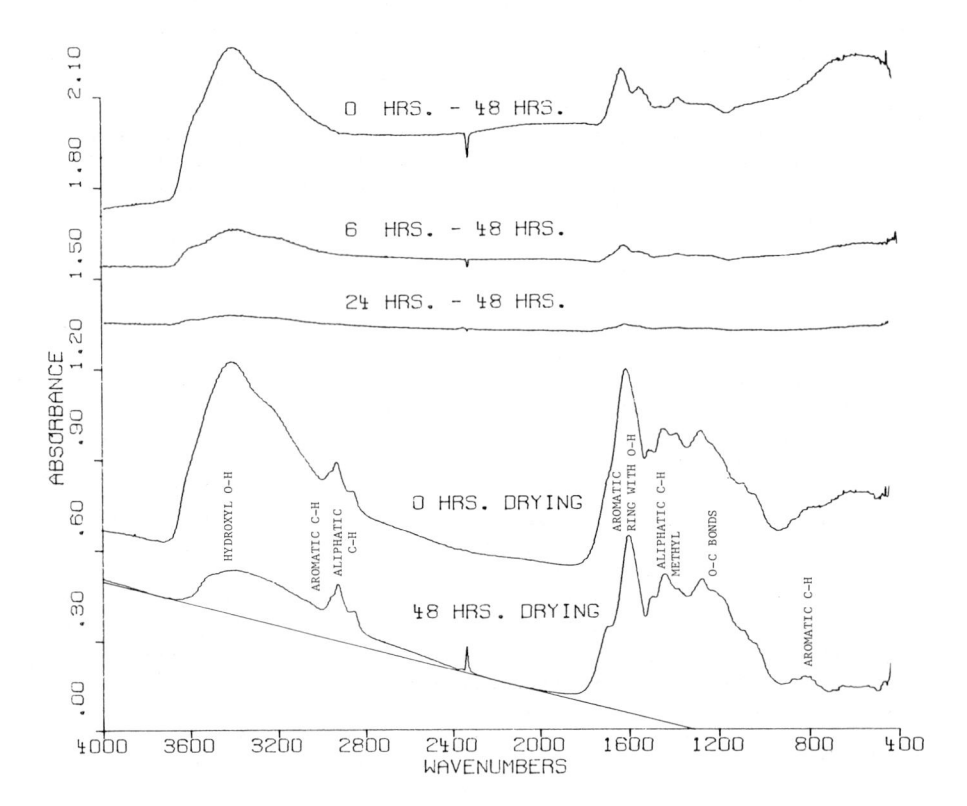

Figure 1. FTIR spectra of a North Dakota lignite showing the effects of drying a KBr pellet sample up to 48 hours.

Drying Pellets to Remove Water The problem of the KBr-H$_2$O bands is discussed in Refs. (1, 3, 13 and 26). They appear only after grinding the KBr. To remove them, investigators have tried heating the pellets with variable success. Friedel reported work of Tschamler which suggested that the complete removal of the KBr-H$_2$O bands occurred only when the pellets were measured at 175°C (3). Upon cooling, the bands reappeared, although at reduced intensity. Osawa and Shih dried pellets for 2 days at 60°C and found residual absorption attributable to water (27). Roberts reported that drying at 100 to 110°C diminished or eliminated the KBr-H$_2$O bands (28). Solomon and Carangelo achieved good results by drying at 110°C for 48 hours (13). For a pure KBr pellet, this procedure reduced the intensity of the 3400 cm^{-1} band from .06 absorbance units in the undried pellet to .01 absorbance units. For coal, these conditions are a good choice since 104° to 110°C is the specified drying temperature for determining coal moisture.

The effectiveness of heating the pellets depends on the pellet preparation procedure which incorporates the water and on the drying conditions which removes it. In considering the results of Ref. (13) it must be considered that the coals were first dried and ground and then mixed with the KBr using a 30 second grind so that available moisture from the coal is minimized and long grinding of the KBr does not occur.

The results of drying a coal pellet are illustrated in Fig. 1. The pellet was dried in vacuum at 105°C for 6, 12, 24, and 48 hours. The figure shows the spectra of the undried pellets, the spectra of the pellets dried for 48 hours and the difference between all the spectra and the 48 hour spectra. The difference spectra show the following features: 1) There are large water peaks in the coal pellets which are larger than in a KBr pellet without coal. The size of the peak increases with decreasing coal rank. 2) Drying for 48 hours substantially reduces the water absorptions at 3450, 1640 and 600 cm^{-1}. Longer times and higher temperatures produce little additional change in the spectra. 3) The drying also reduces the peak intensities at 1560 and 1360 cm^{-1}. 4) Drying produces no change in the peak intensities associated with aliphatic or aromatic hydrogen.

It is reasonable to assume that the spectra obtained after 48 hours of drying (at which point the spectra stop changing) are "moisture free" in the same sense that coal dried at 110°C until it stops loosing weight is defined to be "moisture free". The residue peak intensity contributed by the KBr is typically less than 10% of the intensity contributed by the coal (13).

Beer's Law To verify Beer's law, samples of coal were prepared using approximately 0.5, 1.0 and 1.5 mg of coal in each 300 mg pellet. To compare absorbances of the samples it is necessary to separate the effect of the absorption from that of scattering caused by the coal particles or by imperfections in the KBr pellet. Empirically, this scattering can be represented reasonably well by a straight line. The appropriateness of this straight line scattering correction is discussed in (13). To make the scattering correction, a line which is tangent to the spectrum near 3800 and 2000 cm^{-1} is subtracted from the spectrum over the region in which the base line has positive values of absorbance.

The scattering correction was applied to the spectra of the coal considered in Fig. 1. The corrected spectra have also been scaled to 1 mg (DAF) per cm^2 by multiplying by

$$f = \frac{100}{(100 - \%ASH)} \cdot \frac{A}{W_s \cdot W_p/(W_{KBr} + W_s)}$$

where A is the pellet area, W_s, W_{KBr} and W_p are the weights in mg of the coal sample, the KBr and the pellet, and % ASH is the weight percent ash determined by ASTM procedures. The results are presented in Fig. 2. The scaled spectra show some minor variations but in general are reproduceable within 5%. The most noticeable differences are in the KBr-H$_2$O regions especially in the .5 mg samples which, when scaled, make the KBr-H$_2$O more prominent. The near equality of the spectra for the three different sample weights indicates that Beer's law applies. Earlier studies reached the same conclusion (27, 29 - 31).

Photoacoustic Spectroscopy A more recent measurement technique for the analysis of fuels, particularly solid samples, is the photoacoustic (PAS) measurement (32). This technique allows IR spectra of solid samples to be obtained essentially without preparation. PAS should be considered whenever physical sample preparation is difficult or when an artifact of preparation is suspected to occur.

The photoacoustic effect in solids involves a thermal transfer from the solid sample to the surrounding gas. Energy absorbed from the IR radiation by the solid heats the surface of the solid, which in turn conductively heats a boundary layer of gas next to the solid surface. Thus, sound is generated at the frequency of the IR modulation and can be detected with a microphone. This signal can be processed in the same way as the modulated IR. Because of the different transformations in the forms of the energy, the PAS signals are much lower than in direct IR absorption measurements. Long data collection times are, therefore, required.

Diffuse Reflectance Spectroscopy A third technique for the study of solids by FT-IR is diffuse reflectance spectroscopy. The technique which was recently described by Fuller and Griffiths, (33, 34) allows good quality spectra to be obtained on neat powdered samples. The technique requires fine grinding of the samples so, in this aspect, it is more restrictive than PAS but is substantially faster than PAS so is applicable to following reactions. Fig. 3 compares spectra for a Pittsburgh seam coal using the three techniques.

Diffuse reflectance spectroscopy is particularly useful in studying reactions where the KBr pellet would restrict access to the coal. As an example, consider the exchange of deuterium for

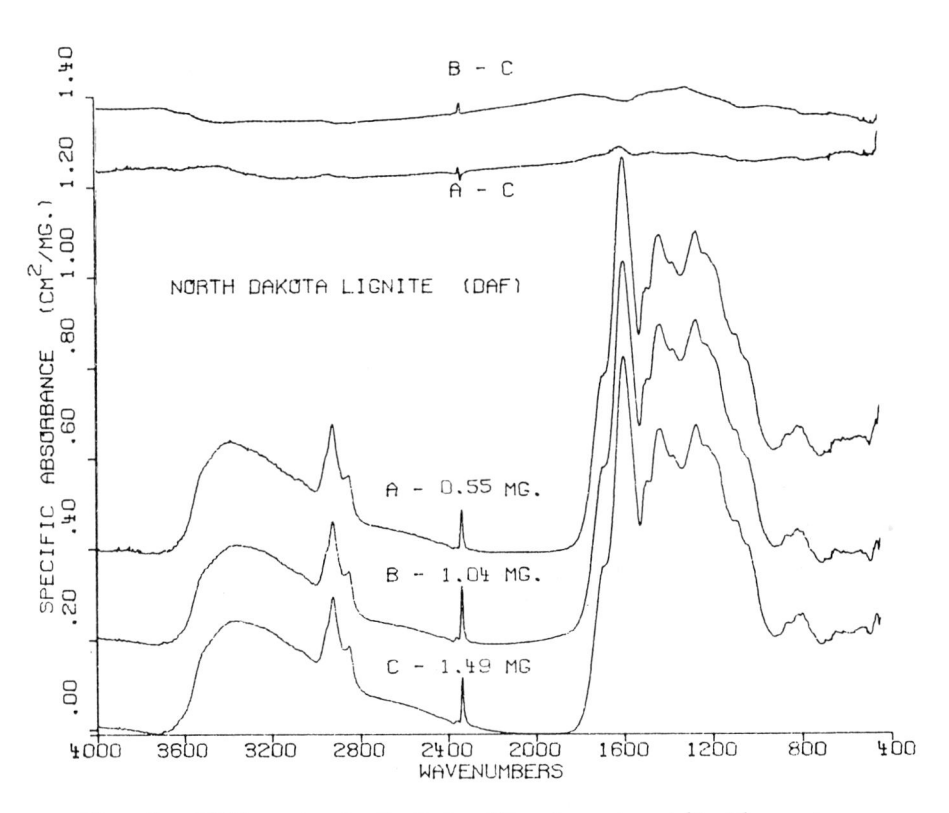

4. SOLOMON ET AL. *Applications of FTIR in Fuel Science* 83

Figure 2. FTIR spectra of a lignite for different sample weights. The spectra are normalized to 1 mg/cm^2.

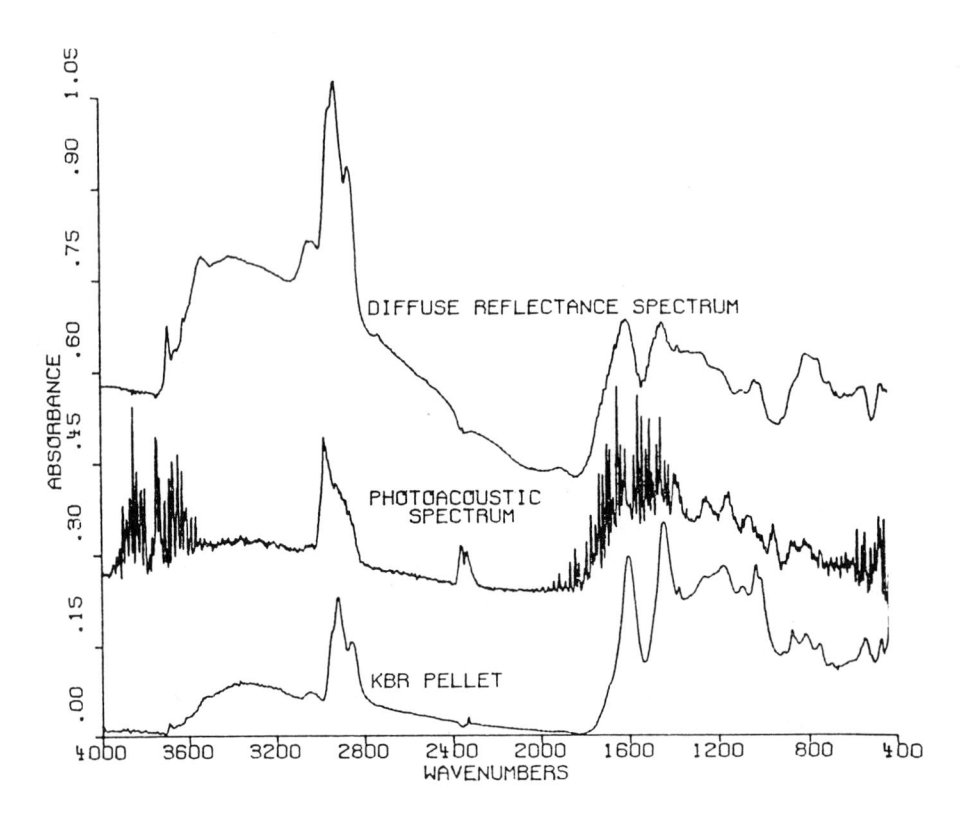

Figure 3. Spectra of a Pittsburgh seam coal by three techniques.

hydrogen on coal hydroxyl groups. This exchange may be accomplished in a few minutes by exposing powdered coal to D_2O vapor. Fig. 4 shows the diffuse reflectance spectrum of a dried finely ground coal powder, the spectrum of the same powder after exposure to D_2O vapor and then after exposure to room air. The upper and lower spectra are almost identical. The middle spectrum shows a decrease in the OH region (3400 wavenumbers) and an increase in the OD region (2600 wavenumbers). The spectra were obtained in about two minutes, permitting the kinetics of the exchange to be followed. Such deuterated coal samples have been used to follow the chemical reactions of the hydroxyl groups in pyrolysis by following the occurrence of the deuterium in the products (35).

<u>Mineral Analysis and Spectral Corrections</u> The FT-IR spectra are obtained in digital form so that corrections for particle scattering and mineral content may be easily made. A typical correction sequence is illustrated in Fig. 5. The upper curve in Fig. 5 is the uncorrected spectrum of a dried coal. It has a slope from 1000 to 4000 cm^{-1} due to particle scattering and contributions from the mineral components near 3,600, 1,000 and 450 cm^{-1}. Identification of the minerals may be made by reference to several previous studies (36 - 38). The bottom spectrum in Fig. 5a has had the mineral peaks removed by subtracting appropriate amounts of the reference spectra shown in Fig. 5b to produce a smooth spectrum in the region of the mineral peaks. Similar procedures for mineral analysis using FT-IR have recently been reported by Painter et al. (7, 8). The spectrum also has a straight line scattering correction and has been scaled to give the absorbance for 1 mg/cm^2 of dmmf coal.

A more accurate way of obtaining the mineral components is illustrated in Fig. 6 which is the spectrum and mineral analysis for the low temperature ash of the coal in Fig. 5.

<u>Spectral Synthesis</u> Much previous work has been done to identify the functional groups responsible for the observed peaks. Extensive references may be found in Lowry, (1), and van Krevelen (2). A convenient way to obtain areas under the absorbance peaks is to use a curve analysis program to synthesize the IR spectra (12). The synthesis is accomplished by adding absorption peaks with Gaussian or Lorenzian shapes and variable position, width, and height as shown in Fig. 7. The peaks are separated according to the indentified functional group. It has been determined that most of the fuel related products which were studied (coal, tar, char, oil shale and tar sand) could be synthesized by varying only the magnitudes of a set of peaks whose widths and positions were held constant. Experimentally, the best shape for most of the peaks is a Gaussian. The fits obtained for six coals of varying rank are illustrated in Fig. 8. The fits to other products and model compounds are shown in Fig. 9. For a typical simulation for coals and chars the absolute value of the area in the difference between the spectrum and the simulation is typically less than 3% of the spectrum. For tars, recycle solvents and oil shales this

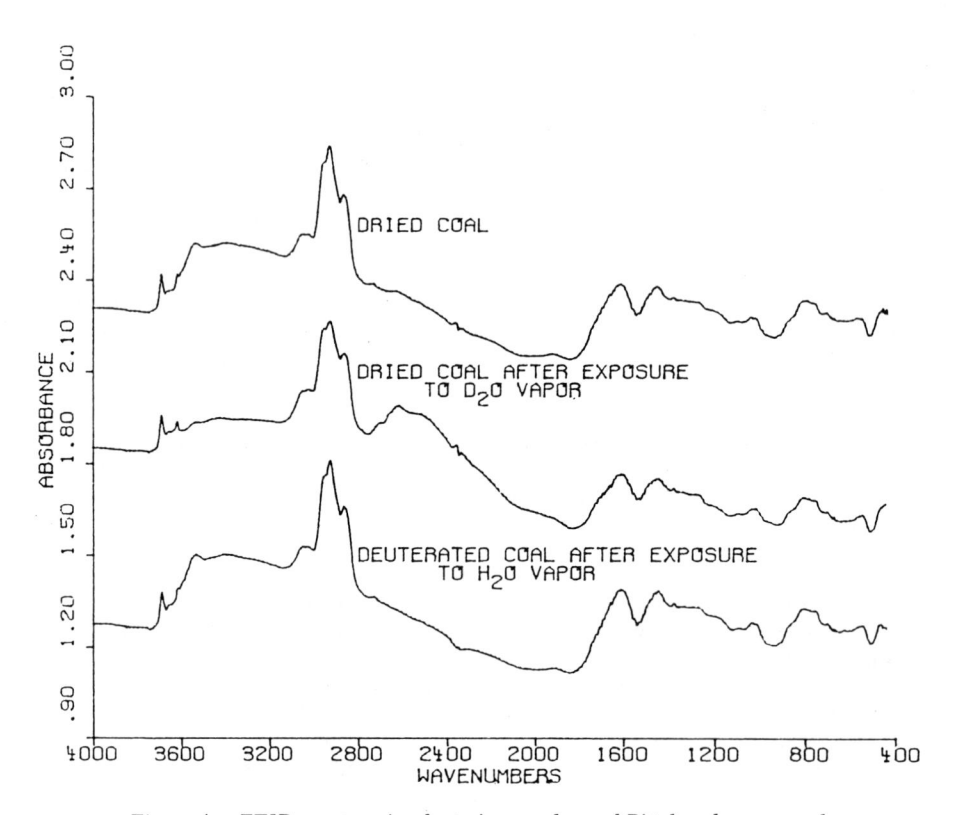

Figure 4. FTIR spectra of a deuterium-exchanged Pittsburgh seam coal.

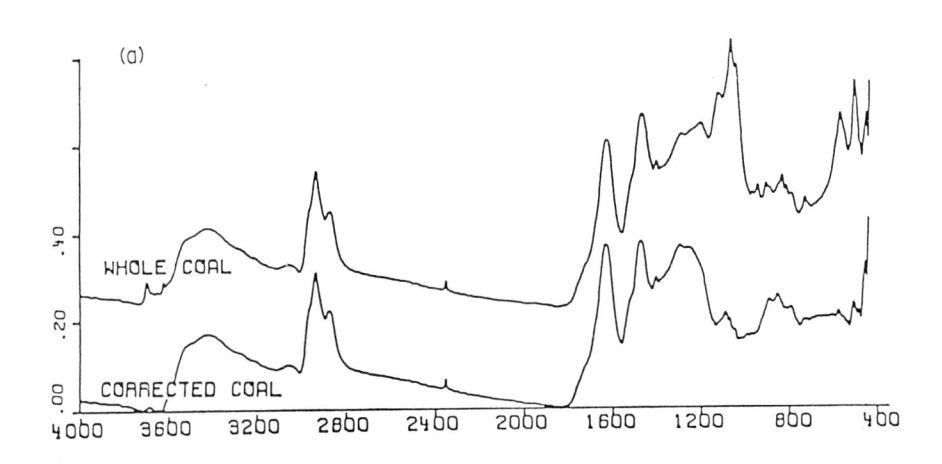

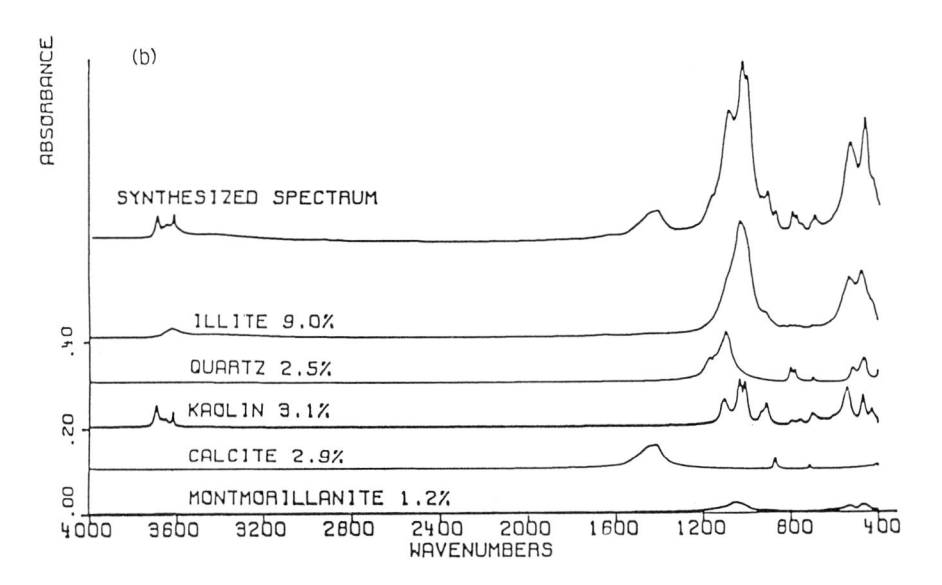

Figure 5. Correction of coal spectrum for scattering and minerals. Key: a, correction of coal spectrum; b, determination of mineral spectrum by addition of reference spectra.

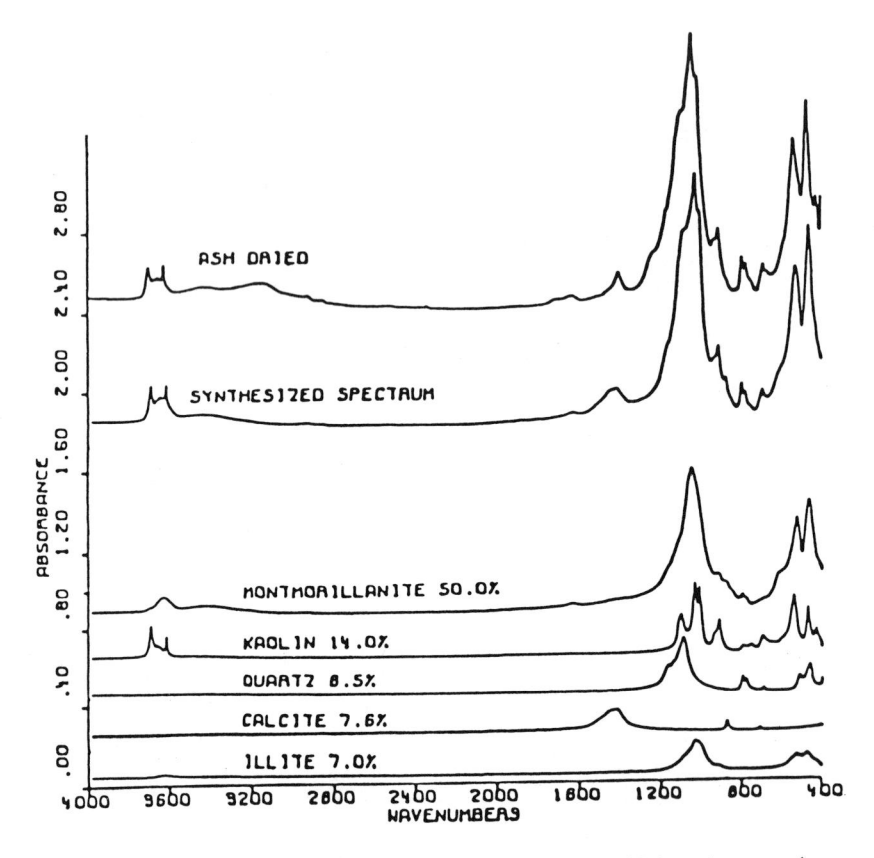

Figure 6. Synthesis of low temperature ash spectrum by addition of spectra from mineral library.

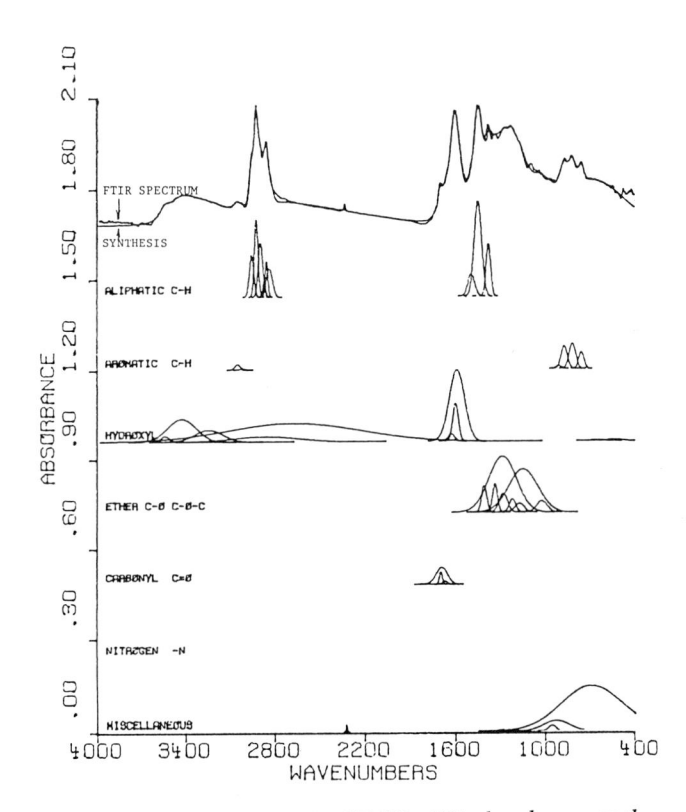

Figure 7. Spectral synthesis of FTIR of Pittsburgh seam coal.

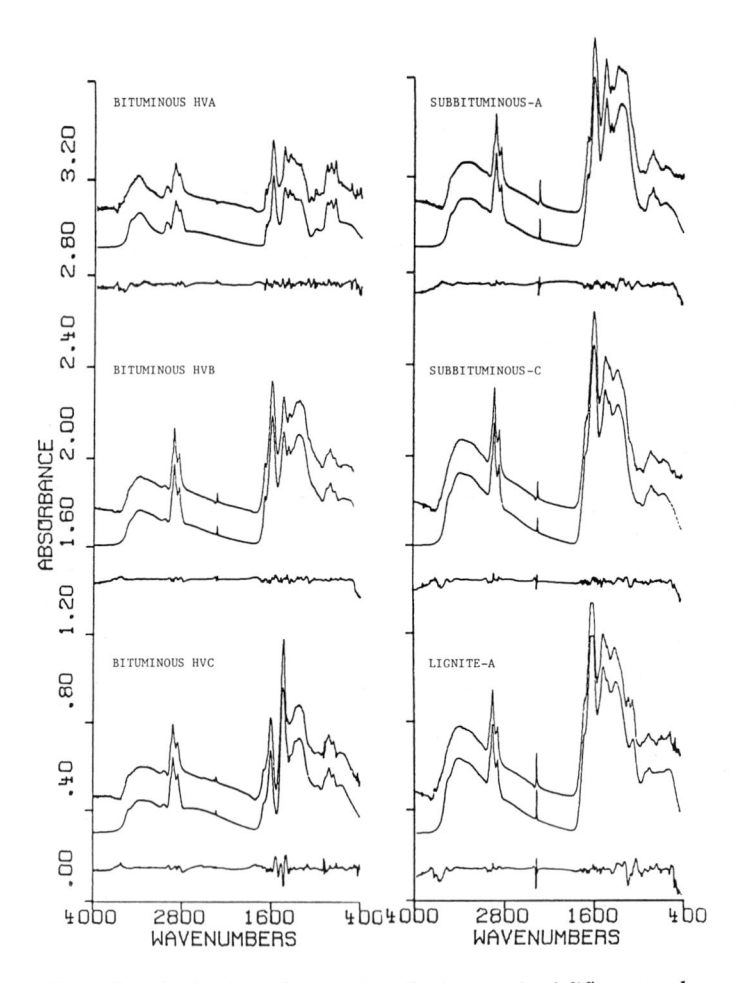

Figure 8. Application of spectral synthesis to coals of different rank.

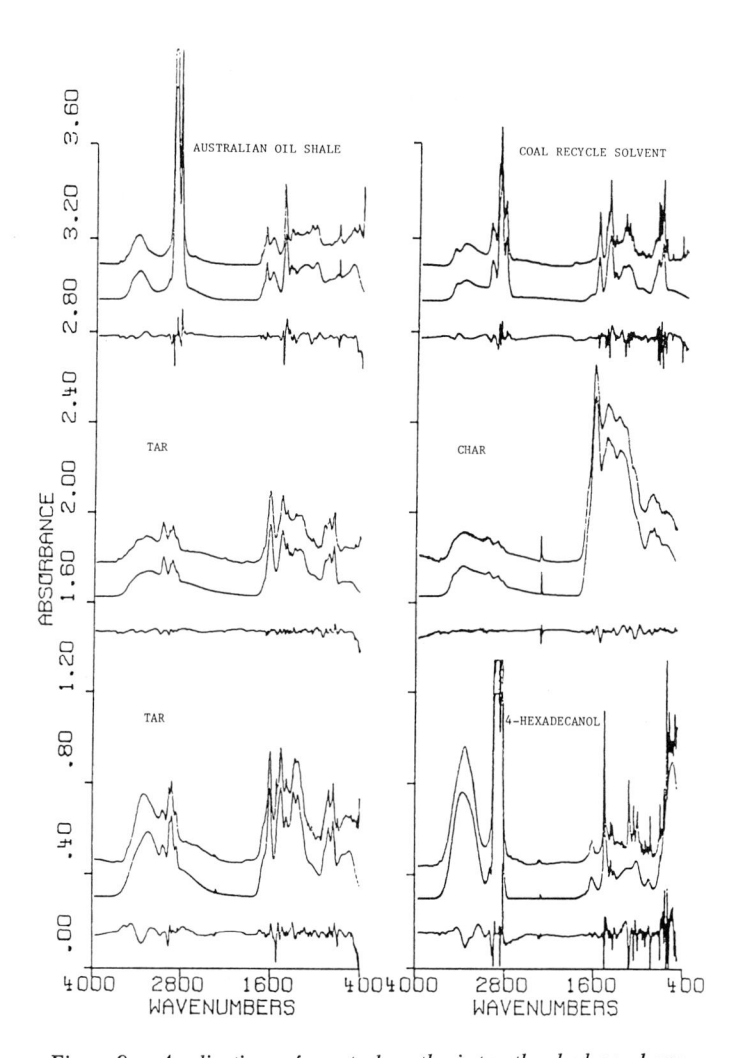

4. SOLOMON ET AL. *Applications of FTIR in Fuel Science* 91

Figure 9. Applications of spectral synthesis to other hydrocarbons.

value increases to 5% but was as high as 14% in one case. Optimum use of such a fitting procedure requires a set of calibration constants which relates the distribution of peaks in the synthesis to the absorptivities of functional groups in the sample. This is the subject of a continuing investigation.

The Hydroxyl Concentration The infrared measurement of the hydroxyl concentration of coals prepared in KBr pellets was discussed by Osawa and Shih (27). These authors used the absorption of the O-H stretch at 3450 cm^{-1}. Two major problems in making this determination are the overlap of a band due to water in the coal or water absorbed in preparing the KBr pellet and scattering which also affects this region of the spectrum. To overcome these problems Osawa and Shih did the following: 1) obtained spectra of coal in KBr pellets after drying the pellets to reduce the water absorption; 2) obtained spectra at several sample sizes and determined "specific extinction coefficients" (more commonly called absorptivity) from the slopes of absorbance vs sample size and 3) used a linear base line correction for scattering. Applying these techniques, they found a linear relation between the absorptivity at 3450 cm^{-1} and the hydroxyl content of the coal determined chemically. Many of Osawa and Shih's suggestions were subsequently used in measurements made by Solomon (12). Painter and co-workers (26) have recently questioned several aspects of these procedures. For the hydroxyl determination these authors have suggested an alternative orginally used by Durie and Sternhell (29), which employs acetylation of the coal and analysis of the difference between spectra of raw and acetylated coal. This procedure has advantages in avoiding the water problem and in distinguishing different types of OH groups but is time consuming. The determination of hydroxyl directly from the absorbance in the OH stretch region would be much simpler so it is worthwhile to evaluate its accuracy. This was done recently by Solomon and Carangelo (13). The results are summarized below.

As seen in the coal spectra the hydroxyl O-H stretch absorbance appears to be a broad band stretching from 3600 to 2000 cm^{-1}. The broadness of the band has been attributed to hydrogen bonding. It is important to be sure that the band shape is real and not an artifact of the scattering correction. Several observations indicate that the band shape is real. 1) Similar band shapes were observed by Friedel and Queiser (39) using thin sections of coal which don't exhibit scattering and by Brown (4) using nujol mulls. 2) Similar band shapes are also observed in non scattering spectra of vacuum distilled coal tar melted onto a KBr blank (12). 3) Spectra for coal may also be obtained by diffuse reflectance spectroscopy (33, 34) and by photoacoustic spectroscopy (32) which don't require KBr. Spectra of a Pittsburgh seam coal taken using these methods are compared in Fig. 3. All three spectra show the broad hydroxyl absorbance. 4) It is possible to change the shape of the hydroxyl absorbances by allowing the hydrogen to exchange with deuterium. This can be done by exposing the coal to D_2O vapor. Fig. 4 shows the diffuse reflection spectrum of a dry coal and the coal after exposure for 1/2 hour to D_2O. The broad OH absorption peak is drastically reduced and the O-D peak at 2600 cm^{-1} appears. 5) The broad peak is also drastically reduced by acetylation (Durie and Sternhell (29)) and by alkylation (Liota (40)).

To verify that the broadness in the peak is due to hydrogen bonding, spectra of a tar sample coated on a KBr pellet were obtained at elevated temperatures (12). As expected for hydrogen bonding, the absorption for bonded OH groups (3200 - 2400 cm^{-1}) decreases while that for free OH increases. The change in the high temperature spectrum was not permanent as the room temperature spectrum returned after cooling.

Osawa and Shih found a good linear relationship between specific extinction coefficients and hydroxyl concentrations determined using an acetylation method. The correlation was especially good when considering only Japanese coals. In the work of Solomon and Carangelo (13) the correlation was made using samples from the Pennsylvania State University coal bank. Hydroxyl concentrations were determined chemically by Yarzab, Abdel-Baset and Given (41). The correlation was determined for the absorbance at 3200 cm^{-1} rather than 3450 used by Osawa and Shih (27) to avoid problems of residual KBr water left after drying. The broad OH absorbance in coal has an intensity at 3200 similar to that at 3450 while the intensity due to water in KBr is close to zero. The absorbances at 3200 cm^{-1} for three coals are plotted as functions of coal amount in Fig. 10. The variation of absorbance is linear in sample amount. The slope of the line gives a(3200), the absorptivity at 3200 cm^{-1} in (abs. units/cm^2/mg). The correlation of hydroxyl content and absorptivity is illustrated in Fig. 11.

The hydroxyl content computed for 46 coals using the correlation of Fig. 11 is illustrated in Fig. 12 as a function of carbon content. The data are for coals from the Exxon coal sample bank which were analyzed recently. The Exxon coals which were sealed in nitrogen when prepared were analyzed immediately within days after breaking the seal. Except for the coals with high hydroxyl content, the Exxon coals appear to lie in a tight band which is slightly lower than the average values for the Penn State coals. The results are, however, different from the chemical determinations of Osawa and Shih (27) and of Blom (42).

The results indicate that hydroxyl oxygen concentrations in coal can be determined by FT-IR with an accuracy of about 10% compared to the chemically determined standards. The consistency of the chemical determination of OH should, however, be improved.

<u>Aliphatic and Aromatic Hydrogen</u> A calibration for the aliphatic peaks near 2900 cm^{-1} and the aromatic peaks near 800 cm^{-1} was obtained by Solomon (12). The objective was to determine the values of the integral absorptivities a'(al) and a'(ar) (in abs. units cm^{-1}/mg/cm^2), which relate peak areas (in abs. units cm^{-1}) to the hydrogen concentration (in mg/cm^2), i.e.,

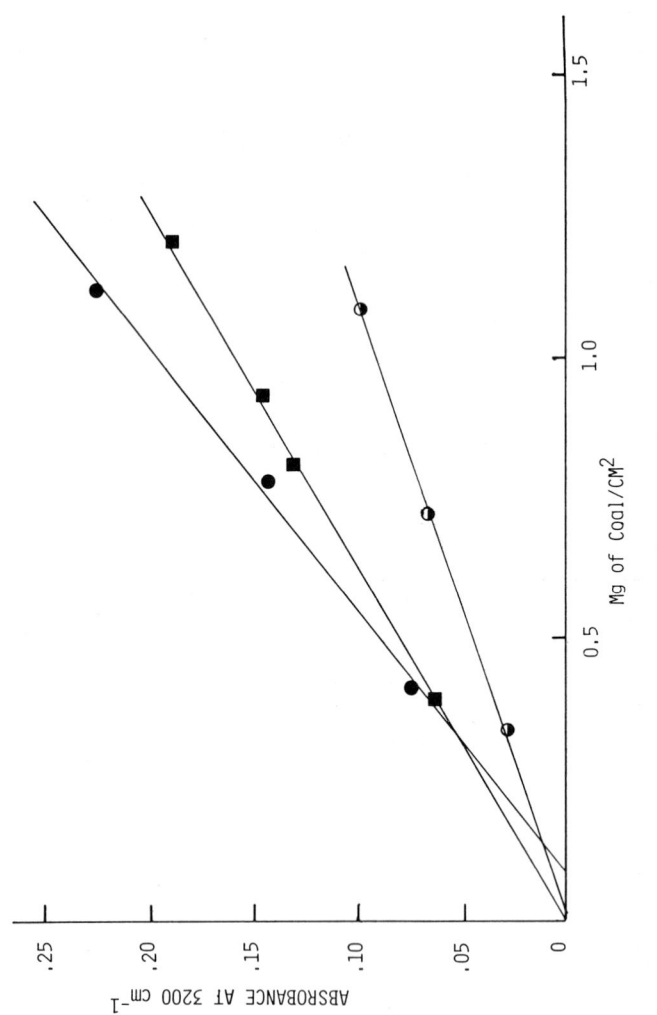

Figure 10. Variation of absorbance at 3200 cm^{-1} with sample amount. Key: ●, lignite; ■, sub-bituminous; and ◐, bituminous. (Reproduced, with permission, from Ref. 13.)

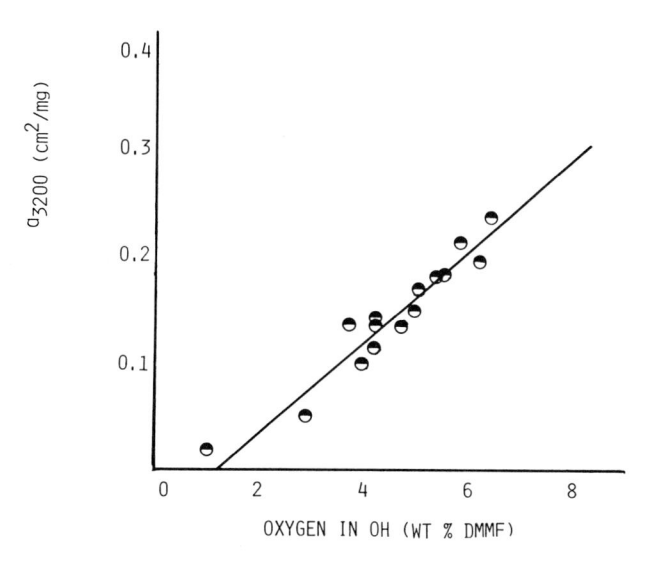

Figure 11. Variation of absorptivity with hydroxyl concentration. (Reproduced, with permission, from Ref. 13.)

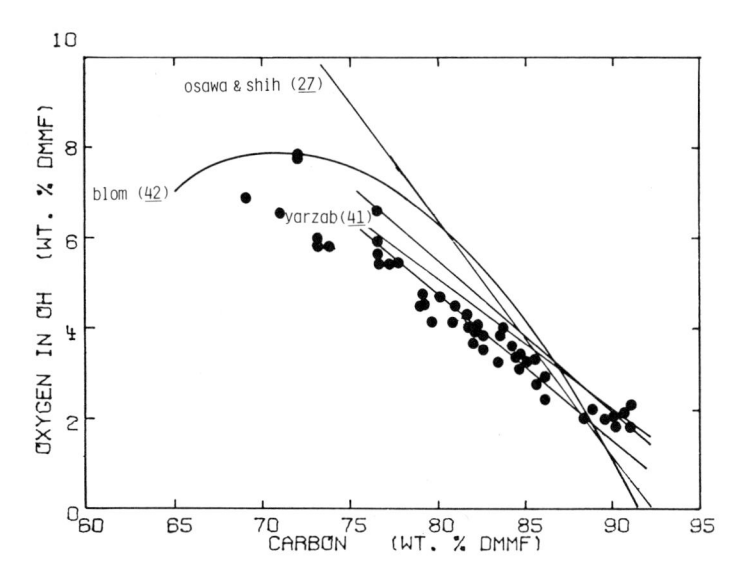

Figure 12. Variation of hydroxyl concentration determined by FTIR with carbon concentration. (Reproduced, with permission, from Ref. 13.)

$$A(al) = a'(al) H(al)$$

and

$$A(ar) = a'(ar) H(ar)$$

where $A(al)$ is the area under the aliphatic peaks, $A(ar)$ is the area under the aromatic peaks and $H(al)$ and $H(ar)$ the aliphatic (or hydroaromatic) and aromatic hydrogen concentrations in the sample, respectively. The equation for total hydrogen concentration $H(total) = H(al) + H(ar) + H(hydroxyl)$ may be combined with the above equations to yield,

$$\frac{1}{a'(al)}\left[\frac{A(al)}{H(total) - H(hydroxyl)}\right] = 1 - \frac{1}{a'(ar)}\left[\frac{A(ar)}{H(total) - H(hydroxyl)}\right]$$

where $H(hydroxyl)$ is the percent hydrogen in hydroxyl groups. If $a'(al)$ and $a'(ar)$ are constant for all products than plotting one term in the square brackets against the other should yield a straight line with intercepts $a'(al)$ and $a'(ar)$. In Ref. (12) the above analysis was performed for a group of coals and model compounds. Peak areas were determined using a spectral synthesis routine employing 26 Gaussian peaks as described above. Hydroxyl concentrations were determined from the infrared spectra using the absorptivity of Osawa and Shih (27). The results indicate $a'(al)$ and $a'(ar)$ to be constant for most coal products as well as many model compounds containing aromatic rings. Exceptions were long aliphatic chain model compounds, and two high aliphatic coals and a low temperature tar. For these samples, it appears that longer aliphatic chains result in a higher absorbance per hydrogen atom. The integral absorptivities obtained were

$$a'(al) = 900 \text{ and } a'(ar) = 800$$

The above analysis was applied to a wider group of coals (including 44 coals from the Exxon Library and a number of coals from the Pennsylvania State University coal bank) and several sets of chars, tars and recycle solvents. Several improvements were also made in the spectral synthesis routine.

The results are shown in Fig. 13. Fig. 13a compares results for bituminous coals and their chars and tars. For this group of materials the integral absorptivities, determined from the regression analysis are

$$a'(al) = 746$$
$$a'(ar) = 686$$

Fig. 13b contains results for subbituminous coals and lignites and

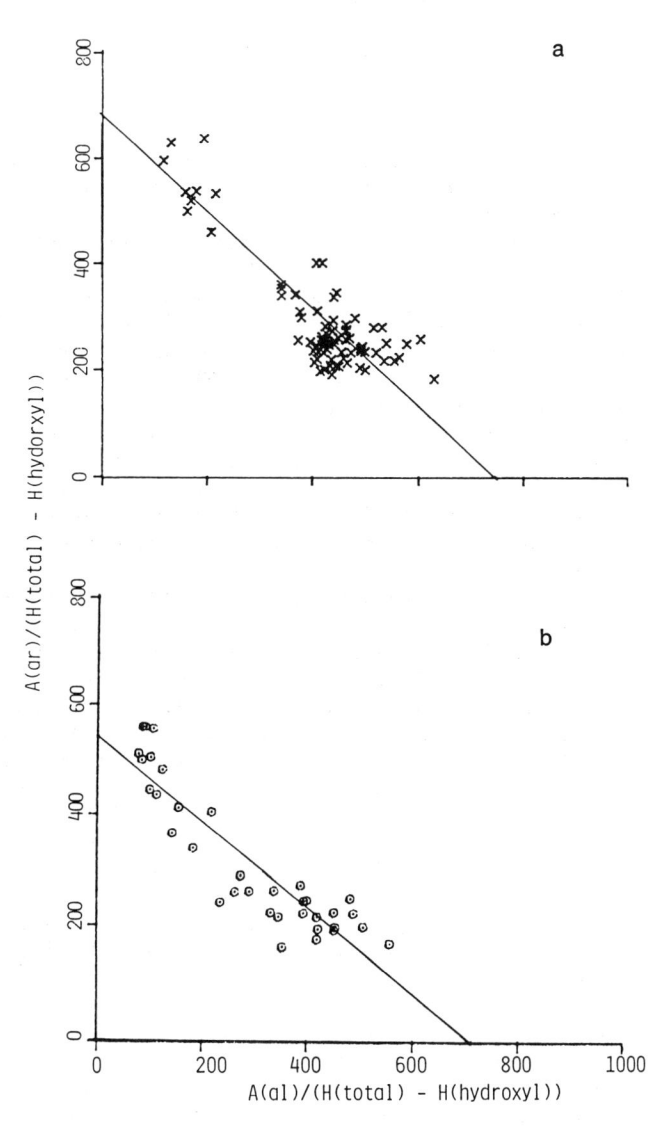

Figure 13. Regression analysis to determine aromatic and aliphatic absorptivities. Key: a, bituminous coals and products; and b, lignite and subbituminous coals and products.

their chars. The data lie in a band which is lower than for the bituminous (Fig. 13a) indicating some variations in absorptivity with rank. The integral absorptivities are

$$a'(al) = 710$$
$$\text{and}$$
$$a'(ar) = 541$$

The integral absorptivities are lower than those previously obtained by Solomon (12) even though the bituminous group contains many of the coals originally used. The major reason for the difference is variations in the choice of component peaks to use for the spectral synthesis. A parallel problem is the choice of integration limits and the base line when direct integration is used to obtain peak areas. Also, the previous work contained only one lignite and hydrogen values were obtained on vacuum dried samples by an elemental analyzer. These values were typically lower than those obtained by ASTM procedurds. In reanalyzing these coals the ASTM hydrogen values were used for consistency.

In view of the possible variations in absorptivity with the choice of peaks, integration limits or base line, it is essential to use the same methods for determining peak areas for both the calibration of absorptivities and the determination of hydrogen concentrations.

Fig. 14 compares values of H(al) + H(ar) determined using the integral absorptivities for the bituminous coal (a'(al) = 746 and a'(ar) = 686) and for the subbituminous coals and lignites (a'(al) = 710 and a'(ar) = 541) with H(total) - H(hydroxyl).

The integral absorptivities for aliphatic and aromatic hydrogen appear accurate for most coals. As in Ref. (12) exceptions are seen for some materials with high aliphatic or hydroaromatic concentrations (above 4 to 5%). These include alginite coals, some tars, oil shales, oils and coal liquefaction recycle solvents. These compounds fall below the parity line. The aliphatic line shapes for these materials appears to be higher and narrower and the absorptivities are larger than those for most of the coals and chars considered above. Model compounds containing long aliphatic chains or having extensive hydroaromaticity show similar shapes and have similar high absorptivities (47).

The aromatic hydrogen values computed, using the above absorptivities, are in agreement with those for coals of similar rank determined by van Krevelen (43) and by Mazumdar, et al. (44) from pyrolysis measurements but are larger (by about a factor of 2) than those determined by Brown (4) using IR techniques for the peaks near 3100 cm^{-1}.

Determination of Carbon Functional Group Concentrations   The concentration of aromatic and aliphatic carbons may be obtained using some simple assumptions. The stoichiometry of the aliphatic portion of the sample can be estimated and C(ar) can be calculated using a method suggested by Brown and Ladner (14). The method determines C(ar) by difference.

$$C(ar) = C(total) - C(al)$$

$$= C(total) - (12/x) H(al)$$

Where C(total) is the total carbon concentration and x is a parameter which describes the stoichiometry of the aliphatic material CH_x. The accuracy of the method has recently been verified by Retcofsky et al. (45) for coal derived liquids. These authors obtained excellent agreement between values of C(ar) determined directly from ^{13}C cross polarization NMR and C(ar) computed from the Brown-Ladner relation using x = 2 and H(al) determined from ^{1}H-NMR. For coals, a value of x = 1.8 is reasonable. This is based on estimates of the aliphatic stoichiometry made by several investigators, a summary of which appears in Ref. (15).

<u>Comparision with NMR</u> To check the accuracy of the determination of aromatic and aliphatic carbons, eighteen coals for which quantitative FT-IR data were obtained were also studied by ^{13}C NMR by Bernard Gerstein et al., (46) at Ames Laboratory. The ratios of aromatic to total carbon obtained by the two methods are compared in Fig. 15. There is good agreement between results of the two methods.

<u>Model Compounds</u>

A library of 156 model compounds of interest in coal has been created under EPRI and DOE sponsored programs (47, 48). Quantitative FT-IR spectra were obtained in duplicate for each compound. The compounds which have known molecular structure were used to obtain integral absorptivities for functional groups of interest. In working with coal and coal liquids which produce broad infrared absorptions because of their inhomogeneity, it is preferable to work with integrated areas rather than peak heights. For each compound, the integral absorptivity a'(i) in Abs. units $cm^{-1}/(mg/cm^2)$ was determined which relates the integrated area A(i) in absorbance units times wavenumber to the concentration of the absorbing species C(i) in the sample in mg/cm^2 by the equation

$$A(i) = a'(i)C(i)$$

Table I summarizes the absorbing groups and the regions of the spectrum considered and provides a summary of the average integral absorptivities with 90% confidence limits. A number of conclusions may be drawn from these data.

<u>Aliphatic and Aromatic Hydrogen</u> The average integral absorptivities for H(al) and H(ar) are

$$a'(al) = 963 \pm 109 \quad \text{and} \quad a'(ar) = 768 \pm 85$$

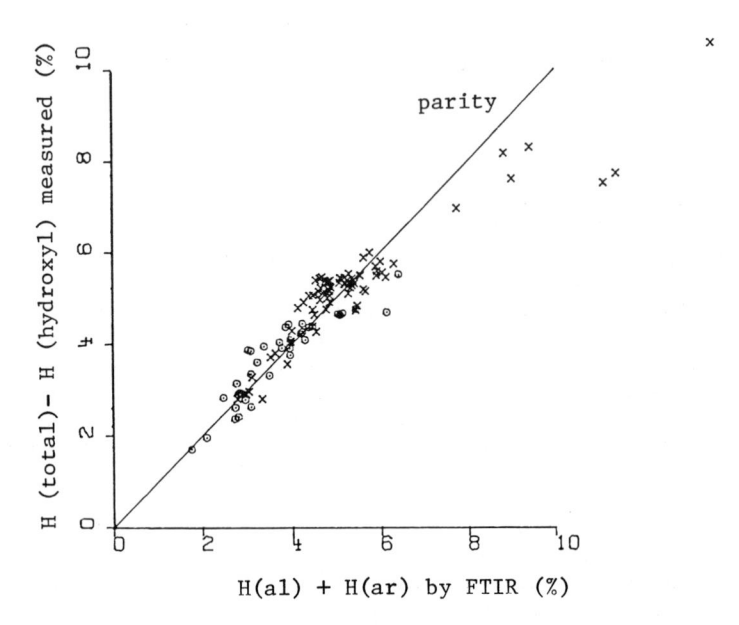

Figure 14. Comparison of hydrogen concentrations determined chemically with those determined by FTIR. Key: ⊙, *subbituminous coals, lignites, and products; and* ×, *bituminous coals and products.*

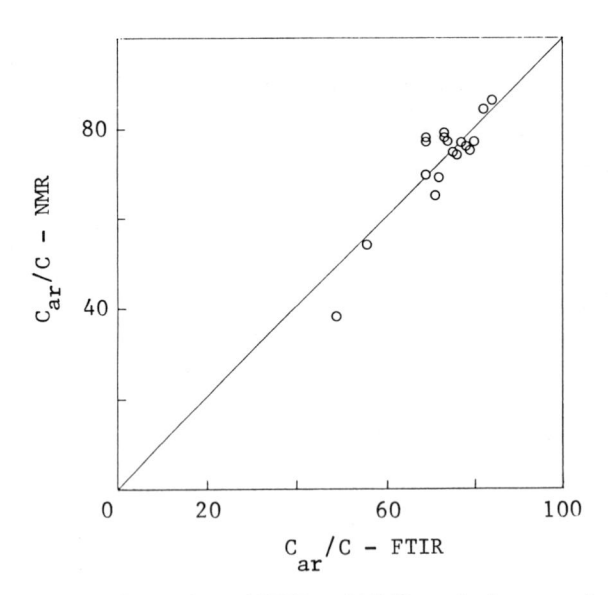

Figure 15. Comparison of FTIR and NMR results for aromatic carbon.

TABLE I

INTEGRATION LIMITS AND MEAN EXTINCTION COEFFICIENTS FOR ABSORBING GROUPS OF 156 MODEL COMPOUNDS

REGION	TITLE	BASELINE LIMITS	INTEGRATION LIMITS	INTEGRAL ABSORPTIVITY
1	H(al)	2600 TO 3120	2800 TO 3000	963.189 +/- 108.975
2	H(ar)	2600 TO 3120	3000 TO 3120	242.693 +/- 31.5441
3	H(ar) WAG	665 TO 925	680 TO 920	768.354 +/- 85.9592
4	N-Hx	3140 TO 3600	3140 TO 3600	7443.55 +/- 4793.54
5	O-C(al)	NONE	1000 TO 1200	1078.05 +/- 266.587
6	O-C(ar)	NONE	1160 TO 1320	1181.72 +/- 285.379
7	C=C ar Rng	NONE	1525 TO 1700	57.9966 +/- 17.0062
8	H-C (str)	2600 TO 3120	2800 TO 3000	9876.80 +/- 6586.99
9	H2-C= str	2600 TO 3120	2800 TO 3000	1338.76 +/- 188.994
10	H3-C- str	2600 TO 3120	2800 TO 3000	3524.42 +/- 1679.72
11	H3-C (wag)	1360 TO 1400	1360 TO 1400	69.6250 +/- 22.3508
12	H-O-	NONE	3200 TO 3600	22797.5 +/- 3982.39
13	O=C	NONE	1640 TO 1760	603.513 +/- 209.718
14	H-S- str	2500 TO 2600	2500 TO 2600	568.171
15	=N- rng	NONE	NONE	
16	C(al)-C	NONE	NONE	
17	N other	NONE	NONE	
18	-S-	NONE	NONE	

The average aromatic absorptivity for model compounds (a'(ar) = 768) are higher than those derived for bituminous coals and products ('(ar) = 686) and for lignite and subbituminous coals and products (a'(ar) = 541). The comparison is similar for the average aliphatic absorptivity for model compounds a'(al) = 900 and those derived for bituminous (a'(al) = 746) and lignite and subbituminous (a'(al) = 710) coals and products. Examination of the data show that compounds with CH_2 groups (either aliphatic or hydroaromatic) removed from aromatic carbons have the highest integral absorptivities (1060 $\pm$ 147) in the 2900 cm^{-1} region compared with methyl groups (753 $\pm$ 159) or CH_2 groups attached to aromatic carbons (778 $\pm$ 213). The latter values are in better agreement with most of the coals while the former agrees with the products with high aliphatic or hydroaromatic concentrations.

The 1600 cm^{-1} Absorption The absorption at 1600 cm^{-1} is characteristic of an aromatic ring stretching mode. The average absorptivity of a' = 7.4 for pure hydrocarbon species is quite low and would not explain the large absorption in coals and coal liquids. The presence of hydroxyls attached to the ring enhances this absorption. The average integral absorptivity goes up to a' = 41.6. The largest absorptivity is, however, obtained by the combination of an attached hydroxyl and a nitrogen in the ring a' = 187. This combination also produces strong absorption in the region between 2800 cm^{-1} and 2000 cm^{-1} characteristic of a strong hydrogen bonding shift in the OH absorption discussed below.

Ring Nitrogen Compounds In the absence of hydroxyls, or other hydrogens which can bond to the nitrogen, the spectra for ring nitrogen compounds are similar to their hydrocarbon analogs. Fig. 16 compares such a pair. Substantial changes occur however, when hydrogen bonding occurs. As an example, Fig. 17 compares Napthol with 4-hydroxy quinoline. The 1600 cm^{-1} absorption becomes enormous and the free O-H absorption at 3450 is shifted to 2800 and increased in intensity and width.

Hydrogen Bonding O-H absorption 2800 to 2000 cm^{-1} Coals and derived liquids produce a substantial absorption below 2800 cm^{-1} which is believed to be due to a shift in the O-H absorbance due to hydrogen bonding. Examination of the model compound spectra shows the absorptions are all small for the region 2800 cm^{-1} to 2000 cm^{-1} except for 4-5 dihydroxy phenanthrene which has neighboring OH groups and for compounds where hydrogen bonding to a ring nitrogen is possible (47). The average integral absorptivity for compounds with OH but without nitrogen is a' = 5400. With nitrogen the average is a' = 16,000. This interesting observation indicates the possible importance of nitrogen in hydrogen bonding and suggests further study.

Library Search Routines To aid in the identification of hydrocarbons, computer programs are being developed for comparing unknown spectra to libraries of known compounds. As an example, a sample of 2-hexadecanol was compared to a library of over 100 compounds. The library search identified 2-hexadecanol as the

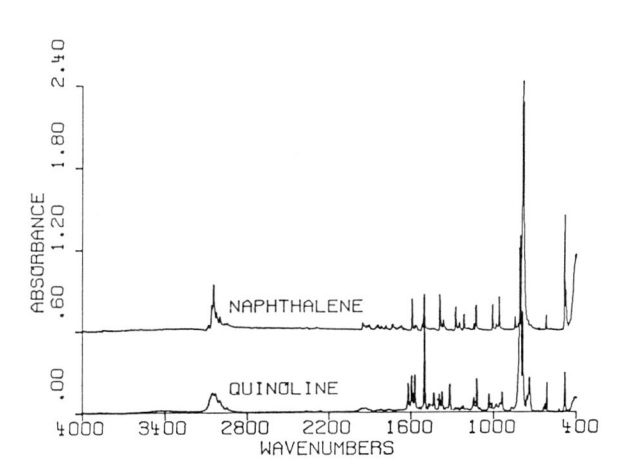

Figure 16. Comparison of the FTIR spectra of naphthalene and quinoline. (Reproduced, with permission, from Ref. 47.)

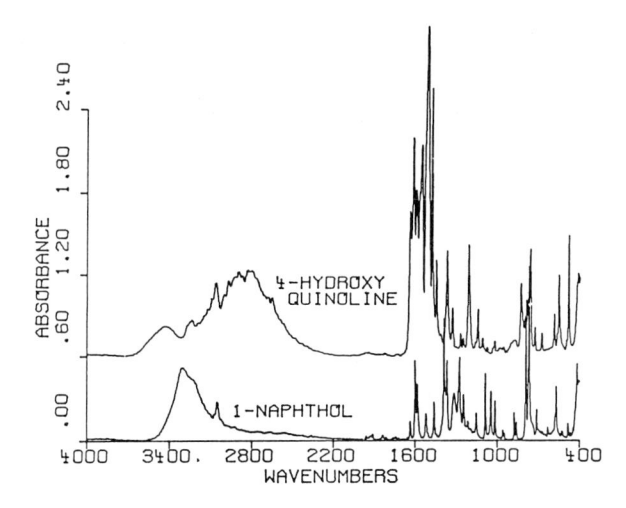

Figure 17. Comparison of the FTIR spectra of 4-hydroxyquinoline and 1-naphthol. (Reproduced, with permission, from Ref. 47.)

best fit but also identified the other spectra in Fig. 18 which have similarities in structure to 2-hexadecanol as close fits.

Such routines work well for pure compounds but not for mixtures. To overcome the problems of identifying components of a mixture, a different routine was developed (47). In this program, a model compound is compared with an unknown mixture in the following way. The important peaks in the known are identified and a comparision with the unknown is made to obtain a ratio of the intensity of the peaks in the unknown to the peaks in the known. It is assumed that the known is included as a component of the unknown when more than 50% of the known peaks appear in the unknown with a consistent ratio. This ratio is the percentage of the known component.

As an example of this procedure, a spectrum obtained by mixing the spectra of 3 model compounds was used as an unknown. The three spectra and the spectrum of the mixture are presented in Fig. 19. The program picked the three components as the top choices and did a good job of picking the concentrations.

Liquids

Quantitative Analysis Quantitative analysis of liquids may be obtained by using a cell of known thickness and determining the sample density. A cell of 10 to 20 microns is adequate for most hydrocarbon liquids of interest. The cell width may be calibrated using interference fringes. Results for a coal recycle solvent are illustrated in Fig. 20 (plotted in abs. units/mg per 1.33 cm^2). A KBr pellet was prepared for the same solvent for comparisons, Fig. 21. The two spectra agree within 5%. The pellet spectrum has an extra peak near 3500 wavenumbers due to water absorbed by the KBr.

LC/FT-IR Liquid chromatographic separation of materials with FT-IR detection is a good method to characterize heavy hydrocarbon products. The optical design and computational power of an FT-IR can be utilized in an integrated LC/FT-IR system for routine analysis of extremely complex samples. An integrated LC/FT-IR system has been described by Vidrine and Mattson (49). A simple flow cell with a very small internal volume can be used for detection so there is no peak-spreading and other detectors can be used simultaneously. Chromatographic fractions of interest can be easily collected for further analysis. In operating the system, a complete FT-IR spectrum can be obtained every 100 milliseconds as the fractions enter the cell. The FT-IR spectra so obtained may subsequently be compared with a library of model compounds. Application of LC/FT-IR to coal liquids has been described by Brown et al. (50, 51).

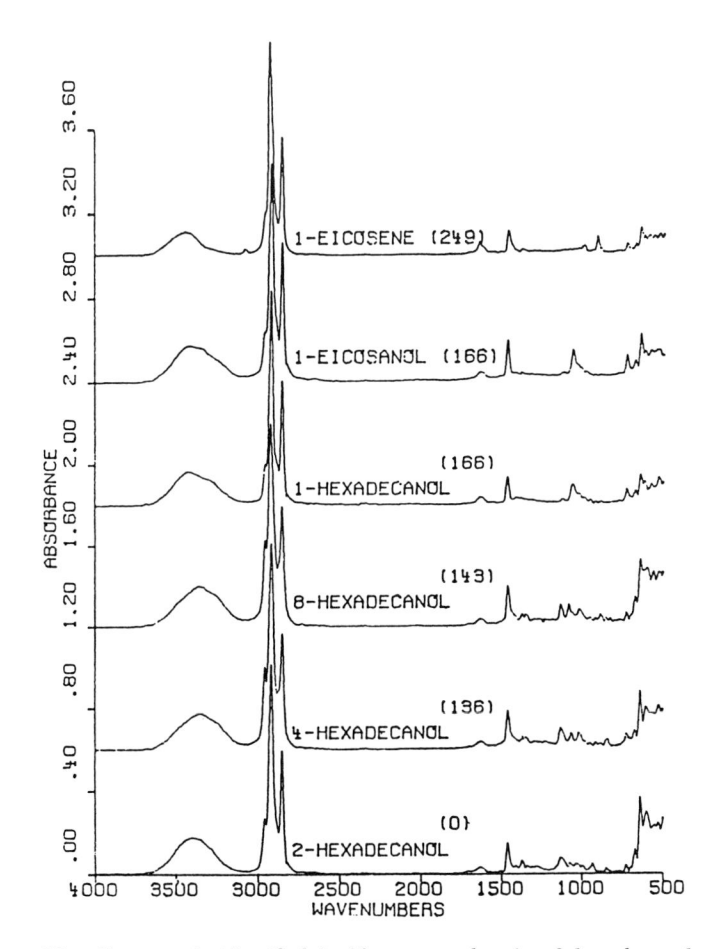

Figure 18. Compounds identified in library search using 2-hexadecanol as the unknown. (Reproduced, with permission, from Ref. 47.)

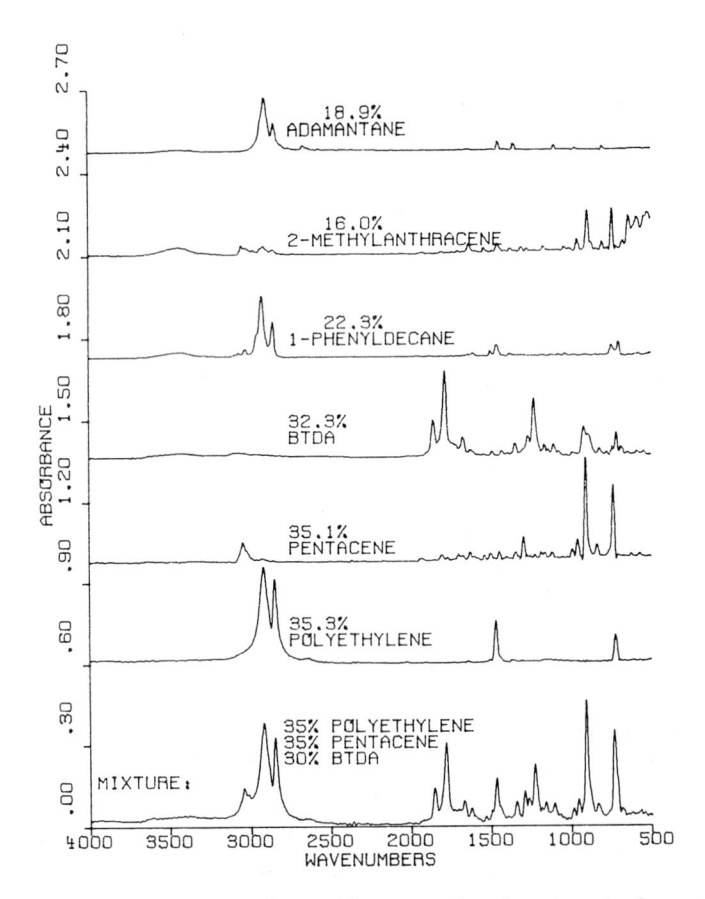

Figure 19. Compounds identified in library search using the mixed spectra of 35% polyethylene, 35% pentacene, and 30% BTDA as the unknown. (Reproduced, with permission, from Ref. 47.)

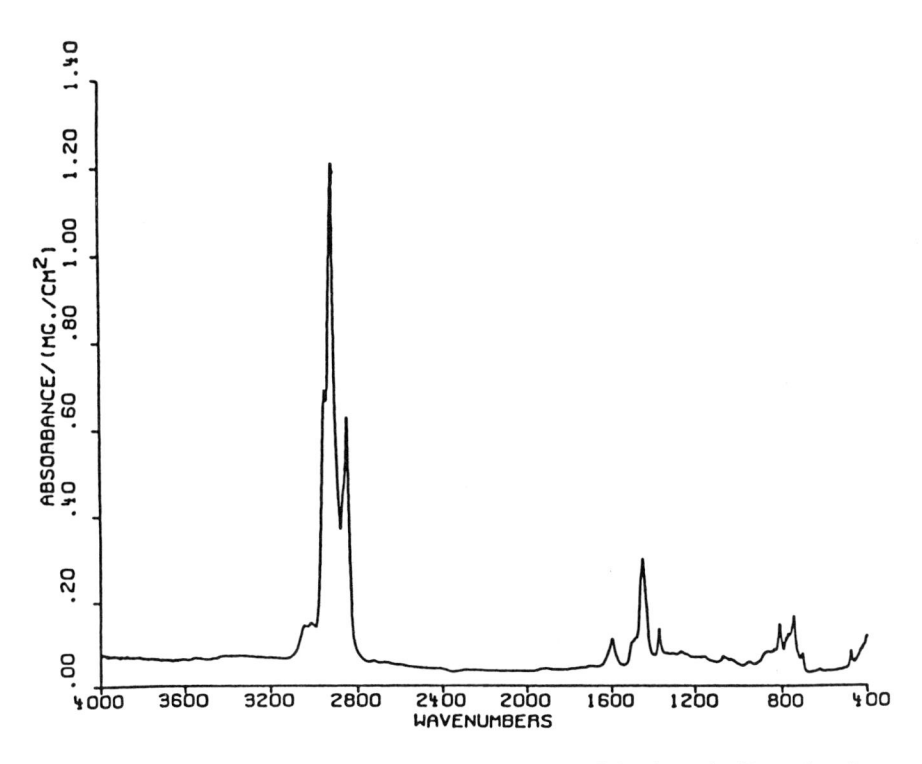

Figure 20. Liquid cell spectrum of recycle solvent (18.9 microns). (Reproduced, with permission, from Ref. 47.)

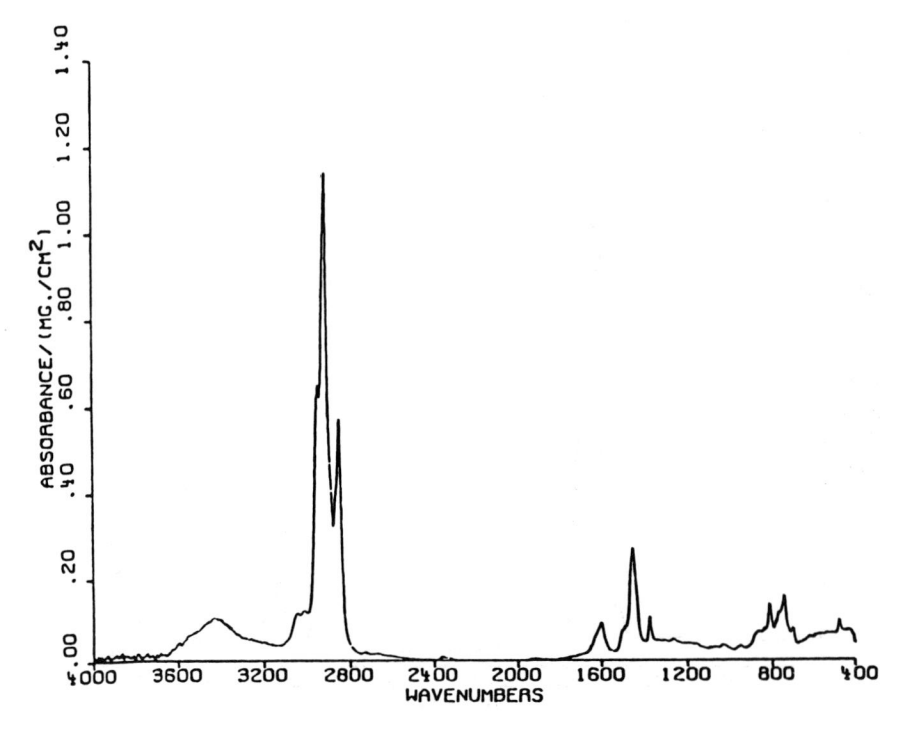

Figure 21. KBr pellet spectrum of recycle solvent. (Reproduced, with permission, from Ref. 47.)

Gases

Quantitative Analysis Quantitative measurements of complicated gas mixtures may be made with the FT-IR without using chromatographic separation. An example is the spectrum in Fig. 22 from coal pyrolysis gas at 0.5 wavenumber resolution. The figure compares spectra for the pyrolysis gas in a room temperature cell collected after injecting coal into a furnace at 800 and 1200°C. The top pair of spectra show the region between 3500 and 2800 cm^{-1}. The 1200°C spectrum shows HCN, C_2H_2, and CH_4. The 800° spectrum shows less methane and little HCN or C_2H_2 but significant amounts of ethane and heavy paraffins (indicated by the broad background). The region between 2600 and 1900 cm^{-1} shows the CO_2 and CO. The region between 1800 and 1200 cm^{-1} shows a water and methane. The region between 1200 and 500 cm^{-1} shows olefins, acetylene, HCN and CO_2.

Calibration of the FT-IR can be made using pure gases or prepared gas mixtures. Care must be exercised because most of the gases of interest show a marked increase in absorbance with dilution. The explanation for this effect is that the absorption lines for these gases are extremely sharp and for moderate concentrations all the infrared energy is absorbed in the line center in a path shorter than the absorption cell. The instrument resolution is substantially broader than the line width so the lines do not appear to be truncated. Dilution of the gas broadens the line, reducing the absorbance at line center so that a longer path contributes to the absorptivity, thus increasing the average absorbance. This effect makes calibration of these gases in the pyrolysis gas mixture difficult. A solution is to dilute the mixture to a fixed pressure at which calibrations are made.

Kinetic Measurements The Nicolet 7199 has the capability to obtain and store a complete low resolution spectrum every 80 msec. This capability may be used to study time dependent events such as pyrolysis, polymerizations, surface adsorption, etc. As an example consider the application of rapid FT-IR in the study of coal pyrolysis. The experimental apparatus consists of a heated grid pyrolyzer in a large gas cell with KBr windows to permit infrared analysis. The coal is evenly distributed between the folds of a metal screen and a current is passed through the screen to heat the coal. Coal temperatures of 1800°C and heating rates in excess of 10,000°C/sec for the higher temperatures can be achieved.

A typical set of spectra showing the evolution of gases during a 10 sec pyrolysis is illustrated in Fig. 23. The low resolution analysis can determine CO, CO_2, H_2O, CH_4, SO_2, CS_2, HCN, C_2H_2, C_2H_4, C_3H_6, benzene, COS, and heavy paraffins and olefins.

The time-temperature evolution for each species can be determined from such scans as indicated in Fig. 24 which shows the CO_2 yield as a function of time at several temperatures.

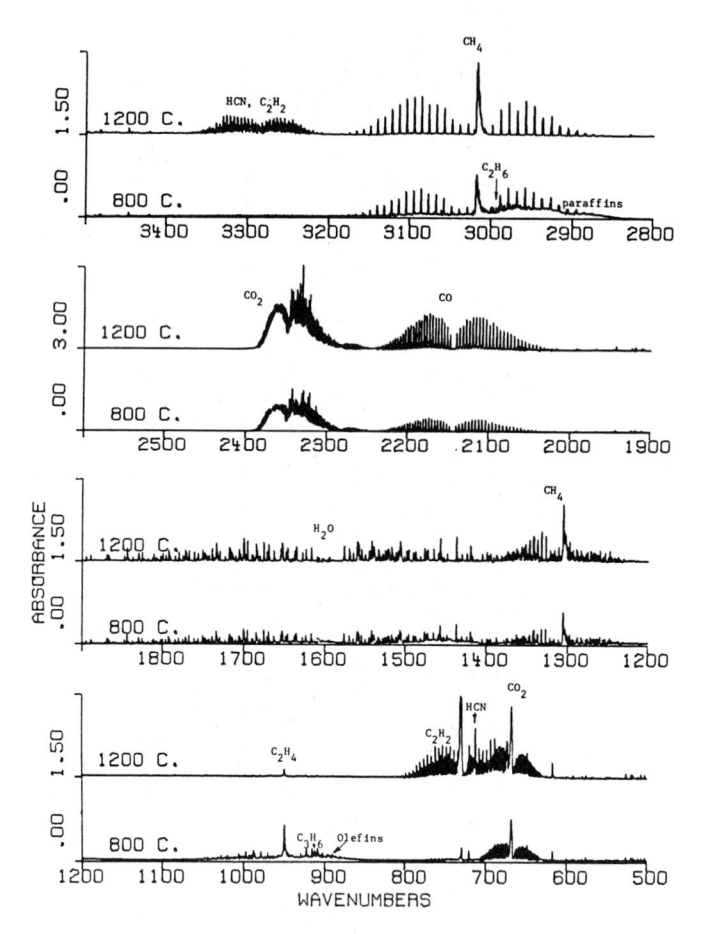

Figure 22. High-resolution spectra of gases from coal pyrolyzed at 1200°C and 800°C. Spectra are obtained in a room temperature cell. (Reproduced, with permission, from Ref. 25.)

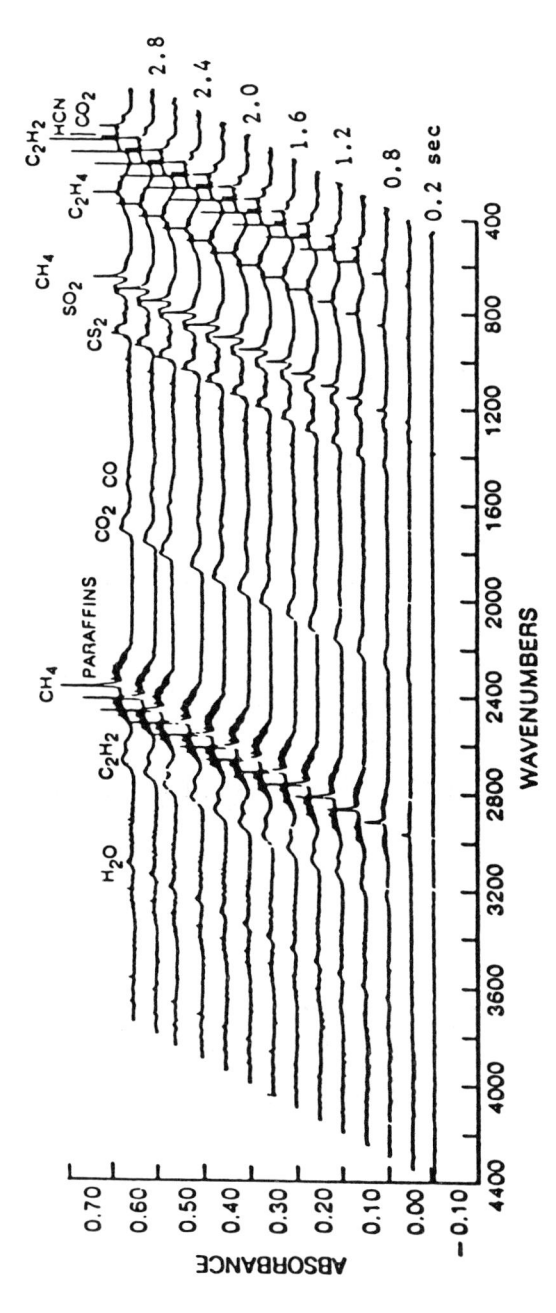

Figure 23. *FTIR spectra taken during coal pyrolysis showing gas evolution.*

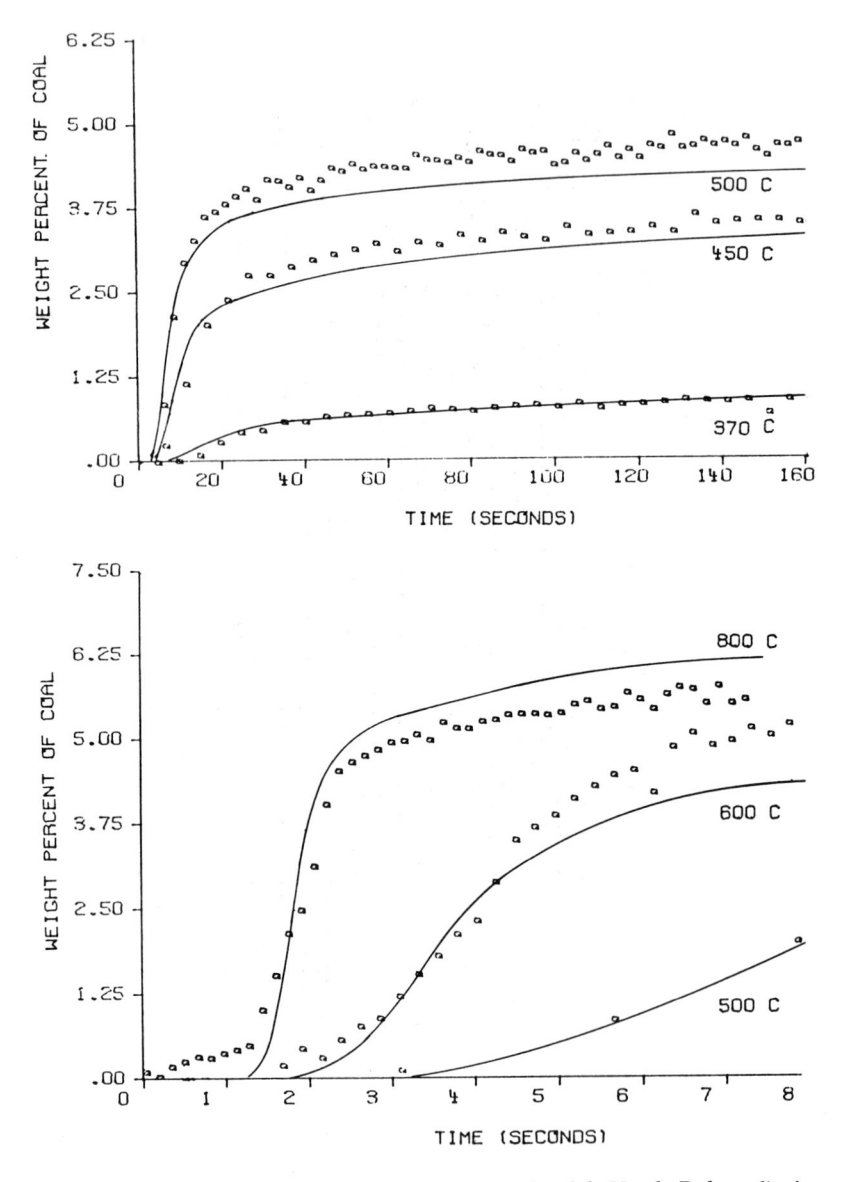

Figure 24. CO_2 yield vs. time from pyrolysis of Beulah North Dakota lignite. Symbols are data and lines are theory using distributed rate kinetics. (Reproduced, with permission, from Ref. 22.)

In Situ Analysis in a Furnace Successful in situ measurement of coal pyrolysis have recently been made (24, 25). In one application (24), small samples of coal were rapidly injected into a preheated zone which is traversed by the IR beam. With this apparatus spectra have been obtained for gases at temperatures up to 1000°C. In the second application, a gas stream of predetermined composition is heated during transit through a bed of alumina chips maintained at furnace temperature. The gas stream then enters a test section, maintained at the same temperature, where coal is introduced through a water cooled injector. After a variable residence time the reacting stream crosses the FT-IR beam and is quenched in a water cooled collector. FT-IR spectra have been obtained at temperatures up to 1200°C. The input gas stream and effluent stream can also be analyzed by routing the stream through an infrared cell. The spectra presented in Fig. 22 were of the effluent gas in the external cell.

Figure 25 shows the in-situ gas analysis. There is an acceptable noise level and no drastic effects from the particle scattering. The analyses are for the coal injector at positions from 5 to 66 cm above the optical port. The species which can easily be seen are CO, CO_2, H_2O, CH_4, C_2H_2, C_2H_4, and heavy paraffins. Additional species could be observed through the use of software signal enhancement technique. FT-IR spectra obtained directly within the hot furnace allows the observation of heavy products such as tar which don't appear in the gas phase at room temperature and provide a means to determine whether reactions occur during the quenching and sampling of the gas stream.

Temperature measurements may also be obtained using intensity ratios of absorption lines. As the temperature of a gas changes, the populations in its higher energy levels increases. This generally means the absorbance intensities became more uniform. The effect is illustrated in Fig. 26 which compares the hot and cold CO spectra. The energy is clearly shifted from the central lines toward the wings. Such a shift could be used for temperature measurement.

GC/FT-IR Complete GC/FT-IR systems are now available from the major FT-IR manufacturers. The system requirements, techniques and applications have recently been reviewed by Erickson (52). Spectra can be obtained as the components are eluted from the column, stored and identified using model compound libraries.

Applications

Coal Structure and Thermal Decomposition The study of thermal decomposition is important since all coal conversion processes (combustion, liquefaction and gasification) are initiated by this step. In addition, it appears that many of the coal structural elements are preserved in the heavy molecular

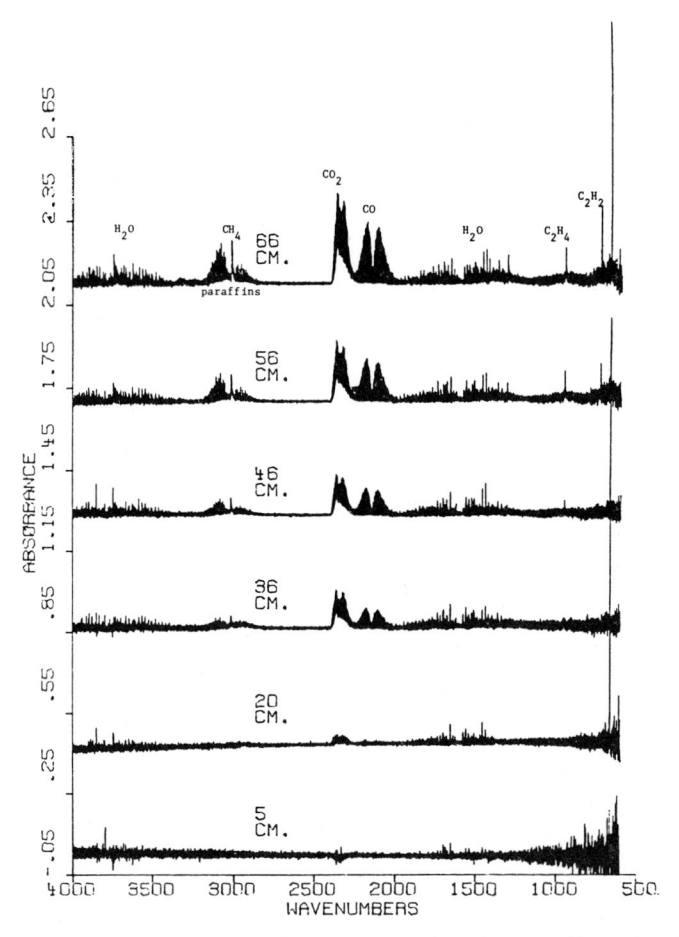

Figure 25. In situ FTIR spectra of pyrolysis gas in a furnace. (Reproduced, with permission, from Ref. 25.)

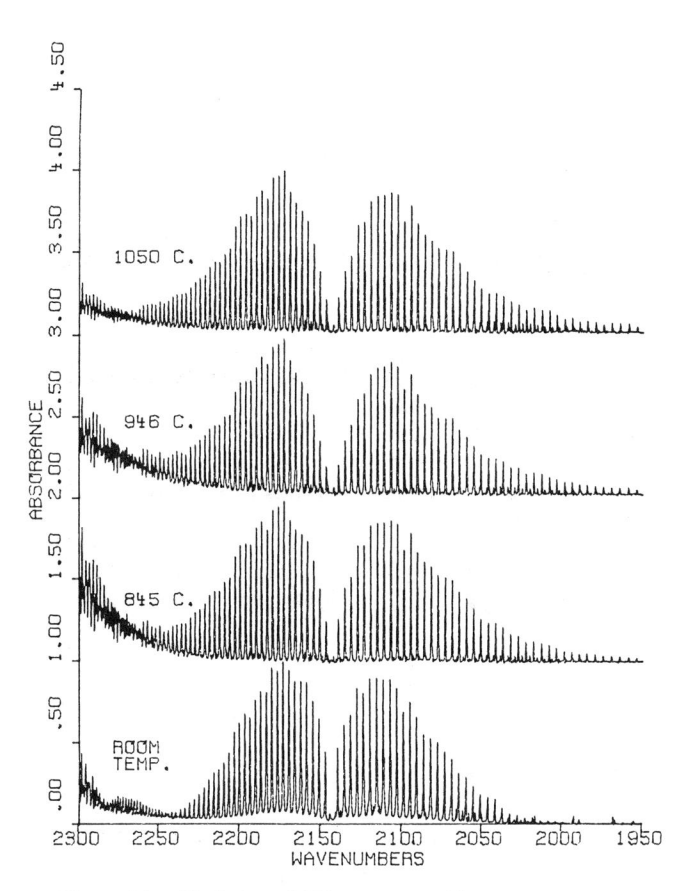

Figure 26. Variation of CO spectrum with gas temperature.

weight products (tar) released in thermal decomposition so analysis of the products can supply important clues to the structure of the parent coal. FT-IR analysis of the solid products, char and tar, produced at successive stages of thermal decomposition provides information on the changes occurring in the chemical bonds. Rapid infrared scans taken during the thermal decomposition provide data on the amount, composition and rate of evolution of light gas species. By comparing the rate of evolution with the rate of change in the chemical bonds, a general kinetic model was developed which relates the products of thermal decomposition to the organic structure of the parent coal (12, 15, 18, 21, 22).

The model uses a widely accepted view of coal structure which assumes a macromolecule consisting of groups of fused aromatic ring clusters linked by relatively weak aliphatic bridges. The ring clusters contain heteroatoms (oxygen, sulfur and nitrogen) and have a variety of attached functional groups. During thermal decomposition, the weak links are ruptured, releasing the clusters and attached functional groups. These large molecules comprise the coal tar. In vacuum devolatilization in a thin heated grid, the tar molecules may be removed quickly from the bed and undergo few secondary reactions. The evidence for this is the striking similarity between the tar and parent coal which has been observed in elemental analysis, FT-IR spectra and NMR spectra (12, 21, 53-55). The FT-IR spectra for four coals and their vacuum tar are illustrated in Fig. 27. For bituminous coals, the two materials are almost identical except for a higher concentration of aliphatic (hydroaromatic) hydrogen (especially methyl) in the tar. This extra hydrogen is presumably abstracted from the char to stabilize the free radical sites formed when the bridges were broken. Similar arguments were given for pyrolysis of model compounds by Wolfs et al. (56).

Since the abstracted hydrogen is most likely to come from the hydroaromatic portion of the coal, it is reasonable to expect the tar yield to depend on H(al). In Fig. 28, taken from Ref. (17), the tar yield (circles) in vacuum pyrolysis is plotted against H(al) for a number of coals. Also plotted are the yields of heavy hydrocarbons, (i.e., oils and BTX) from hydropyrolysis (squares), (57). Indeed, there is a strong correlation between tar yield and H(al). The observation that the tar is rich in methyl hydrogen but not in aromatic hydrogen when compared to the parent coal has implications concerning the nature of the aliphatic linkages. The result suggests that the bonds which were broken to free the ring clusters were predominantly between two aliphatic carbons, not between an aromatic and an aliphatic.

Simultaneous with the evolution of tar molecules is the competitive cracking of the bridge fragments, attached functional groups and ring clusters to form the light molecules of the gas. The quantity of each gas species depends on the functional group distribution in the original coal. At low temperatures there is very little rearrangement of the aromatic ring structure. There

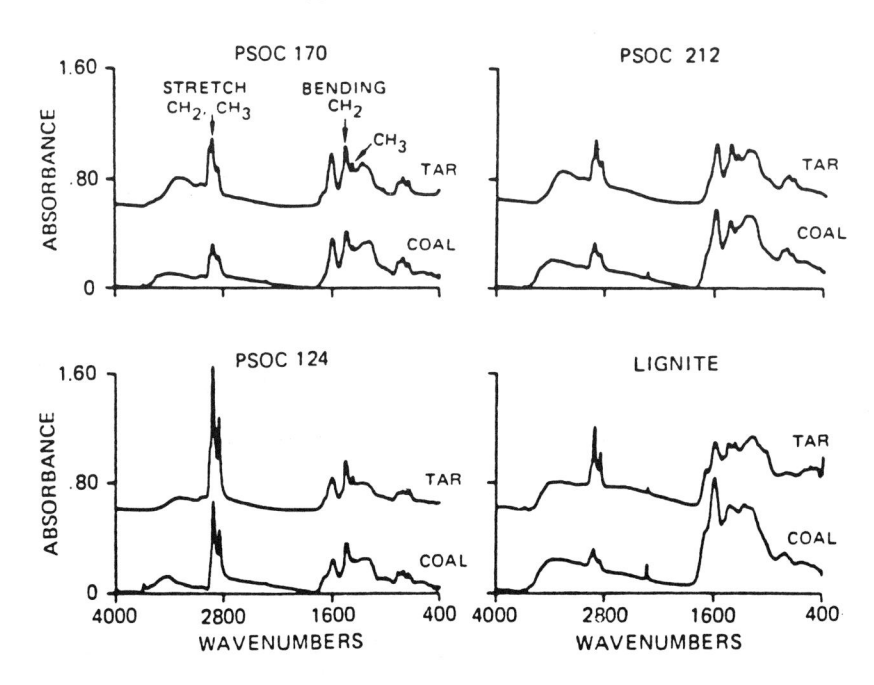

Figure 27. Comparison of FTIR spectra of coals and their vacuum distilled tars. (Reproduced, with permission, from Ref. 12)

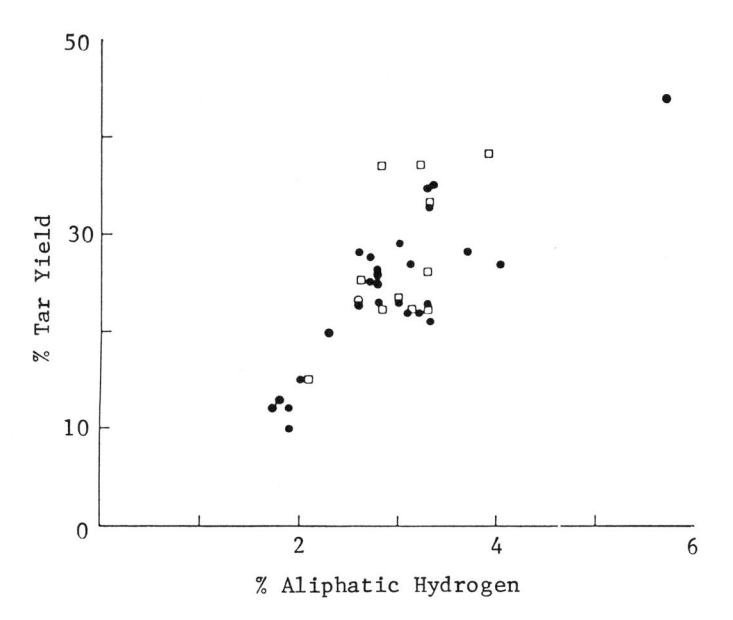

Figure 28. Correlation of tar yield with aliphatic (hydroaromatic) hydrogen. Key: ●, tar yield in vacuum pyrolysis; and □, heavy hydrocarbon yield from hydropyrolysis. (Reproduced, with permission, from Ref. 17.)

is decomposition of the substituted groups and aliphatic (or hydroaromatic) structures resulting in CO_2, H_2O, hydrocarbon gases and some CO release from the carboxyl, hydroxyl, aliphatic and weakly bound ether groups, respectively. At high temperature there is breaking and rearrangement of the aromatic rings. In this process, H_2 is released from the aromatic hydrogen and HCN and additional CO are released from ring nitrogen and tightly bound ether linkages. The evolution of each species is characterized by rate constants which are insensitive to coal rank. The differences between coals are due to differences in the mix of sources in the coal for the evolved species.

The pyrolysis chemistry may be followed by examining the infrared spectra of chars and tars. The spectra for a series of chars is illustrated in Fig. 29. The chars were produced by devolatilizing coal in an entrained flow reactor at the indicated temperature and reaction distance. The results are similar to those observed by Brown (58) and Oelert (59). The rapid reduction of the aliphatic and hydroxyl peaks corresponds to the high rate formation of tar, hydrocarbon gases and H_2O. The aromatic and C-O peaks remain to high temperature. The spectra for tars produced at various temperatures are illustrated in Fig. 30. The spectra indicate pyrolysis of the tar at high temperatures. There is a decrease in hydroxyl and aliphatic groups and an increase in aromatic hydrogen, especially the peak associated with 3 and 4 adjacent hydrogens at 750 cm^{-1}.

Relation Between Coal Structure and Proximate Analysis Fixed Carbon One of the most commonly used determinations of the thermal decomposition yield has been proximate analysis volatile matter determination in which weight loss is measured after rapid heating of the coal to 950°C. An early attempt to correlate the "fixed carbon" residue (FC) with structural properties was discussed by van Krevelen, (43). Based on experiments with model compounds it was predicted that the aromatic carbon fraction should remain in the FC while the remainder of the coal should be split off completely forming the volatile matter (VM). FC should therefore be roughly equal to the aromatic carbon content of the coal, C(ar). The correlation was verified for several coals using values of C(ar) determined from densimetric constitution analysis.

In a recent study, measurements were made for a number of coals to verify van Krevelen's correlation (15). Values of C(ar), computed using the Brown-Ladner relation with values of H(al) determined by FT-IR are plotted in Fig. 31 against the ASTM determined values for FC. Also plotted are data for the coals reported by van Krevelen. The values of C(ar) for similar rank coals determined by the two methods are in agreement. Values for C(ar) show a systematic variation with FC but, as noted by van Krevelen, are typically higher than FC. The data seem to contradict the observation that volatile tars have appreciable aromatic contents (see Figs. 27 and 30). The reasons for this correlation are discussed in (15).

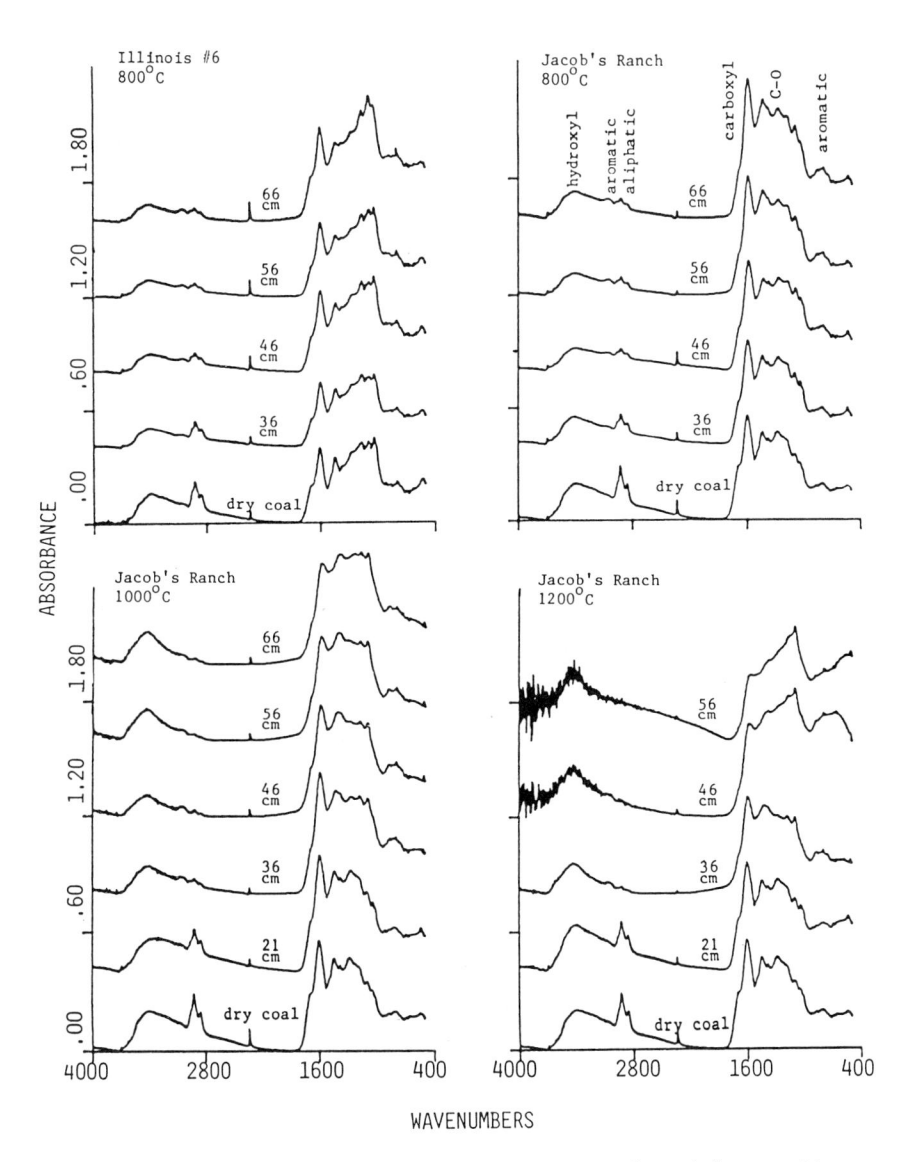

Figure 29. FTIR spectra of chars at successive stages in thermal decomposition. Chars were produced in an entrained flow reactor at 800°C, 1000°C, and 1200°C by reaction over the indicated distances.

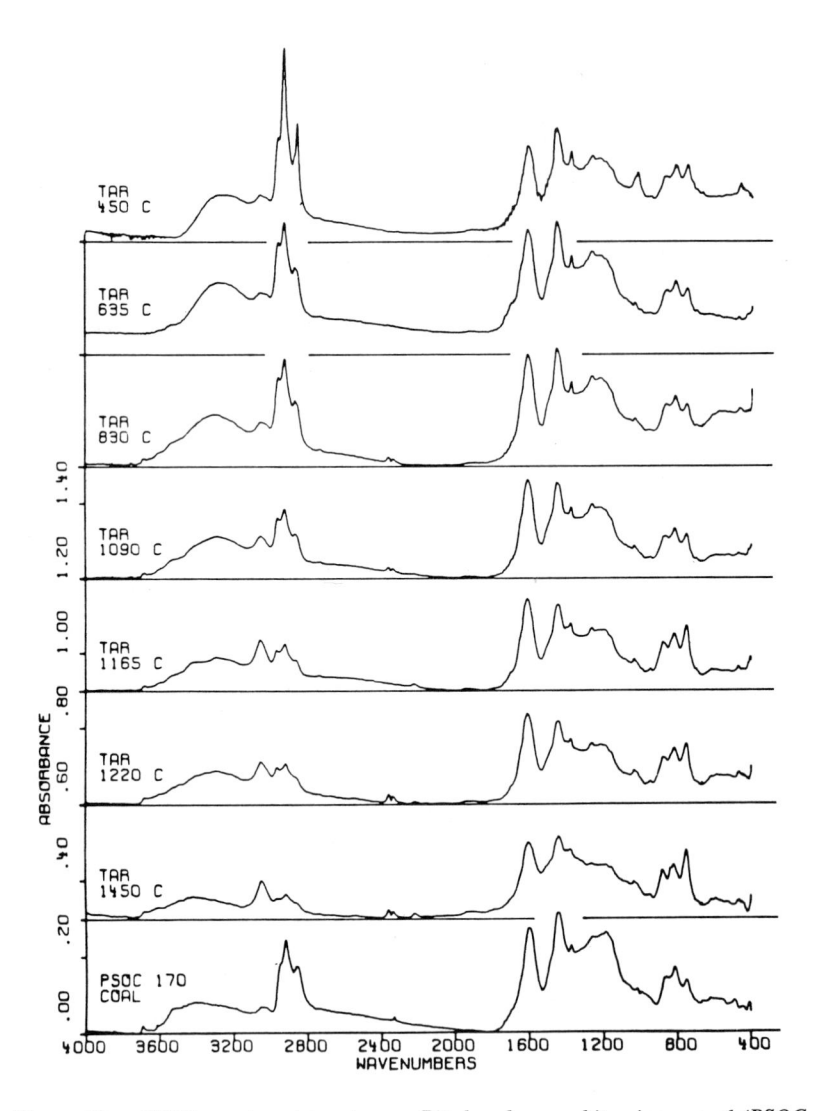

Figure 30. FTIR spectra of tars from a Pittsburgh seam bituminous coal (PSOC 170). Tars from 10 s vacuum pyrolysis at the indicated temperature.

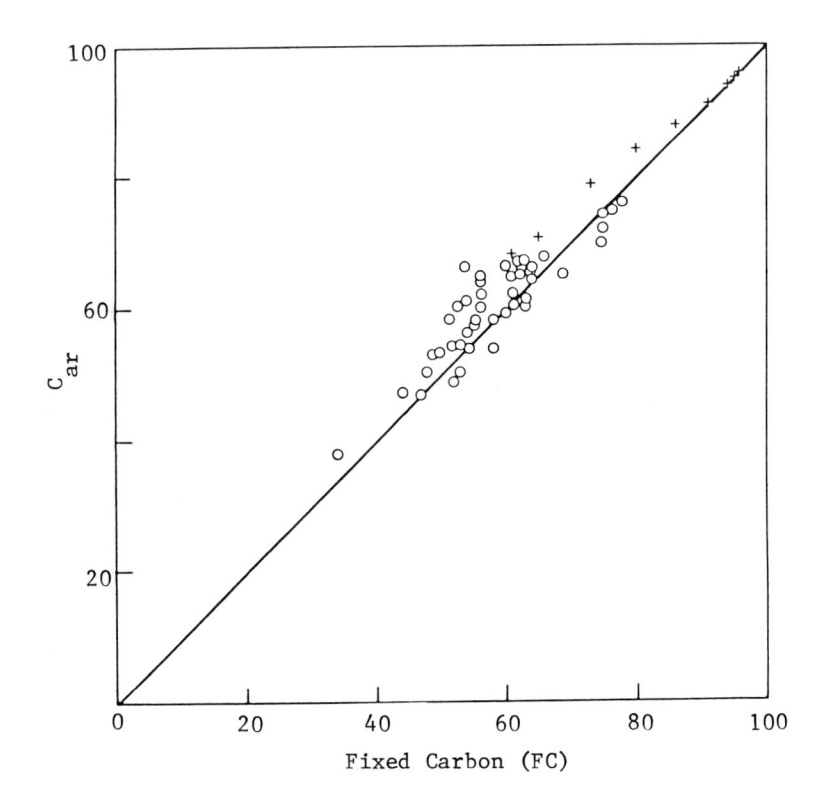

Figure 31. Correlation of proximate analysis fixed carbon with aromatic carbon C(ar). Key: +, data obtained by van Krevelen using densimetric constitution analysis (43); and ○, data obtained by Solomon using FTIR and the Brown–Ladner relation (15). (Reproduced, with permission, from Ref. 15.)

Oil Shales   Among the methods used for characterizing oil shale, those applied to the raw shales are most useful because they are not influenced by possible changes in the organic structure produced by the extraction procedure. Recently, Maciel et al., (60, 61) demonstrated the usefulness of the ^{13}C cross polarization NMR technique for oil shale measurements. The method was applied to a number of oil shales to determine the ratio of aliphatic to aromatic carbon. From these measurements, it was found that there exists an approximate rectilinear correlation between the "apparent" aliphatic carbon content of the oil shale and the oil yield as determined by Fischer assay.

In a recent study by Solomon and Miknis, (16) several raw oil shales from the group studied by Maciel et al., were analyzed by FT-IR using the same techniques employed for analyzing coals. The results are presented in Fig. 32. The figure shows the aliphatic content computed in two ways. First, it can be determined directly as the sum of H(al) + C(al) (squares) where H(al) and C(al) are determined from FT-IR and ^{13}C NMR, respectively. Second, it can be determined from H(al) using an assumed stoichiometry of CH_2 i.e. %C-H = 7H(al). The agreement of the two determinations of aliphatic C-H suggests that the assumed stoichiometry is reasonable. The results of Fig. 32 show that the weight percent oil yields are only slightly less than the weight percent aliphatic C-H. An attractive explanation of the result is that under conditions of the Fischer assay there is almost complete conversion of the aliphatic components but little of the aromatics.

Liquefaction   The ability to do quantitative structural analysis makes FT-IR a good tool for coal liquefaction research. As an example, Fig. 33 compares the feedstock and the product in a liquefaction process (47). The feedstock consists of 33.3% coal plus 66.6% recycle solvent. The combined spectrum is compared with the spectrum of the product. The lowest spectrum is the difference between the feedstock and the product. The product is lower in oxygen groups but higher in aromatic hydrogen.

Effect of Hydrodesulfurization on Coal Structure   Several processes have been developed to reduce the concentration of sulfur in coal prior to its utilization. Besides assessing the sulfur cleaning potential of these processes, it is important to determine what changes have been made in the overall properties of the coal.

Figure 34 illustrates the results for a raw and chemically cleaned coal (62). A substantial decrease is observed in the aliphatic peak (2900 cm^{-1}) and substantial increases are observed in the carbonyl peak (1700 cm^{-1}), the hydroxyl peak (3500-2400 cm^{-1}) and the ether peak (1250 cm^{-1}).

According to the pyrolysis model discussed above, the changes in the organic structure which are indicated in the FT-IR spectra will result in substantial changes in the distribution of volatile products. A comparison of the gas yields as determined by FT-IR for a raw and treated coal are shown in Fig. 35. As expected, the

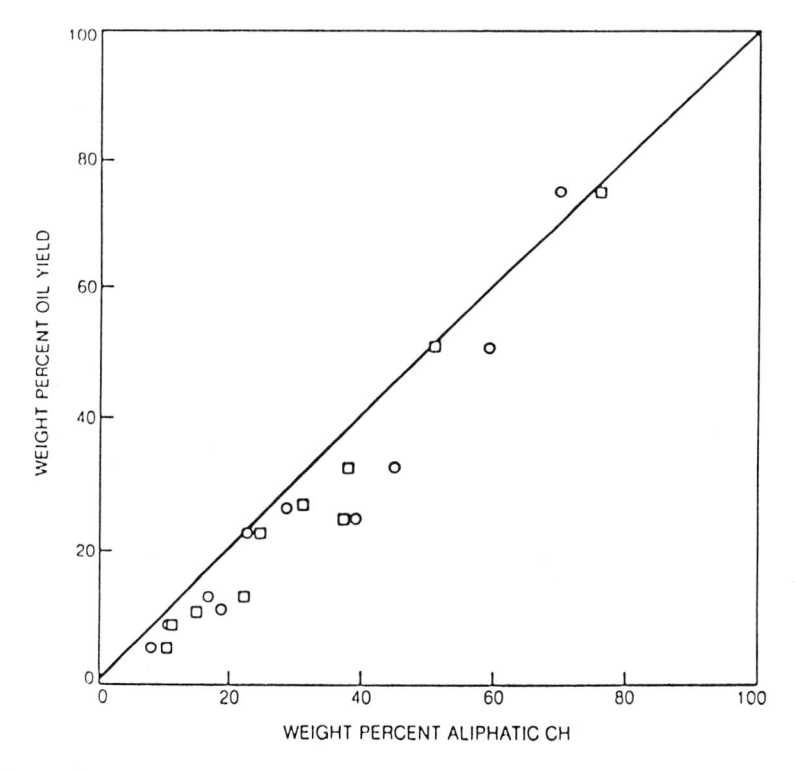

Figure 32. Correlation of Fischer assay oil yield with aliphatic hydrocarbon concentration. Key: ○, %CH=7H(al); and □, %CH=H(al) + C(al). (Reproduced, with permission, from Ref. 16.)

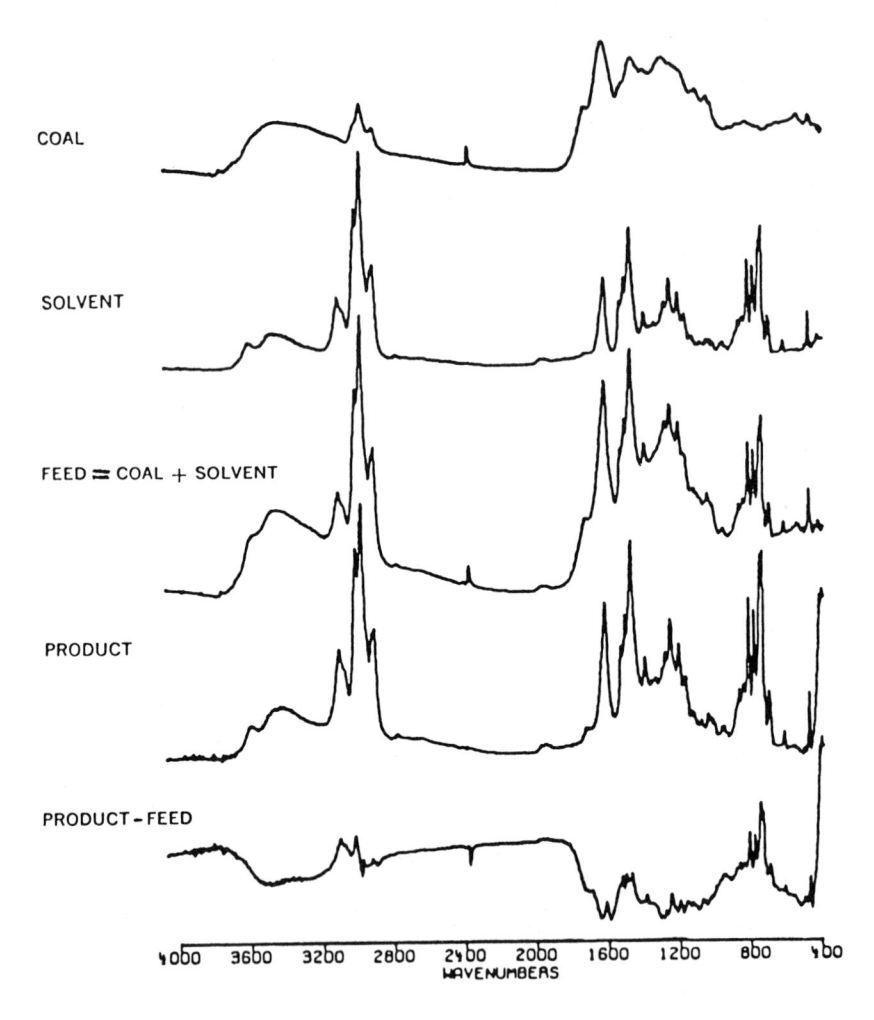

Figure 33. FTIR spectra of liquefaction feedstock and products. (Reproduced, with permission, from Ref. 47.)

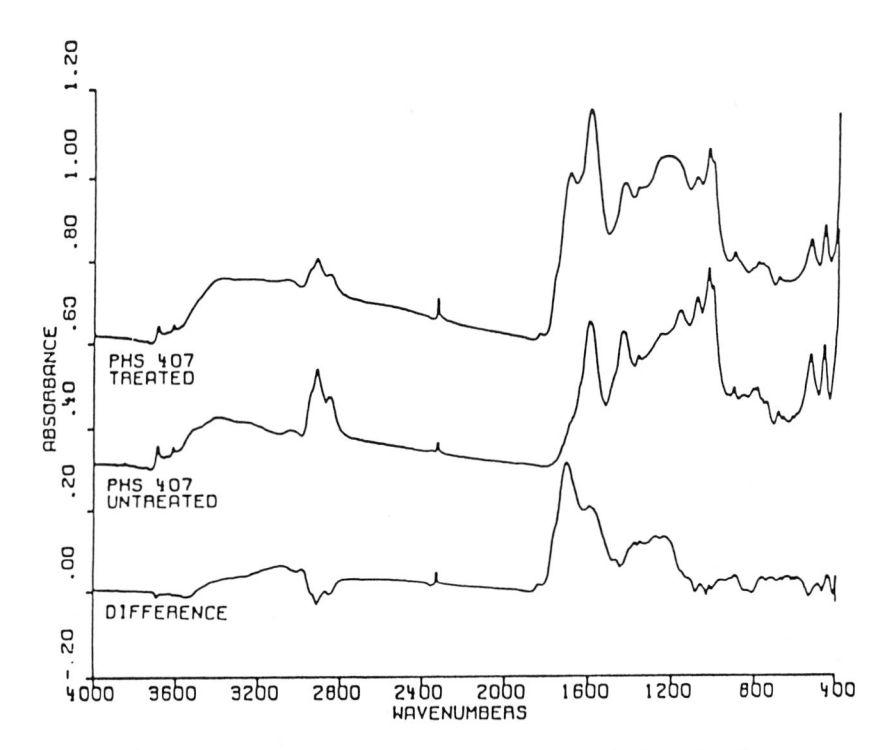

Figure 34. *FTIR spectra of a raw and chemically cleaned coal.*

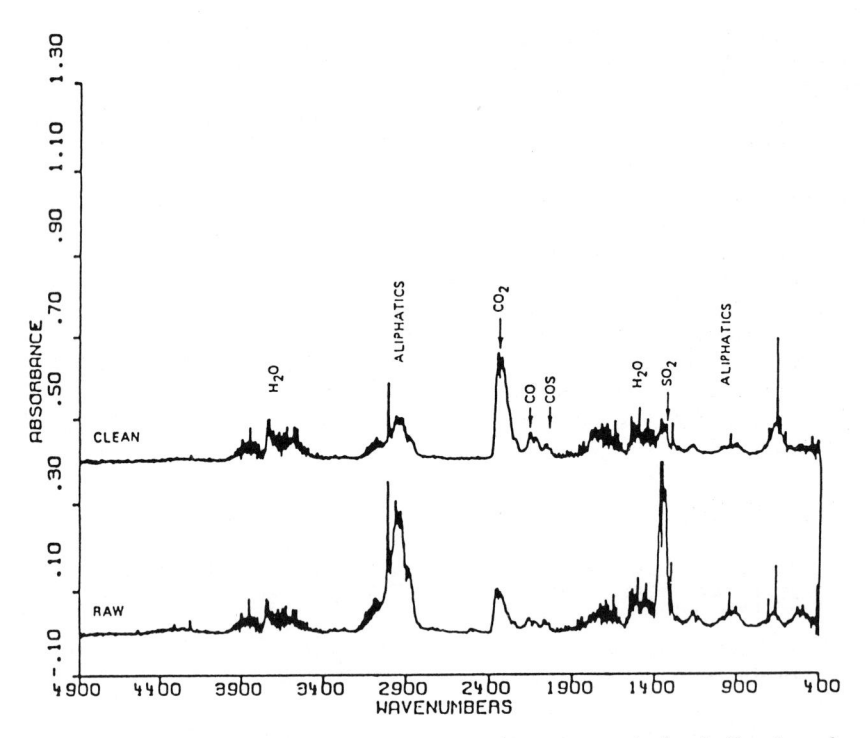

Figure 35. FTIR spectra of pyrolysis gases from raw and chemically cleaned coals.

cleaned coal shows reduced levels of hydrocarbon gases, and increased levels of oxygen containing species, H_2O, CO and CO_2. The level of SO_2 is sharply reduced while the level of COS is relatively unchanged. These differences are related to the change which the treatment imposes on the sources for these products. The tar yield for the cleaned coal is also reduced in agreement with the relation between tar yield and aliphatic hydrogen which was discussed above (see Fig. 28).

Acknowledgement

The authors would like to acknowledge the contribution of several colleagues and their institutions. Warren Vidrine of Nicolet Instrument Corporation provided his assistance and laboratory to obtain the photoacoustic attenuation and diffuse reflectance spectra. Richard Neavel and the Exxon Corporation supplied samples from the Exxon sample bank and allowed the publication of the hydroxyl data. Professor Douglas Smoot of Brigham Young University supplied the samples on which the drying and Beer's Law studies were performed and has allowed the use of data obtained for him. Much of the early FT-IR work was done at United Technologies Reserach Center under DOE Contract ET-78-C-01-316. More recent work has been done at Advanced Fuel Research, Inc. under EPRI contracts RP 1604-2 and RP 1654-8 and DOE Contract DE-AC01-81-FE05122. Support for some of the pyrolysis/FTIR studies was received from the Grand Forks Energy Technology Center.

Literature Cited

1. Lowry. H. H., Chemistry of Coal Utilization, Supp. Volume, Wiley, NY. (1963)

2. van Krevelen, D. W., Coal, Elsevier Publishing Co., Amsterdam. (1961)

3. Friedel, R. A., in Applied Infrared Spectroscopy, p. 312, edited by D. N. Kendall, Reinhold, NY. (1966)

4. Brown, J. K., J. Chem. Soc., 744 (1955)

5. Brooks, J. D., Durie, R. A. and Sternhell, S., Aust. Journal, Appl. Sci. 9, 63 (1958)

6. Friedel, R. A. and Retcofsky, H., Proceedings of the Fifth Carbon Conference, Volume II, 149, Pergamon Press. (1963)

7. Painter, P. C., Coleman, M. M., Jenkins, R. G., Whang, P. W. and Walker, P. L., Jr., Fuel 57, 337 (1978)

8. Painter, P. C., Coleman, M. M., Jenkins, R. G., and Walker, P. L., Jr., Fuel 57, 125 (1978)

9. Painter, P. C., Snyder, R. W., Pearson, D. E. and Kwong, J., Fuel 59, 282 (1980)

10. Painter, P. C. and Coleman, M. M., Fuel 58. 301 (1979)

11. Painter, P. C., Yamada, Y., Jenkins. R. G., Coleman, M. M. and Walker, P. L., Jr., Fuel 58, 293 (1979)

12. Solomon, P. R., ACS Division of Fuel Chem. Preprints, 24, #2, 184 (1979) and Advances in Chemistry Series, Vol. 192, "Coal Structure", 95 (1981)

13. P. R. Solomon and R. M. Carangelo, FTIR Analysis of Coal: I. Techniques and Determination of Hydroxyl Concentrations. (submitted to Fuel)

14. Brown, J. K. and Ladner, W. R., Fuel 39, 87 (1960)

15. Solomon, P. R., Fuel 60, 3 (1981)

16. Solomon, P. R. and Miknis, F. P., Fuel 59, 893 (1980)

17. Solomon, P. R., Hobbs, R. H., Hamblen, D. G., Chen, W., La Cava, A. and Graff, R. A., Fuel 60, 342 (1981)

18. Solomon, P. R., Coal Structure and Thermal Decomposition, in "New Approaches in Coal Chemistry" ACS Symposium Series. 169 pg 61 (1981)

19. Erickson, M. D., Frazier, S. E. and Sparacino, C. M., Fuel 60, 263 (1981)

20. P. R. Solomon, ACS Div. of Fuel Chemistry Preprints, 24, #3, 154 (1979)

21. Solomon, P. R. and Hamblen, D. G., Understanding Coal Using Thermal Decomposition and Fourier Transform Infrared Spectroscopy, Presented at the Conference on the Chemistry and Physics of Coal Utilization, Morgantown, W. Va., June 2-4. (1980)

22. Coal Pyrolysis, P. R. Solomon, D. G. Hamblen, R. M. Carangelo, AICHE, Symposium on Coal Pyrolysis. (Nov., 1981)

23. Characterization of Coal and Coal Thermal Decomposition, P. R. Solomon, Chapter III of EPA Monograph on Coal Combustion, (In Press).

24. J. D. Freihaut, P. R. Solomon, D. J. Seery, ACS Div. of Fuel Chem. Preprints 25, #4, 161 (1980)

25. Solomon, P. R. and Hamblen, D. G., ACS Division of Fuel Chemistry Preprints 27, (to be published)

26. Painter, P. C., Snyder, R. W., Starsinic, M., Coleman, M. M. and Kuehn, D. J. and Davis, A., Concerning the Application of FTIR to the Study of Coal, A Critical Assessment of Band Assignments and The Applications of Spectral Analysis Program. The Pennsylvania State University Report, (1980) and presentation at ACS 181st National Meeting, Atlanta, GA, March. (1981)

27. Osawa, Y. and Shih, J. W., Fuel 50, 53 (1971)

28. Roberts, G., Anal. Chem. 29, 911 (1957)

29. Durie, R. A. and Sternhell, S., Aust. J. of Chem. 12, 205 (1959)

30. Kirkland, J. J., Anal. Chemistry 27, 1537 (1955)

31. Fujii, S., Osawa, Y. and Sugimura, H., Fuel 49, 68 (1970)

32. Vidrine, D. W., Appl. Spectroscopy 34, 314 (1980)

33. Fuller, M. P. and Griffiths, P. R., Applied Spectroscopy 34, 533 (1980)

34. Fuller, M. P. and Griffiths, P. R., Anal. Chem. 50, 1906 (1978)

35. Solomon, P. R., Hamblen, D. G. and Carangelo, R. M., Proceedings of the International Conference on Coal Science, pg. 719, Dusseldorf, Germany. (September, 1981)

36. Estep, P. A., Kovach, J. J., Hiser, A. L. and Karr, C., Jr., Spectrometry of Fuels, Plenum Press, New York, pp. 228 (1970),

37. Huggins, C. W., Green, T. E. and Turner, T. L., U.S. bureau of Mines, Report of Investigation 7781, (1973),

38. van der Marel, H. W. and Beutelspacher, H., Atlas of Infrared Spectroscopy of Clay Minerals and Their Admixtures, Elsevier, New York. (1976)

39. Friedel, R. A. and Queiser, J. A., Anal. Chem. 28, 22 (1956)

40. Liotta, R., Fuel 58, 724 (1979)

41. Yarzab, R. F., Abdel-Baset, Z. and Given, P. H., Geochim. et Cosmochim., Acta, 43, 281 (1979)

42. Blom, L., Edelhausen, L. and Van Krevelen, D. W., Fuel 38, 537 (1959)

43. van Krevelen, D. W., and Schuyer, J., Coal Science, Elsevier, Amsterdam. (1957)

44. Mazumdar, B. K., Chakrabartty, S. K. and Lahiri, A., Fuel 41, 129 (1962)

45. Retcotsky, H. L., Schweighardt, F. K. and Hough, M., Chem. 49, 585 (1977)

46. Gerstein, B. C., Murphy, P. D., Ryan, L. M., and Solomon, P. R., A Study of Carbon and Hydrogen Aromaticity in Coals by High Resolution Solid State Nuclear Magnetic Resonance and Fourier Transform IR Spectroscopy. (to be published)

47. P. R. Solomon and R. M. Carangelo, Characterization of Wyoming Subbituminous Coals and Liquefaction Products by Fourier Transform Infrared Spectrometry, EPRI Final Report for Contract #1604-2. (Sept., 1981)

48. Solomon, P. R., Monthly Topical Report, Investigation of the Devolatilization of Coal under Combustion Conditions, Dept. of Energy, Contract ET-78-C-01-3167. (November, 1979)

49. Vidrine, D. W. and Mattson, D. R., Applied Spectroscopy 32, 502 (1978)

50. Brown, R. S., Hausler, D. W. and Taylor, L. T., Anal. Chemistry 53, 197 (1981)

51. Brown, R. S., Hausler, D. W. and Taylor, L. T., Reprint Anal. Chemistry 52, 1511 (1980)

52. Erickson, M. D., Applied Spectroscopy Reviews 15, 261 (1979)

53. Brown, J. K., Dryden, I. G. C., Dunevein, D. H., Joy, W. K., and Pankhurst, K. S., J. Inst. Fuels 31, 259 (1958)

54. P. R. Solomon and Mereidth B. Colket, Fuel 57, 749 (1978)

55. Orning, A. A. and Greifer, B., Fuel 35, 381 (1956),

56. Wolfs, P. M. J., van Krevelen, D. W. and Waterman, H. I., Fuel 39, 25 (1960)

57. Chen, W., La Cava, A., and Graff, R. A., ACS Div. of Fuel Chemistry Preprints 24, #3, 94 (1979)

58. Brown, J. K., J. Chem. Soc., 752 (1955)

59. Oelert, H. H., Fuel 47, 433 (1968),

60. Maciel, G. E., Bartuska, V. J. and Miknis, F. P., Fuel 57, 505 (1978)

61. Maciel, G. E., Bartuska, V. J. and Miknis, F. P., Fuel 58, 155 (1979)

62. The samples were supplied by Sid Friedman of the Pittsburgh Energy Technology Center.

RECEIVED September 7, 1982

Chemistry and Structure of Coals
Diffuse Reflectance IR Fourier Transform (DRIFT) Spectroscopy of Air Oxidation

N. R. SMYRL and E. L. FULLER, JR.

Union Carbide Corporation, Oak Ridge Y-12 Plant, Nuclear Division, Oak Ridge, TN 37830

Diffuse reflectance infrared Fourier transform (DRIFT) spectroscopy has been proven to be an excellent means of characterizing coals and related materials. This report is devoted to the evaluation of the technique as a method for *in situ* monitoring of the chemical structural changes wrought in reactions of coal with fluid phases. This technique does not require a supporting medium (matrix) which can contain chemical artifacts which inherently serve as a barrier for access to the solid coal. The rapid response of the Fourier transform infrared technique is further beneficial for kinetic studies related to combustion, liquefaction, gasification, pyrolyses, etc. Experimental equipment and techniques are described for studies over wide ranges of pressure (10^{-5} Pa to ca 1.5 x 10^2 kPa) and temperature (298°K to 800°K).

Fourier transform infrared (FTIR) spectroscopy has been used extensively in the past several years in the study of coal and its related products (1-13). The majority of this work has been done by transmission spectroscopy utilizing the KBr pellet technique. In relation to coal, the method has several drawbacks including variable H_2O content in the KBr, high scattering background, difficulty in reproducing background for accurate subtraction studies, the unknown effect of pressure in fabricating the pellet, and the inability to perform *in situ* studies. Diffuse reflectance (DR) and photoacoustic (PA) detection as infrared (IR) sampling methods both possess the potential to alleviate most of the above mentioned problems; although, neither of these techniques is completely without its own pitfalls.

In the field of IR spectroscopy, DR has only recently experienced a wave of renewed interest when coupled with FTIR instrumentation (14-20). There are also recent reports by Hattori, et al, (21-24) of the design of an emissionless DR-IR grating

0097-6156/82/0205-0133$06.00/0
© 1982 American Chemical Society

spectrometer along with its application to catalytic reactions at elevated temperatures and to adsorbed species. We have recently reported the development of a high-vacuum DR-IR sample cell having a temperature controlled sample stage for use with our FTIR equipment (25). The capabilities of this cell were demonstrated by monitoring the in situ reactions of LiH and LiOH with H_2O and CO_2.

Photoacoustic detectors for use with FTIR spectrometers have been developed and are presently available for a number of commercial instruments. Since PA-FTIR represents a very recent area of interest, the published literature is, therefore, rather limited (26-34). The photoacoustic and diffuse reflectance methods are complimentary and both have been shown to be applicable to coal analysis (16,19,29,31,34).

Our intention in this report is to demonstrate the utility of diffuse reflectance infrared Fourier transform (DRIFT) spectroscopy for coal analysis, particularly in relation to monitoring the in situ oxidation of coal, and to compare its relative merits to those of the KBr pellet and PA sampling techniques.

Experimental

The DR equipment used for the present studies consisted of a Model DRA-SID accessory designed for Digilab FTS-15 side-focus FTIR spectrometers by Harrick Scientific Corporation. The ultra-high-vacuum (UHV) DR cell was a modified version of a cell also developed by Harrick Scientific. A detailed description of the DR cell and the accessory are given in a previous report (25).

The coal samples utilized in the present work were a sub-bituminous coal and a bituminous coal obtained from freshly opened mine faces in the Wyoming Wyodak mine and Pennsylvania Bruceton mine, respectively. The samples were stored under argon prior to initiation of any experimental work. The chemical analyses for these particular coal samples are given in Table 1. The samples were prepared for the DR studies by grinding in a ball mill under an argon atmosphere to pass a 200 mesh screen.

The Wyodak sample was utilized for the oxidation studies. A powdered sample of this coal was placed in the DR cell in the neat form (no supporting matrix medium). The pressure was slowly reduced from atmospheric pressure (~ 100 kPa) to 1 Pa followed by a gradual increase in temperature from 27°C to 191°C in order to follow the desorption of moisture. The sample was then heated to 393°C and oxidized in 2.7 kPa of air for ~ 24 hrs. Spectra were periodically obtained throughout this sequence of events.

The spectra were scanned at 2 cm^{-1} resolution on a Digilab FTS-15C Fourier transform infrared spectrometer as the signal average of 100 interferograms. A few selected spectra were obtained at a higher signal to noise ratio by scanning for a longer period. The spectra are displayed in the ordinate format of [- log (R_s/R_o)] which is referred to in this paper as reflectance (by analogy to absorbance in normal transmission work) where R_s is the reflection of the sample and R_o is the reflection of the reference. The reflection from a mirror mounted at 45° in the DR accessory was used for the reference which had only the effect of vertically shifting the overall baseline when compared to a material such as KCl for a reference standard. The mirror reference was preferred since it could be assured that there would be no moisture contribution to the spectra.

Results and Discussion

In order to perform in situ reaction studies and to monitor sorption-desorption phenomena, it is very beneficial to have the capability to view a solid material as a powder in the neat form. Both DR and PA infrared sampling methods possess this ability. Although DR can sometimes yield highly distorted band structure when viewing highly crystalline neat materials (16,25), DR-IR would still appear to have the edge over PA-IR for in situ powder reaction studies. The reason for this is the fact that the PA method requires a diluent gas for sound propagation which is required for detection and, therefore, cannot be used to acquire spectra under vacuum conditions. Photoacoustic IR is not, however, restricted to only powder samples and therein lies one of the principal advantages of the PA method.

As Figure 1 demonstrates, DRIFT spectroscopy can be utilized for the analysis of powdered coal without the necessity of dilution in a supporting medium. This figure also demonstrates the aiblity to follow the moisture desorption process as the Wyodak material is both evacuated and heated. Curve A of Figure 1 exhibits a rather broad band extending from ~ 3700 cm^{-1} to 2100 cm^{-1} which can be attributed to the O-H stretch of H_2O and various hydroxy containing constituents in the coal in a variety of hydrogen bonded environments. Superimposed on this broad band are the aromatic and aliphatic C-H stretching bands of the organic coal constituents. As the head space is evacuated and the sample is heated, the broad O-H band is observed to decrease in intensity and become more structured with several distinct peaks being resolved at 3540, 3390, 3290 cm^{-1} as noted in Curves B and C. Curve B is the DRIFT spectrum at room temperature and ~ 1 Pa pressure while Curve C is the spectrum of the material at the same pressure heated to 191°C. Most of the changes occur on initial evacuation with very little occurring in the heating stage as noted in Curves D and E which represent the respective difference spectra (A-B) and (B-C). (These difference spectra were obtained by subtraction of the C-H stretching bands which presumes that no volatile organic fraction is lost in the evacuation and heating process.)

The major portion of the changes appear to result from moisture desorption. One of the problems observed in KBr pellet transmission work is the uncertainty associated with drying the

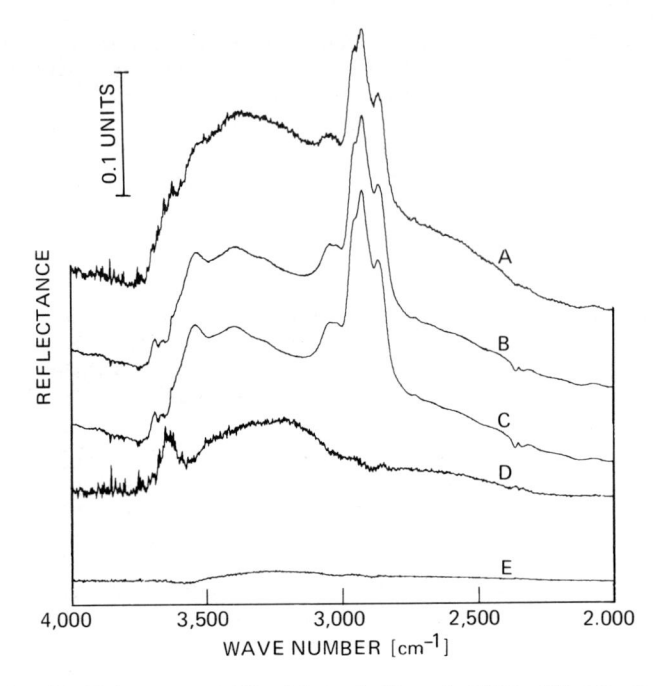

Figure 1. DRIFT spectra of Wyodak coal. Key: A, 27°C, 100 kPa; B, 27°C, 1 Pa; C, 191°C, 1 Pa; D, difference spectrum (A − B); and E, difference spectrum (B − C).

Table 1. Chemical analysis of coal samples.

Analysis	Wyodak Composition (wt %)	Bruceton Composition (wt %)
C	72.5	82.5
H	5.45	5.5
O (by difference)	20.5	7.7
N	1.01	1.3
S	0.53	3.0
Moisture	28.9	1.6
Ash	5.98	14.3

KBr and the possible contribution of the moisture from this material to the O-H band in the coal spectra (10). The potential value of DR for obtaining better understanding of the relative contributions to the O-H band from the water and the organic constituents is apparent.

Another desirable characteristic which is demonstrated in the DRIFT spectra of coal is the relatively flat baselines. The advantage from this particular aspect is apparent when compared to the rather severe sloping baselines which appear in the spectra of KBr pellets of coal due to light scattering.

FTIR has been used successfully in the qualitative and quantitative analysis of mineral matter in coal (1,5,13). The success of these methods relied heavily on computer subtraction techniques. Figure 2 illustrates that similar techniques may be applicable to DRIFT spectra. Curve B in Figure 2 represents the DRIFT spectrum of kaolin, a common mineral component found in coal, diluted in KCl at the 2% level. The absorption contributions due to kaolin in the Bruceton coal can be removed by subtraction of Curve B from Curve A to yield Curve C. The 3390 cm^{-1} band of kaolin was used to determine the proper degree of subtraction. It is necessary in this case to dilute the kaolin in KCl to avoid problems arising from specular reflection. Figure 2 was included for illustrative purposes only, and no quantitative work has been attempted in the present study.

Painter, et al, have studied coal oxidation both for samples obtained from various areas in a vein exploration audit through a high volatile coking coal seam(8) and in laboratory studies at low temperature (100°C) (9). It was concluded from this work that a variety of carbonyl, carboxylic acid, and carboxylate containing products are formed in the oxidation process. Rockley and Devlin(29) have compared the surfaces of freshly cleaved coal with aged surfaces by PA-FTIR and have also noted carbonyl species on the aged surfaces which were attributed to oxidation. Our in situ studies at a somewhat higher temperature (393°C) essentially parallel the observations by Painter's group, although, there were some differences that are noted in the following discussion.

The DRIFT spectra of an essentially unoxidized state of the Wyodak coal (Curve B) is compared to that of the same sample after 1 hour of oxidation in 2.7 kPa of air at 393°C (Curve B) in Figure 3. The most notable differences in these spectra are decreased absorption of the aliphatic C-H stretching bands at ~ 2900 cm^{-1} and the appearance of a carbonyl stretching band at ~ 1700 cm^{-1}. These features are accentuated in the difference spectrum (Curve C). Figures 4 and 5, which represent the series of difference spectra for the 3900-2500 cm^{-1} and 2000-1200 cm^{-1} regions, respectively, measured at successive time intervals in the oxidation, illustrate in considerably more detail the changes which are occurring. The negative peaks are indicative of species or functional groups which are lost and the positive peaks to those which are formed. Since the study was done in situ, there was little

138   COAL AND COAL PRODUCTS

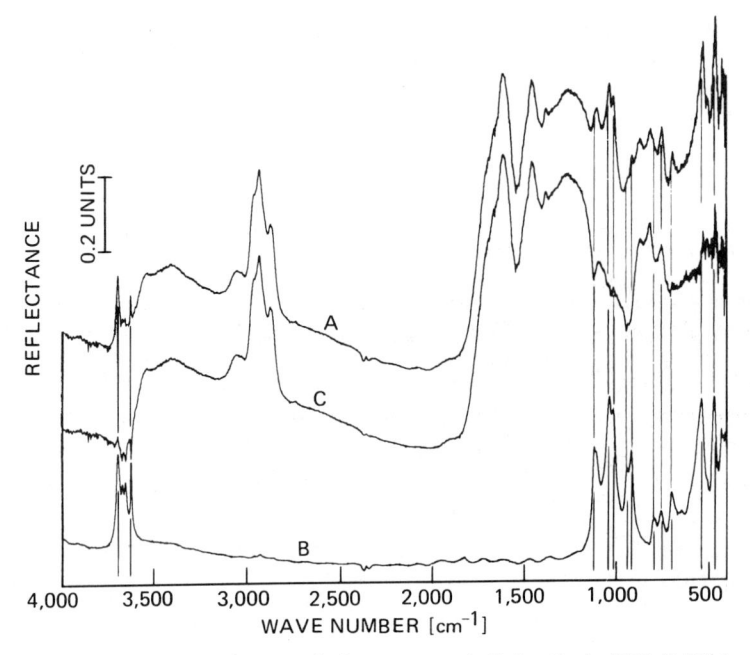

Figure 2.   DRIFT spectra. Key: A, Bruceton coal; B, kaolin in KCl (2.0%); and C, difference spectrum (A − 0.7404 × B).

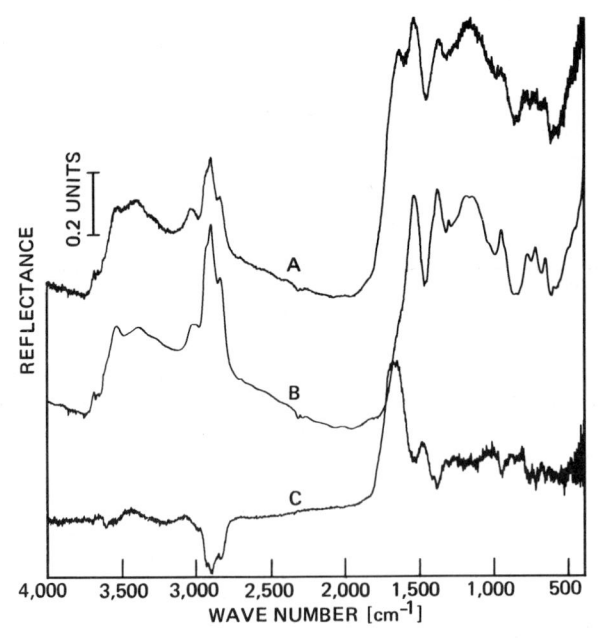

Figure 3.   DRIFT spectra of Wyodak coal. Key: A, sample after ∼ 24 h of oxidation in 2.7 kPa of air at 393°C; B, dried unoxidized sample; and C, difference spectrum (A − B).

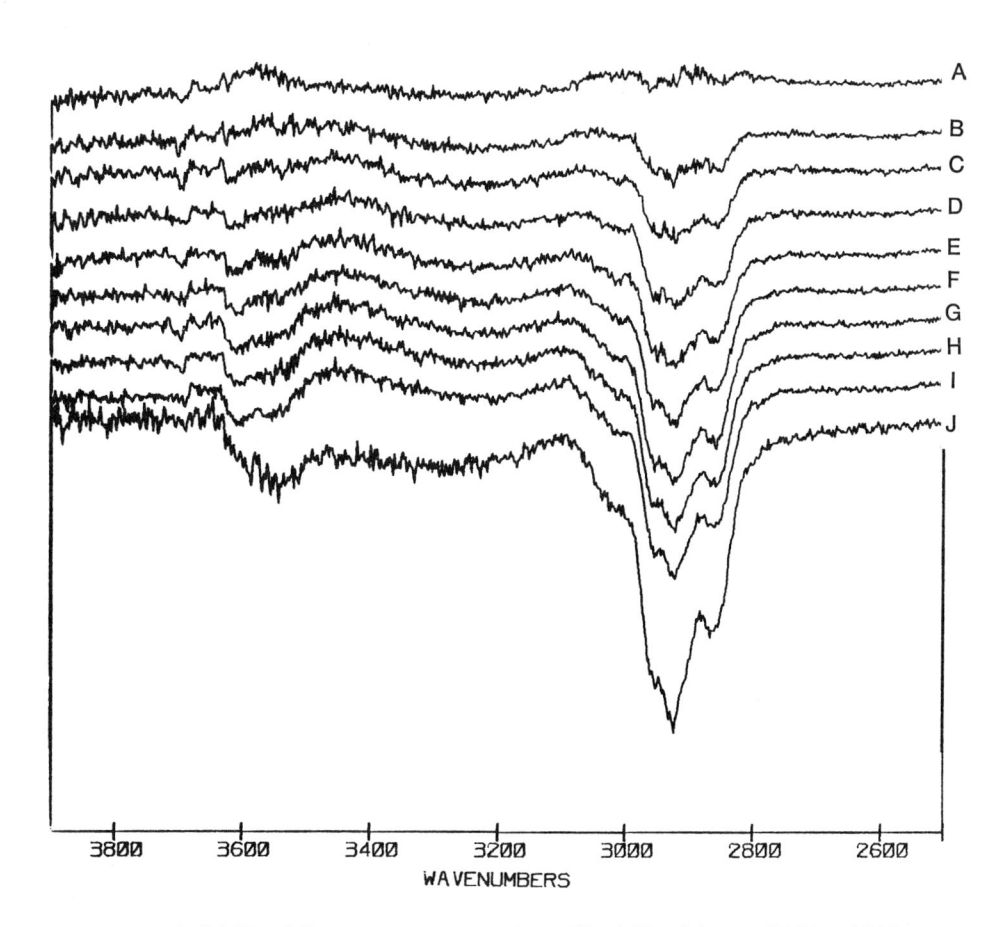

Figure 4. DRIFT difference spectra of in situ oxidized Wyodak coal (3990 − 2500 cm^{-1}). Key to time elapsed: A, 1 min; B, 5 min; C, 15 min; D, 30 min; E, 45 min; F, 1 h; G, 1.5 h; H, 1.75 h; I, 2 h; and J, 19.5 h.

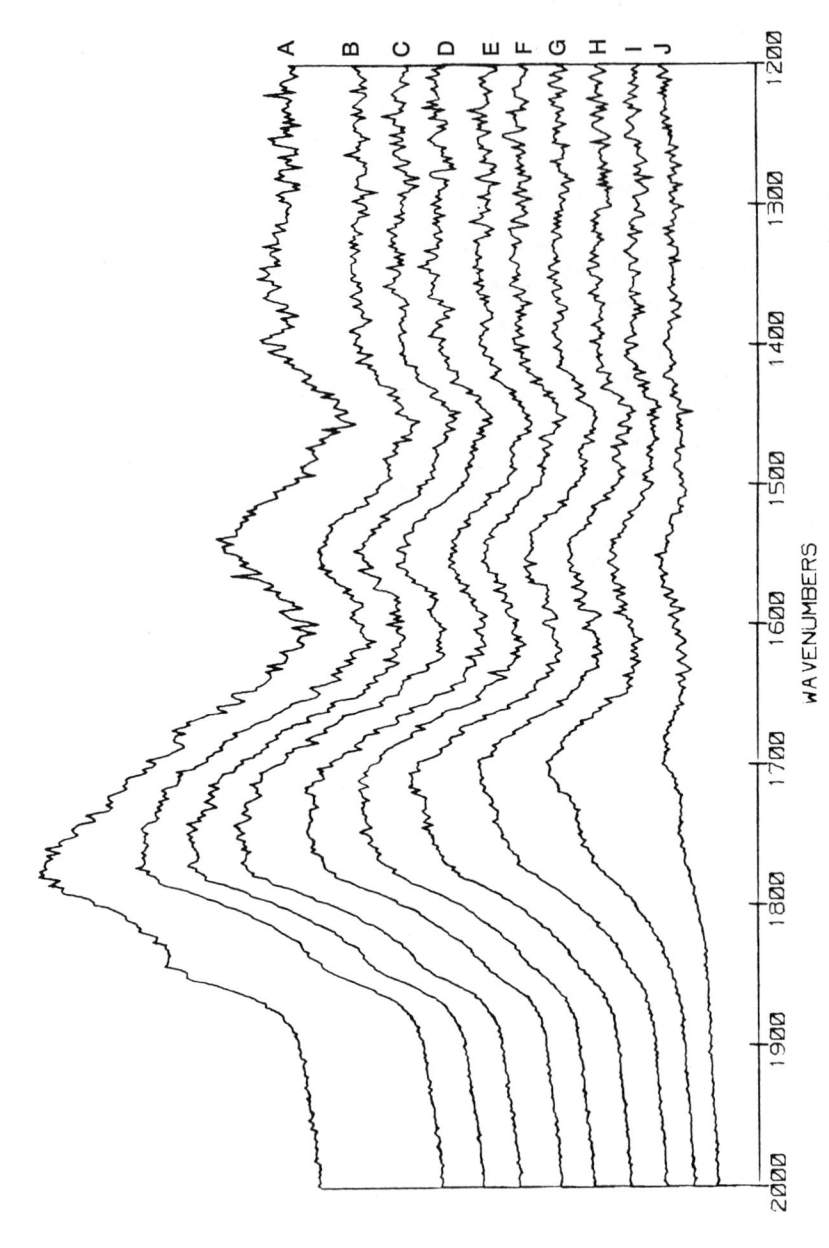

Figure 5. *DRIFT difference spectra of in situ oxidized Wyodak coal (2000–1200 cm^{-1}). Key to time elapsed: A, 1 min; B, 5 min; C, 15 min; D, 30 min; E, 45 min; F, 1 h; G, 1.5 h; H, 1.75 h; I, 2 h; and J, 19.5 h.*

doubt about the proper weighting factor for the subtraction (1:1). Problems associated with variations between samples, matrix affects, transport, and handling, etc. are nonexistent. The changes due to oxidation are evident since the mathematical differences need not be adjusted for extraneous factors.

The loss of aliphatic C-H functional groups is a measure of the degree of oxidation as noted by the increasing negative features at 2950, 2925, and 2865 cm^{-1} (Figure 4). The loss of aliphatic C-H stretching intensity has been interpreted as oxidative attack at aliphatic sites (primarily methylene groups in the benzylic position) in coal (9). There also appears to be a slight loss in aromatic C-H stretching intensity at 3035 cm^{-1} with oxidation. However, by comparing the spectrum of the unoxidized coal to that of the final oxidized state in Figure 6, the aliphatic and aromatic C-H stretching bands are observed to be roughly the same intensity in the final state indicating that oxidation has had only a minor effect on the aromatic portion of the coal. It should be noted at this point that the difference in Curves A and B of Figure 6 is represented by Curve J in both Figures 4 and 5.

The formation of certain oxidation products are noted in the series of difference spectra displayed in Figure 5. The complex band structure from ~ 1650 cm^{-1} to ~ 1875 cm^{-1} represents a broad range of carbonyl species formed during oxidation. There are at least four distinct bands which can be distinguished at various stages in the oxidation. There is also a band noted at lower wavenumbers (1550 cm^{-1}).

Painter, et al, assigned a band at 1575 cm^{-1} to a carboxylate salt. The bands observed in the present study at 1550 cm^{-1} might correspond to the frequency observed for the antisymmetric O-C-O stretching in certain carboxylate salts (35). We would, however, tend to discount carboxylate salt formation during oxidation due to lack of mobility of both the organic and mineral phases of the coal at these lower temperatures. At this particular time we have no straightforward alternative suggestion for the origin of this band. The slight increasing negative feature at 1455 cm^{-1} should also be noted. This feature is due to the loss of absorption for the methylene bending band which should accompany the corresponding loss of methylene C-H stretching absorption as previously noted.

In Figure 5 a band at 1705 cm^{-1} is observed which appears to predominate in the early stages of oxidation (note in particular Curve B). In the low temperature oxidation work of Painter, et al (9), in a higher ranked coal, a band near this value, attributed to aryl alkyl ketones, predominated in the spectra of both the early and latter stages of oxidation. A prominent shoulder is also noted in Curve B of Figure 5 at ~ 1745 cm^{-1}. This band along with a band observed at 1775 cm^{-1}, which is the dominant feature in the latter stages of oxidation (see Curve J), are perhaps due to a variety of ester type functional groups (9,10,35). The observation of a prominent shoulder at ~ 1845 cm^{-1} in Curve J might

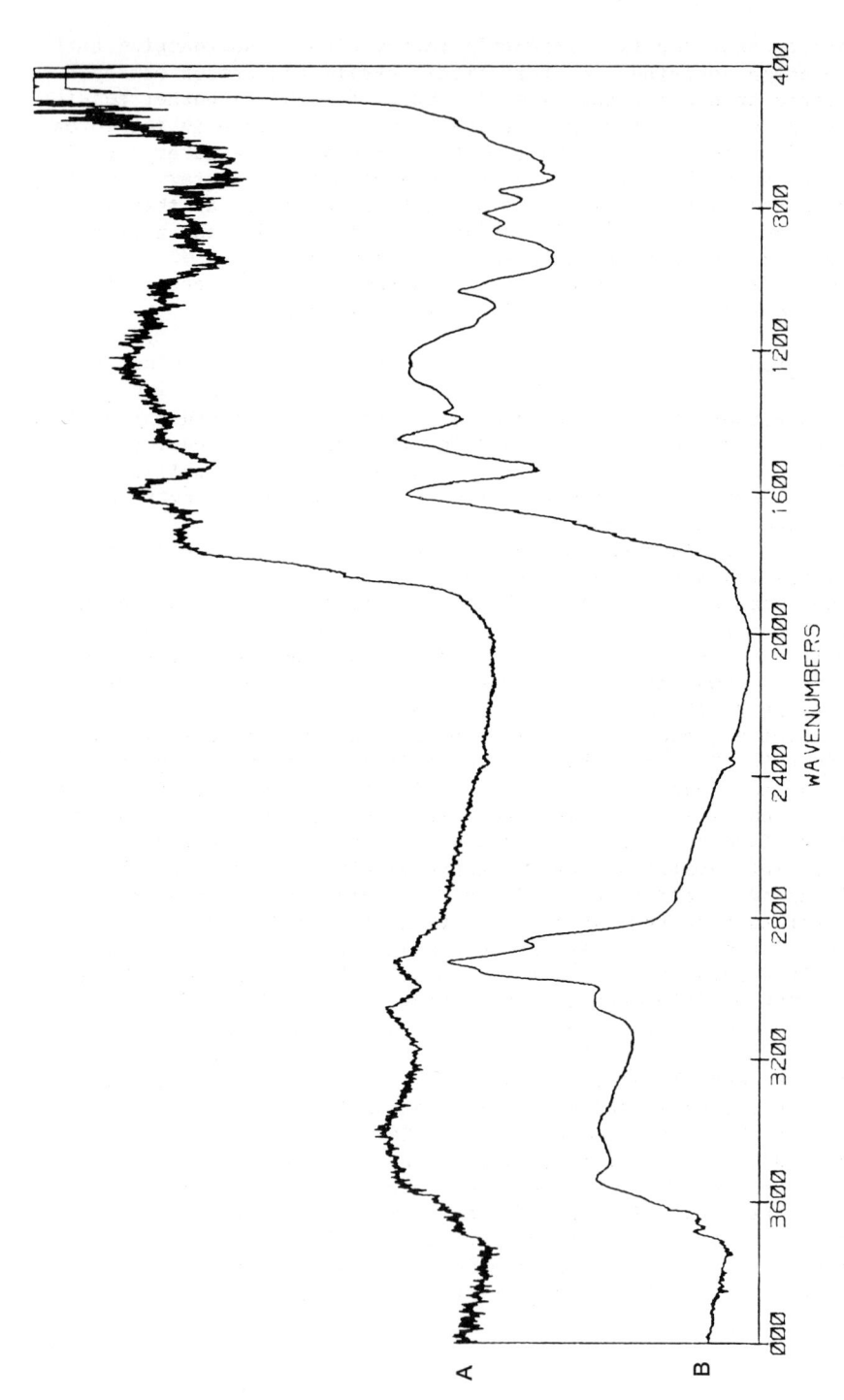

Figure 6. DRIFT spectra of Wyodak coal. Key: A, unoxidized; and B, final oxidized state.

indicate that the 1775-1750 cm^{-1} region might also contain an absorption contribution from certain cyclic anhydrides attached to aromatic or unsaturated ring structures (10,35). Another possible explanation for the origin of the 1845 cm^{-1} feature is the formation of organic carbonates (35,36). Certainly, the interpretation of the carbonyl region of the difference spectra is far from being straightforward; however, a general trend does appear throughout these spectra which would indicate a logical progression of oxidative products that might be expected from increasing reaction (36,37). The effect is manifest in two aspects: (1) enhanced absorption at (2) higher wavenumbers.

Conclusion

A number of distinct advantages over the standard KBr pellet technique has been demonstrated for the analysis of coal using DRIFT spectroscopy. The principle advantage is the ability to monitor various reaction processes in situ where temperature and fluid phase environment can be accurately controlled. Specifically, data was presented describing moisture desorption and intermediate temperature air oxidation of a powdered subbituminous coal. In comparison to its companion method, PA-IR, DRIFT spectroscopy would appear to be the technique of choice for the study of such reaction processes involving powdered samples since the temperature and environment of the sample are more conveniently controlled. Also PA-IR in general requires longer data acquisition times than DRIFT to produce a similar quality S/N ratio (34). No effort has been made in this report to treat in any way the quantitative aspects which most surely at some point must be considered. Most quantitative work involving DR spectra has utilized the Kubelka-Munk Equation to mathematically treat the data. This Equation seems to apply mainly to species in highly reflecting matrices at low dilution. Therefore, it remains to be determined what treatment may be required for DR spectral data obtained from neat materials such as coal.

It should be emphasized that the technique has been shown to work very well in defining the organic intermediates for air oxidation of coal. The progressive dehydrogenation and subsequent oxygenation of the solid substrate are quite well defined; thus we have the capability to elucidate, in real time, the intermediate states ("activated states", "surface complexes", etc.) (38) that exist in the controlled combustion the of powder where the final states are carbon dioxide, carbon monoxide, and water.

Future studies will evaluate the kinetics of combustion and the relative contribution of these oxidized entities in the global analyses related to relevant parameters: (1) coal rank, (2) particle size, (3) gas phase transport, (4) catalytic adducts, (5) coal porosity, (6) temperature, (7) pressure, etc.

Acknowledgment

This work was carried out at the Oak Ridge Y-12 Plant, operated for the U.S. Department of Energy by the Union Carbide Corporation, Nuclear Division, under U. S. Government Contract W-7405-eng-26.

Literature Cited

1. Painter, P. C., Coleman, M. M., Jenkins, R. G., and Walker, P. L., Jr., Fuel, 57, 125 (1978).
2. Painter, P. C., Coleman, M. M., Jenkins, R. G., Whang, P. W., and Walker, P. L., Jr., Fuel, 57, 337 (1978).
3. Painter, P. C., Snyder, R. W., Youtcheff, J., Given, P. H., Gong, H., and Suhr, N., Fuel, 59, 364 (1980).
4. Painter, P. C., Stepusin, S., Snyder, R. W., and Davis, A., Appl. Spectrosc., 35, 102 (1981).
5. Painter, P. C., Youtcheff, J., and Given, P. H., Fuel, 59, 523 (1980).
6. Painter, P. C., Coleman, M. M., Fuel, 58, 301 (1979).
7. Painter, P. C., Yamada, Y., Jenkins, R. G., Coleman, M. M., and Walker, P. L., Jr., Fuel, 58, 293 (1979).
8. Painter, P. C., Snyder, R. W., Pearson, D. E. Pearson, and Kwong, J., Fuel, 59, 282 (1980).
9. Painter, P. C., Coleman, M. M., Snyder, R. W., Mahajan, O., Komatsu, M., and Walker, P. L., Jr., Appl. Spectrosc., 35, 106 (1981).
10. Painter, P. C., Snyder, R. W., Starsinic, M., Coleman, M. M., Kuehn, D. W., and Davis, A., Appl. Spectrosc., 35, 475 (1981).
11. Solomon, P. R., and Colket, M. B., Fuel, 57, 749 (1978).
12. Solomon, P. R., Fuel, 60, 3 (1981).
13. Solomon, P. R., Relation Between Coal Structure and Thermal Decomposition Products, in "Coal Structure: Advances in Chemistry Series", Gorbaty, M. L., and Ouchi, K., Eds., American Chemical Society, Washington, D. C., Vol 192, Chap. 7, p 95 (1981).
14. Willey, R. R., Appl. Spectrosc., 30, 593 (1976).
15. Fuller, M. P., and Griffiths, P. R., Anal. Chem., 50, 1906 (1978).
16. Fuller, M. P., and Griffiths, P. R., Am. Lab., 10 (10), 69 (1978).
17. Kuehl, D., and Griffiths, P. R., J. Chromatogr. Sci., 17, 471 (1979).
18. Fuller, M. P., and Griffiths, P. R., Appl. Spectrosc., 34, 533 (1980).
19. Krishnan, K., Hill, S. L., and Brown, R. H., Am. Lab., 12 (3), 104 (1980).
20. Kaiser, M. A., and Chase, D. B., Anal. Chem., 52, 1849 (1980).

21. Niwa, M., Hattori, T., Takahashi, M., Shirai, K., Watanabe, M., and Murakami, Y., Anal. Chem., 51, 46 (1979).
22. Hattori, T., Shirai, K., Niwa, M., and Murakami, Y., React. Kinet. Catal. Lett., 15, 193 (1980).
23. Hattori, T., Shirai, K., Niwa, M., and Murakami, Y., Anal. Chem., 53, 1130 (1981).
24. Hattori, T., Shirai, K., Niwa, M., and Murakami, Y., Bull. Chem. Soc. Jpn., 54, 1964 (1981).
25. Smyrl, N. R., Fuller, E. L., Jr., and Powell, G. L., Appl. Spectrosc., In Press (1982).
26. Rockley, M. G., Chem. Phys. Lett., 68, 455 (1979).
27. Rockley, M. G., Chem. Phys. Lett., 75, 370 (1980).
28. Rockley, M. G., Appl. Spectrosc., 34, 405 (1980).
29. Rockley, M. G., and Devlin, J. P., Appl. Spectrosc., 34, 407 (1980).
30. Rockley, M. G., Davis, D. M., and Richardson, H. H., Science, 210, 918 (1980).
31. Vidrine, D. W., Appl. Spectrosc., 34, 314 (1980).
32. Royce, B. S. H., Enos, J., and Teng, Y. C., Bull. Am. Phys. Soc., 25, 408 (1980).
33. Laufer, G., Juncke, J. T., Royce, B. S. H., and Teng, Y. C., Appl. Phys. Lett., 37, 517 (1980).
34. Krishnan, K., Appl. Spectrosc., 35, 549 (1981).
35. Colthup, N. B., Daley, L. H., and S. E. Wiberly, Introduction to Infrared and Raman Spectroscopy, Academic Press, New York (1964).
36. Valbarth, A., Problems in Oxygen Stiochiometry in Analyses of Coal and Materials, in "Analytical Methods for Coal and Coal Products", Karr, C., Ed., Academic Press, New York (1979).
37. Berkowitz, N., An Introduction to Coal Technology, Academic Press, New York, p 100 (1979).
38. Essenhigh, R. H., Fundamentals of Coal Combustion in "Chemistry of Coal Utilization", Elliot, M. A., Ed., John Wiley and Sons, New York, 1981.

RECEIVED July 15, 1982

Comprehensive Elemental Analysis of Coal and Fly Ash

R. A. NADKARNI

Exxon Research and Engineering Company,
Analytical Research Laboratory, Baytown, TX 77520

> All major ash elements and some trace elements are determined in coal or fly ash by inductively coupled plasma emission spectrometry. Parr oxygen bomb combustion followed by ion selective electrode, X-ray fluorescence or atomic absorption spectrometric measurements are used to determine halogens, sulfur, nitrogen, mercury, arsenic, selenium, and phosphorus. Hydride generation-atomic absorption spectrometry is used to determine traces of As, Se, Sn, Sb, Te, Pb, and Bi. Spectrophotometric determinations are used for gallium and germanium.

With the increased emphasis on the development of synthetic fuels to supplement the depleting natural petroleum resources, coal is coming into prominence as a viable alternative fuel source. Since coal is a heterogeneous mixture of many minerals, it is important to have analytical methods to measure the inorganic constituents accurately and thus to be able to follow their path through various stages of coal production and utilization. This paper describes the approach of Exxon's Baytown Research and Development Division to analyzing coal and fly ash samples with some established techniques and some newer improvements that have been incorporated in them. The first paper in this series was published by us earlier ([1]) emphasizing the analysis of major elements.

Experimental

Sample Preparation.

Finely ground (usually -300 mesh) coal samples are ashed at 750°C in a muffle furnace to a constant weight. Alternatively, an RF plasma low temperature asher can be used; however, the time needed for ashing is of the order of a few days. The coal ash or fly ash thus prepared is brought into solution by dissolving in a mixture of aqua regia + HF in Parr Teflon bombs heated at 110°C for two hours. The details are described elsewhere (1).

Inductively Coupled Plasma Emission Spectrometry (ICPES).

The sample prepared by dissolution in a Parr bomb is analyzed by ICPES. We use a Jarrell-Ash Plasma AtomComp 90-975 with 90-55 spectrum shifter for background correction. The details of our instrumentation and the operation have been described elsewhere (2).

Atomic Absorption Spectrophotometer (AAS).

A Varian 475 spectrophotometer and M-65 vapor generation accessory is used for the determination of mercury through cold vapor generation, and for the determination of As, Se, Sb, Bi, Te, Sn, and Pb through hydride generation. The procedure for mercury is described elsewhere (1). In practice, after dissolution and dilution of the sample in a fixed volume, an aliquot of the solution is taken and is treated with enough HCl to proper normality.

As described later, the solutions are treated with additional reagents and allowed to react for desired time periods. The solution is transferred to the M-65 hydride generator accessory. The generator is continuously flushed with nitrogen gas to carry the hydrides to the burner. On dropping a pellet of $NaBH_4$ into the generator, gaseous metal-hydrides are immediately produced and are swept by the nitrogen gas into a heated open-ended quartz tube located on the burner-head in the light path of the AA-475 instrument. The hydrides are decomposed in the heated tube with hydrogen burning quietly at both ends of the tube. Some particles of $NaBH_4$ are physically carried over into the flame, which imparts a bright yellow color to the flame. The atomized elements are measured by the AA instrument, preferably in the peak area mode. After one measurement is completed, the solution in the generator is drained out, and it is ready for the next analysis. Details of this method are published elsewhere (3).

Parr Oxygen Bomb.

A Parr 1901 oxygen bomb apparatus is used with stainless steel combustion capsules. Less than 1 gm of coal sample is mixed with a few drops of white oil in the cup. Five mL of water is placed in the bottom of the bomb. The assembled and closed bomb is pressurized to 30 atm of oxygen. After a few seconds of combustion (<10) the bomb is allowed to cool and then carefully opened. The inside walls of the bomb are washed with water and combined with 5 mL water in the bomb. If necessary, the solution is filtered, and diluted to 50 mL. Aliquots of this solution are analyzed for halogens, mercury, sulfur, nitrogen, arsenic, and selenium by different analytical techniques. Details of this method are published elsewhere (4).

Miscellaneous Instrumentation.

Halogens are determined by an Orion 901 ion analyzer using 94-09 fluoride, 94-17 chloride, 94-35 bromide, or 94-53 iodide electrodes. These determinations can be carried out sequentially on 1 aliquot of sample in the order chloride, bromide, iodide, and fluoride. Five M KNO_3 is used as the ionization buffer for the first three ions, while Orion "total ionic strength adjustor buffer" is used for fluoride analysis. All determinations are done by the standard additions technique.

Colorimetric measurements are carried out using a Varian 634 UV-VIS spectrophotometer. Gallium and germanium are the only elements determined in this fashion.

Other miscellaneous instrumentation used in this scheme of complete coal analysis include: Phillips PW-1212 X-ray Fluorescence analyzer for sulfur, Antek-771 chemiluminescent analyzer for nitrogen, and Texas Nuclear 9500 14-MeV neutron generator for oxygen.

Results and Discussion

ICPES Analysis.

We have described our basic procedure for the "ash elements" analysis by ICPES earlier (1), and we will not describe the details here. Even though our ICPES system is equipped with 34 element channels, normally data on 20-25 elements can be obtained with good precision and accuracy in a coal ash solution matrix. Some additional trace elements have been determined with a computer controlled scanning monochromator ICPES by Floyd et al (5), however, our ICPES is equipped with fixed array of exit slits on a polychromator and some trace elements cannot be determined with reliability due to a combination of their low concentration in the coal or fly ash, and the background interference from major elemental lines in the determination of weaker trace element lines. A typical example of results obtained by the Parr bomb dissolution procedure, followed by ICPES measurements, is given in Table I. Our results for National Brueau of Standards (NBS) Standard Coal-1632 and Fly Ash-1633 are compared with NBS certified values, or "best" values from the literature, where available. The best values were obtained from over 70 papers in the literature. The obvious outliers were omitted in calculating the average values. The agreement between our results and the "known values" is satisfactory. The average accuracy of our data is ±10%, though some trace elements have

Table I

Elemental Analysis of Coal and Fly Ash
by Inductively Coupled Plasma Emission Spectrometry

Element, ppm	NBS-1632 Coal Present(a)	Found	NBS-1633 Fly Ash Present(a)	Found
Al, %	1.78	1.71	12.5	12.7
Ba	342	219	2700	2100
Be	1.45	1.53	11.9	18.7
Ca, %	0.41	0.49	4.60	5.21
Co	5.78	5.5	39.3	25
Cr	20.2±5	12	131±2	104
Cu	18+2	15	128±5	142
Fe, %	0.87±0.03	0.72	6.18	5.53
K, %	0.29	0.20	1.68	1.71
Mg, %	0.16	0.12	1.55	1.30
Mn	40±3	39	493±7	483
Na	380	368	3200	3000
Ni	15±1	15.7	98+3	--
P	71; 118	118	880	1004
Si, %	3.38	2.41	20.5	22.7
Ti	960	702	7200	6800
V	35±3	30	214±8	214
Zn	37±4	39	210±20	197

(a) From Nadkarni (1). Values with standard deviations are N.B.S. certified.

larger errors. Many other standards have been analyzed by this procedure with equally good results.

Parr Oxygen Bomb Procedure.

The main advantage of this method is in determining what we call the "volatile" elements such as halogens, sulfur, nitrogen, phosphorus, arsenic, and selenium. Normally these elements have to be determined individually with separate sample preparation in each case. However, once the coal sample is combusted in a Parr oxygen bomb, and the gases produced are trapped in an extractant such as water or diluted Na_2CO_3, all the above elements can be determined in the resultant solution. All the halogens are determined by ion selective electrodes, mercury by Cold Vapor-AAS, nitrogen by Antek chemiluminescent detector, sulfur by XRF, arsenic, phosphorus and some other elements by ICPES. Our results for NBS-1632 and 1632a coals and 1571 Orchard Leaves are given in Table II. We find good agreement between our results and the literature values. We have seen no iodine or selenium in any coal samples we have analyzed, because of their low levels of coal. The only other trace elements we found in the combustion-absorption solutions were boron and lead; however, neither of these elements seem to be quantitatively recovered by this procedure.

Vapor Generation-Atomic Absorption Spectrometry.

Even though the ICPES system is extremely sensitive for most of the trace elements, it is still not possible to determine several critically important elements such as As, Se, Sb, Pb, etc. in a coal or fly ash matrix. Though theoretical sensitivity may be impressive, the presence of large concentrations of the matrix elements, which produce interferences and high background, make the high ICPES sensitivity ineffective. Atomic absorption spectrophotometry with a hydride generation accessory is one possible way of determining this group of elements. In this mode, the elements of interest are converted into gaseous hydrides, which are then fed into the flame, decomposed therein, and the elements are determined by AAS. Mercury, which does not form a hydride, can also be determined in this same fashion by forming gaseous elemental mercury. The detection limits of ICPES, AAS with flame, and AAS with hydride generation are compared in Table III (6) where it can be clearly seen that the last mode of determination is the best. In addition, the matrix is eliminated in the hydride mode but not in flame AAS or ICPES.

Table II

Trace Elements Determination in Coal Using Parr Oxygen Bomb

Element, ppm	Method of Determination (a)	NBS-1632a Coal Pres. (b)	NBS-1632a Coal Found (c) (N=5)	NBS-1632 Coal Pres. (d)	NBS-1632 Coal Found	NBS-1571 Orchard Leaves Pres. (b)	NBS-1571 Orchard Leaves Found (c) (n=3)
Fluorine	ISE	92	84±8	90	71	-	-
Chlorine	ISE	784 (12)	770 ± 48	962	915	690	638 ± 27
Bromine	ISE	45 (13)	43	-	-	-	-
Mercury	CV-AAS	0.13 ± 0.03	0.17 ± 0.02	0.12 ± 0.02	0.12	-	-
Nitrogen,	Antek	1.27 (12)	1.19 ± 0.08	-	-	-	-
Sulfur, %	XRF	1.62 ± 0.03	1.48 ± 0.07	1.35	1.32	0.19	0.27 ± 0.04
Arsenic	ICPES	9.3 ± 1	8.9 ± 1.2	5.9 ± 0.6	5.31	10±2	8.9 ± 2.2
Phosphorus	ICPES	-	85±17	71;118	156	-	-
Boron	ICPES	53 (12)	22±3	40	29	-	-
Lead	ICPES	12.4 ± 0.6	7±1	30±9	19	45±3	31±2

(a) Explanation in the text
(b) From NBS Certificate of Analysis and literature values where available
(c$_n$) Number of replicate analyses; ± values represent one standard deviation from mean value
(d) From Nadkarni (1)

Table III (6)

Detection Limits of Different Spectroscopic Procedures

Element, ppm	ICPES	AAS Flame	AAS Hydride
As	0.04	0.63	0.0008
Bi	0.05	0.044	0.0002
Ge	0.15	0.02	0.004
Pb	0.008	0.017	0.10
Sb	0.20	0.06	0.0005
Se	0.03	0.23	0.0018
Sn	0.30	0.15	0.0005
Te	0.08	0.044	0.0015

The spectral properties of the hydride forming group of elements are given in Table IV. As is evident, the hydride generation-AAS method is extremely sensitive with a fairly large dynamic range. This higher sensitivity is obtained by using the peak area rather than the peak height absorbance mode. Use of electrodeless discharge lamps instead of the hollow cathode lamps also increases the sensitivity several fold in the case of arsenic and selenium. In the case of other elements, the EDLs are not that advantageous over HCLs.

Table IV

AAS Determination of Hydride Forming Elements

Element	Hydride	B.P. °C	Wavelength nm	Slit nm	Sensitivity ng	Range ng
As	AsH_3	-55	193.7; 197.2	1	0.2	30
Bi	BiH_3	-22	233.1	0.2	0.4	60
Ge	GeH_4	-88.5	265.1	1	0.05	--
Pb	PbH_4	--	217.0	1	1.4	300
Sb	SbH_3	-17	217.6; 231.2	0.2	0.2	30
Se	H_2Se	-42	196.0	1	0.6	100
Sn	SnH_4	-52	286.3	0.5	0.8	100
Te	H_2Te	-4	214.3	0.5	0.5	50

Most of these hydrides are generated in strongly acidic solution with some reducing agents such as zinc, magnesium, titanium trichloride or sodium borohydride. The last reagent is now almost universally perferred. It can be used as a dilute solution, which is somewhat unstable, or as solid pellets. Since ng amounts of the hydride forming elements are determined, it is extremely important to keep the working environment clean. Ultra pure reagents only, where available, should be employed, although sometimes we have found even these to be contaminated. Thus, an identical blank must be carried through the entire procedure each time. Variation in the arsenic content of the NaBH$_4$ pellets from 3 ng to 90 ng per pellet has been reported (7). The concentration of acid in solution has pronounced effect on the absorption signal. No effect on absorbance of As, Bi, and Sb was observed on changing the HCl concentration from 1 to 9 N. The selenium response is optimum in the 2.5 to 5 N range of HCl strength. The optimum HCl concentration for the H$_2$Te production is between 2.5 to 3.6 N. The production of SnH$_4$ is markedly dependent on the HCl concentration, the optimum being 0.6 to 1 N. Outside these limits, the absorbance dropped sharply. For the rest of this work 4 N HCl solution strength was used in the analysis of As, Bi, Sb, and Se. Oxidation state of some elements is also critical in analysis. Thus, arsenic and antimony have to be in +3 form, and selenium in +4 state. This is achieved by appreciable amounts of coal matrix--Na, Al, K, Ca, Mg, Fe, Ti, P, Li, Sr, Ba, Ni, Cr, Si, etc.--were found not to interfere with the hydride determination.

Lead is determined by hydride-AAS procedure following the method of Fleming and Ide (8). Tartaric acid and K$_2$Cr$_2$O$_7$ at pH 1.5 to 2 are used for optimum PbH$_4$ production.

Results of using the hydride-AAS method for two NBS standard coals and a fly ash are summarized in Table V. Each sample was run on five replicate aliquots. No bismuth was detected in any sample. No tellurium was detected in coal 1635. There are no literature values available for either of these coals; two literature values available for tellurium in the fly ash are widely different from each other and from our value, pointing to the difficulty in determining this element by most analytical techniques. Agreement of other elements between our results and the NBS values is satisfactory. The overall precision of the results is of the order of ±10%.

Spectrophotometric Determination of Ga and Ge.

The significance of these two elements in coal or fly ash is from the point of view of their recovery from coal ashes or fly ashes. The concentration levels of these two technologically valuable elements in other mineral sources (principally zinc and aluminum ores) is about the same level as in coal or fly ashes (Table VI). Both of these elements are difficult to determine by most analytical techniques. We have chosen to use two sensitive and relatively selective reagents to extract these elements into organic solvents and then spectrophotometrically measure them. The flow scheme of analysis is given in Table VII.

Table V

Determination of Trace Elements in Coal by
Hydride Generation-Atomic Absorption Spectrometry

Element	Coal-1632a Present (a)	Coal-1632a Found (b) (n=5)	Coal-1635 Present (a)	Coal-1635 Found (b) (n=5)	Fly Ash-1633 Present (a)	Fly Ash-1633 Found (b) (n=5)
Arsenic	9.3±1	9.5 ± 0.6	0.42 ± 0.15	0.28 ± 0.02	61±6	64±4
Selenium	2.6±0.7	3.1 ± 0.2	0.9±0.3	0.79 ± 0.07	9.4±0.5	8.8±0.5
Antimony	0.58	0.41 ± 0.15	0.14	0.13 ± 0.01	6.9	6.0±0.6
Tin	-	8.1±1	-	-	10.9	12.7±0.8
Lead	12.4 ± 0.6	12.4 ± 0.4	1.9 ± 0.2	1.5 ± 0.2	70±4	80±10
Tellurium	-	0.50 + 0.05	-	<0.1	2.3; 9.9	0.9±0.1

(a) From N.B.S. Certificate of Analysis or literature values where available.
(b) Number of replicate analyses; ± values represent one standard deviation from the mean value.

Table VI

Gallium and Germanium in Coals

Material	Ga, ppm	Ge, ppm
Western U.S. coals	1-7	1-22
Illinois coals	1-10	1-43
Appalachian coals	3-11	6
Eastern coals	4	13
N. Dakota lignite ashes	10-50	2-7
S. African coals	50	-
Australian coals	1-20	1-30
Indian (Assam) coals	-	0.06-1%
Indian fly ash	-	1.5%

The gallium (III)-malachite green reaction has to be performed in a reducing medium since the colored oxidation product of the dye is also extracted. Titanium chloride is used for this purpose. The color is extracted into the organic solvents from 1 to 9 N HCl solutions; however, the maximum extraction takes place in the range 5 to 6.5 N HCl. The extraction is complete in less than a minute and further shaking has no effect on the maximum extraction. Various complexing agents such as sodium citrate, sodium potassium tartarate and potassium cyanide have no effect on the extraction of the colored complex. Among the organic solvents tested for extraction behavior, CCl_4 and dichloroethane were found to have the lowest extraction capacity for the complex, while chloroform, benzene and ethyl acetate were all found to extract the complex quantitatively. Chloroform was used as the extractant in all further work.

The maximum absorption of the colored gallium-malachite green complex in chloroform was found to be 623 nm. The color once extracted was stable for three hours, after which the absorption generally decreased. The Beer-Lambert law is obeyed at least up to 0.4 ppm gallium.

Jankovsky (9) and Nadkarni et al (10) have studied the influence of foreign ions in the gallium-malachite green reaction and extraction. Only elements that can interfere are silver (by formation of AgCl); Au, Pb, Se and Te (by reduction to metallic state by $TiCl_3$); and tungsten (by hydropyrolysis). However, none of these elements are present in coal or fly ash at levels more than subppm, and hence, no serious interference from these elements is expected. Copper, Mo, Sb and Tl give colored extracts, but even in the presence of 100 to 1000 fold excess their influence on the gallium determination is small. Tin can be tolerated up to 50 fold excess. Thus, for practical purposes of coal or fly ash analysis, no particular interference is expected in the gallium determination.

Table VII

Analytical Scheme for Gallium
and Germanium Determination in Coal

Parameter	Gallium	Germanium
Dissolution	Aqua-Regia + HF in a Parr Bomb	Aqua-Regia + HF in a Parr Bomb
Solution Treatment	$TiCl_3$ + Na·K Tartrate + HCl to 6N Strength	8-9M in HCl + $FeSO_4$
Reagent	Malachite Green	Phenylfluorone
Extractant	Chloroform	Carbon Tetrachloride
Organic Phase Treatment	Wash with 6N HCl + $TiCl_3$	Back extract into water
Wavelength	623 nm	511 nm
Linearity	0-0.3 ppm	0-1.0 ppm
Interferences	None Serious	None after Preparation
Sensitivity	0.05 ppm	0.1 ppm
Reference	(9), (10)	(11)

The most sensitive and selective method for the determination of germanium appears to be a colorimetric procedure based on the reaction of Ge (VI) with phenylfluorone or 9-Phenyl-2,3,7-trihydroxy-6-fluorone (11). This reagent gives fluorescent red color with Ge (IV) solutions. A prior separation of germanium from other metals in the solution is necessary. The best procedure for this is to extract $GeCl_4$ into immiscible organic solvents. $GeCl_4$ is quantitatively extracted into CCl_4 at HCl concentrations of 8-9 M. Equilibrium is attained in 1-2 minutes of shaking. Germanium is readily recovered from the CCl_4 phase by shaking for 1-2 minutes with water. The only element extracted to a large extent (about 70%) is arsenic (III). Since small amounts of arsenic do not interfere with the phenylfluorone method for germanium, its coextraction is not a serious matter. The only other elements extracted by CCl_4 at 8-9 M HCl concentrations are: Se (IV) 0.01%, Sb (V) 0.01%, Sb (III) 0.003%, As (V) 0.002%, Hg (II) 0.003%, and Sn (IV) 0.001%. These elements are completely recovered from the CCl_4 layer by washing it with 9 M HCl. Once the germanium is separated from other metals, it can be determined by the red color it develops with phenylfluorone. This reaction proceeds slowly in dilute (0.2-1.5 M) HCl solutions. About an hour is needed for the maximum color development. The color is stable for several hours, but gradually a flocculent product precipitates out overnight. There are several ions which will interfere in this reaction by either developing color or oxidizing the reagent, but this is of little importance since the germanium will have been separated from all ions by CCl_4 extraction to its tetrachloride prior to the phenylfluorone color development. The only element to accompany germanium is arsenic (III), which is insensitive to phenylfluorone and does not interfere in the germanium determination.

The maximum absorption of the red germanium-phenylfluorone complex was found to be 510-512 nm. The Beer-Lambert law is obeyed up to at least 1 ppm Ge. Results of the determination of gallium and germanium are shown in Table VIII. Our results are mostly in good agreement with the literature values. The eight literature values available for Fly Ash-1633 seem to fall into two groups, four in between 35-49 ppm, and the other four between 58-77 ppm. The relative standard deviation for gallium is about 3% and for germanium ~3% for the fly ash sample, but ~20% for the BCR rock sample.

Table VIII

Spectrophotometric Determination of Gallium and Germanium

Sample	Gallium, ppm Present	Found(e)	Germanium, ppm Present	Found(e)
Coal-1632	6.8(a)	6.11	2.5(c)	2.6
Coal-1632a	8.49(b)	8.15	-	-
Coal-1635	1.05(b)	1.88	-	-
Fly Ash-1633	35-77(c)	49±0.4 (4)	25(c)	21±1 (3)
Rock BCR-1	20(d)	20±0.5 (4)	1.54(d)	1.33±0.3 (5)

(a) From Nadkarni (1)
(b) From NBS Certificate of Analysis
(c) From literature values
(d) From Flanagan (14)
(e) Number of replicate analyses in parentheses; ± values represent one standard deviation from the mean values

Conclusion

With the completion of this work, we now have analytical capabilities for the determination of over 40 elements in coal and fly ash. Table IX lists these capabilities, and the same data are depicted in Figure 1 in the periodic table form. Depending on the analysis desired, the coal or fly ash analytical procedure can be varied. Traditionally, a coal sample is ashed in a muffle furnace at 750°C for preparation of the ash. The same result with better control of trace element losses is achieved by using a low temperature RF plasma asher. The ash from these steps is fused with $Li_2B_4O_7$ on a Claisse Fluxer to prepare a solution for major elements analysis by ICPES. Alternatively, the ash can be dissolved in acids in a Parr bomb and analyzed by AAS or ICPES for the measurement of all major as well as trace metals. Other volatile elements can be determined by Parr oxygen bomb combustion and absorption of the gaseous products in a solvent. From this solution, individual elements can be determined, such as halogens by ISE, sulfur by XRF, nitrogen by Antek and arsenic, selenium or phosphorus by ICPES/AAS. In addition, a few specific elements are determined directly, e.g. oxygen by 14-MeV NAA and mercury by Cold Vapor-AAS. These methods, developed over the last three years, are now being applied routinely for the analysis of trace elements in several hundred coals and similar materials each month in the BARD Analytical Resarch Laboratory.

Table IX

Exxon-Baytown Analytical Methods
for Coal and Fly Ash Analysis

Technique	Sample Preparation	Elements Determined	Detection Limits in Sample ppm
ICPES	Acid Dissolution or Alkali Fusion	Al, As, B, Ba, Be, Ca, Cd, Co, Cr, Cu, Fe, K, Li, Mg, Mn, Mo, Na, Ni, P, Pb, Sb, Si, Sr, Ti, V, Zn, and U.*	0.1-10
AAS	Acid Dissolution - Hydride Generation	As, Bi, Ge, Hg, Pb, Sb, Se, Sn, and Te	0.5
Colorimetry	Dissolution Extraction	Ge and Ge	0.5
INAA	None	Oxygen	0.01%
ISE	Parr Oxygen Bomb Combustion	Halogens	50
Miscellaneous	Parr Oxygen Bomb Combustion	As, S, N, P	10-1000

*Not all of the trace elements can be seen in normal coal or fly ash samples, because of their very low concentrations and/or matrix interferences.

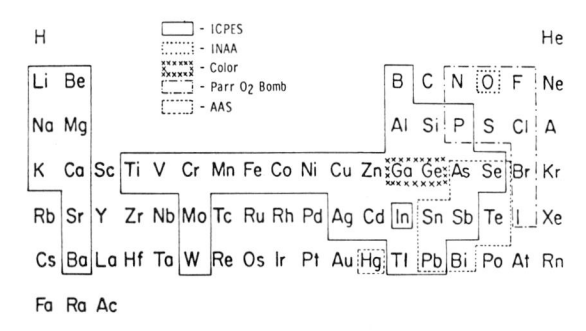

Figure 1. Exxon–Baytown analytical methods for coal and fly ash analysis.

Acknowledgements

The author wishes to express his thanks to R. B. Williams and R. I. Botto for encouragement and enthusiastic support. Able technical helps was provided by H. H. McQuitty, O. Thomas, D. M. Pond, and L. B. Jeffrey during the development of these analytical methods.

Literature Cited

1. Nadkarni, R. A., Anal. Chem. 1980, 51, 929.
2. Botto, R. I., Proc. Intnat. Conf. Dev. At. Plasma Spectrochem. Anal., Puerto Rico, 1980, In Press.
3. Nadkarni, R. A., Anal. Chim. Acta, 1982, In Press.
4. Nadkarni, R. A., Amer. Lab., 1981, 13-8, 22.
5. Floyd, M. A., Fassel, V. A., and D'Silva, A. P., Anal. Chem., 1980, 52, 2168.
6. Robbins, W. B. and Caruso, J. A., Anal. Chem., 1979, 51, 899A.
7. Knudson, E. J. and Christian, G. D., At. Abs. Newslett., 1974, 13, 74.
8. Fleming, H. D. and Ide, R. G., Anal. Chim. Acta., 1976, 83, 67.
9. Jankovsky, J., Talanta, 1959, 2, 29.
10. Nadkarni, R. A., Tejam, B. M. and Haldar, B. C., Radiochem. Radioanal. Lett., 1972, 12, 235.
11. Snadell, E. B., "Colorimetric Determination of Traces of Metals," 3rd Ed., 1959, Interscience Publishers, Inc., N. Y.
12. Failey, M. P., Anderson, D. L., Zoller, W. H., Gordon, G. E. and Lindstrom, R. M., Anal. Chem., 1979, 51, 2209.
13. Gladney, E. S. and Perrin, D. R., Anal. Chem., 1979, 51, 2015.
14. Flanagan, F. J., Geochim. Cosmochim. Acta, 1973, 37, 1189.

RECEIVED April 30, 1982

Application of Inductively Coupled Plasma Atomic Emission Spectrometry (ICP–AES) to Metal Quantitation and Speciation in Synfuels

R. S. BROWN, D. W. HAUSLER[1], J. W. HELLGETH, and L. T. TAYLOR

Virginia Polytechnic Institute and State University, Department of Chemistry, Blacksburg, VA 24061

>Metal analysis in coal derived products has typically been a laborious, time consuming process. Even with multielement emission techniques, typical sample preparations involving destruction of the sample matrix has limited sample throughput while not allowing any subsequent speciation. The direct analysis of coal derived products via inductively coupled plasma atomic emission spectrometry (ICP-AES) in organic solvents without pre-treatment is reported. Several solvents which can be employed with ICP are tabulated along with specific element detection limits. Subsequent analysis by liquid chromatography coupled with ICP detection in several modes is described for model organometallic systems as well as for several coal derived products as a first step toward speciation of organically bound metals in coal derived products.

Highly specific information concerning the chemical nature and concentration of each moiety in coal and coal conversion products is desirable if extensive coal utilization is to be achieved.(1) Considering the high complexity and heterogeneous nature of each sample, multiple, information-specific, chromatographic-spectroscopic analytical methods are required. In solution, where the maximum information is achieveable the analysis problem is further complicated by the fact that most conventional spectroscopic and chromatographic solvents have little affinity for coal and coal liquids.

Within this context, the quantitation and speciation of organically bound trace metals in coal liquefaction soluble products presents a real challenge. Quantitative trace element methods in the solid state on liquefaction

---

[1] Current address: Phillips Petroleum Company, Research and Development Laboratory, Bartlesville, OK 74005

0097-6156/82/0205-0163$06.25/0
© 1982 American Chemical Society

products have been preliminarily explored. Spark source mass spectrometry has been used to survey the distribution and general level of about sixty elements in the feed coal (West Virginia) and products from a single batch of a long-term liquefaction run on the 400 lb coal/day PDU at the Pittsburgh Energy Technology Center.(2) Atomic absorption spectrometry (AAS) was employed for precise analysis of specific elements. In each case, samples were ashed prior to analysis to destroy the organic material. Neutron activation analysis (NAA) has been recently used to obtain information on possible trace element species present in solid SRC I and liquid SRC II products derived from a Western Kentucky coal.(3, 4) Although the detection limits, multielement capability and relative lack of matrix interferences make NAA an attractive analytical procedure, long irradiation times (10 minutes to 8 hours) and counting times (up to 3 weeks) are major undesirable features. Furthermore, "on-line" analysis of separated components is totally impractical from a time/effort point of view.

Quantitation methods in solution have also been minimal and have involved either single element (continuous nebulization) analysis via atomic absorption spectrometry(5) (AAS) or multi-element analysis via inductively coupled plasma atomic emission spectrometry (ICP-AES) both on acid digested aqueous-based matrices.(6) These methods have not proven highly satisfactory because acid digestion risks the loss of volatile elements and introduces, in some cases, a substantial blank value, requires a great amount of time and labor and, in the AAS case, provides for only single element determinations with mandatory background correction. An analytical method which eliminates the sample preparation step and provides for multi-element quantitative analysis in the low ppm range seems highly attractive for coal liquefaction products. In this regard ICP-AES analysis in a pyridine matrix will be reported here. For comparison, the simultaneous determination of 15 different wear metals in lubricating oils dissolved in 4-methyl-2-pentanone appears to be the only previously reported effort at quantitation via ICP-AES in a totally organic matrix.(7)

The need for sample digestion, in most of these reports, in order to achieve a metal analysis results in modification of the chemical nature of the metals in these complex mixtures. Direct analysis in pyridine or any other organic solvent may allow for information regarding the chemical nature of specific entities to be established. Information regarding speciation, however, is not readily ascertained. While certain spectroscopic techniques such as x-ray photoelectron spectroscopy and heavy metal nuclear magnetic resonance spectrometry, are capable of yielding speciation information, these methods normally require a relatively large concentration of individual metal species. A certain degree of success in this area has been recently achieved. EXAFS has been used to characterize titanium species in solid SRC

I and liquid SRC II products and to probe the chemical and structural environment of vanadium in coal and coal liquefaction products.(8) Quantitative aspects of these studies were not considered. In the same vein, an extensive extraction (acidic methanol) and chromatographic procedure has been applied to Daw Mill coal (92 Kg) with the result that 17.8 mg of a mixture of gallium complexes of homologous porphyrins (C_{27}-C_{32}) has been isolated for the first time. Identification was based upon a combination of spectroscopic techniques.(9)

In anticipation of experiments designed to yield quantitative speciation information, we have recently established the capability for multi-element detection in toluene and pyridine matrices.(10, 11) The development of a size exclusion chromatography-ICP-AES interface designed to handle highly volatile organic solvents was reported and tested with mixtures of model organometallics. Multi-dimensional chromatographic fractions containing organically bound metals from a SRC (Wyoming subbituminous coal) have recently been isolated(12) employing this technique as a first step toward metal speciation.

The present work which we believe relates to both quantitation and speciation will (1) extend the number of solvents and chromatographic methods which can be employed using this technique, (2) present metal detection limits in a variety of solvents for both direct aspiration and pumped delivery, (3) discuss trace metal quantitative analysis data obtained on pyridine solutions of several SRC's as a function of processing conditions and coal source and (4) describe the isolation of organically bound metal fractions in a SRC process solvent.

Experimental

The ICP-AES used in this work was an ARL (Sunland, CA) model 137000. The LC interface and instrumental parameters have been described previously(10). An ice bath was used to thermostat the interface with all solvents. Chromatographic equipment consisted of a Waters 6000A (Milford, MA) pump with a Valco (50 and 200 μl loop) injection valve for isocratic separations. Columns were purchased commercially and included a Waters μ-Porasil silica column (3.9mm x 30cm), a Waters normal phase μ-Bondapak CN (3.9 mm x 30 cm) column, a Whatman Polar Amino Cyano column (4.6mm x 75cm) and a Waters 100 Å μ-Styragel size exclusion column (3.9 mm x 30 cm). Solvents were chromatographic grade from Fisher Scientific. Model organometallic compounds were from "in-house" stocks or were purchased from chemical warehouses.

Organo-metal quantitation standards in an oil matrix were obtained from Conostan (Ponca City, OK) at 300 ppm concentration of each element. Calibration standards in the range 0-25 ppm were prepared on a w/w basis in pyridine. Calibration curves were generated from these standards for subsequent metal analysis. Typical root mean square (RMS) concentration error for this curve was ~2% or better for most elements. Pyridine solutions of coal derived material were ~5-7% w/w. In order to provide continuous delivery of pyridine to the nebulizer spray chamber, a standard liquid chromatographic pump in conjunction with an injector equipped with a 3 ml sample loop was employed. To avoid dilution of this sample "plug", air bubbles were placed in front and back of the sample. Flow rates of 0.5 ml/min were employed with signal integration times of 10-30 seconds. With chloroform and heptane as solvents, direct aspiration of standard solutions was employed and detection limits for all solvents were determined for each element by calculating the concentration of analyte necessary to give a signal equal to twice the standard deviation of the background emission at each wavelength. It should be noted that all detection limits reported here are those at the interface and represent the limits for both total unseparated metal and chromatographically separated and detected metal as each comes off the column. We have observed for injected known amounts onto a size exclusion column that a 10-20 fold increase in detection limit due to chromatographic dilution is realized at the interface.

Six SRC samples derived from Kentucky No. 9 coal were obtained from the Southern Services, Inc., Wilsonville, AL pilot plant funded by the Electric Power Research Institute and the U.S. Department of Energy and operated by Catalytic Inc. Information on processing conditions was kindly supplied by Mr. Bill Weber. A process recycle solvent (92-03-035) originating at the SRC-I pilot plant in Wilsonville, AL. was obtained from Mobil Research and Development Corporation Central Research Division, Princeton, NJ.

Results and Discussion

Metal Analysis of Solvent Refined Coals. Initially elemental detection limits were determined in a variety of organic solvents. Chloroform and heptane were chosen because of their desirability as chromatographic solvents. Toluene and pyridine were selected for their tendency to solubilize coal derived products. Detection limits for each metal in chloroform, toluene and heptane were measured by direct aspiration using the nebulizer and spray chamber chromatographic interface described in reference (11), while a pumped delivery system was employed for pyridine. Individual element detection limits are shown in Table I for these solvents along with the analytical wavelengths monitored. These represent compromise emission lines for our polychromator system in order to minimize interferences while maintaining reasonably good detection limits. Background levels for blanks although higher than those of aqueous solutions showed little interference problems from organic emission. Detection limits for each metal are, in general, similar in each organic solvent as well as in MIBK which has been previously investigated.(7) An examination of Table I indicates that water continues to be the superior matrix

TABLE I
ELEMENTAL DETECTION LIMITS AND EMISSION LINES MONITORED IN VARIOUS SOLVENTS[a]

ELEMENT	PYRIDINE[b]	TOLUENE[c]	CHLOROFORM[c]	HEPTANE[c]	MIBK[8]	WATER[15]	WAVELENGTH (Å)
Ag	0.015	0.025	0.013	0.004	0.02	0.007	3280.7
Al	0.060	0.014	0.025	0.025	0.09	0.045	3082.2
B	0.007	--	0.012	0.007	--	--	2497.7
Ba	0.003	0.003	0.003	0.004	--	0.0013	4554.0
Ca	0.067	--	0.673	0.249	--	--	3179.3
Cd	0.004	0.005	0.003	0.005	--	0.0034	2265.0
Cr	0.011	--	0.005	0.009	--	--	2677.2
Cu	0.006	0.017	0.009	0.018	0.006	0.0054	3247.5
Fe	0.009	0.070	0.006	0.008	0.04	0.0062	2599.4
Mg	0.003	0.005	0.001	0.001	0.007	0.00015	2795.5
Mn	0.002	0.002	0.001	0.001	0.01	0.0014	2576.1
Mo	0.074	--	0.025	0.043	--	--	3132.6
Ni	0.038	0.034	0.026	0.096	0.1	0.015	2316.0
Pb	0.079	0.091	0.124	0.258	0.3	0.042	2203.5
Si	0.126	0.197	0.378	0.044	0.07	0.027	2881.6
Sn	0.055	0.041	0.086	0.064	0.03	0.025	1899.8
Ti	0.006	0.003	0.004	0.005	0.03	0.0038	3349.4
V	0.016	0.040	0.025	0.012	0.03	0.0075	2924.0
Zn	0.015	0.005	0.142	0.009	0.04	0.0040	2025.5

[a]All concentrations in ppm
[b]Pumped delivery
[c]Direct aspiration

for metal analysis provided there are no other experimental difficulties.

Detection limits via either direct aspiration or pumped delivery are similar for the same solvent (toluene-pyridine), since nebulizer and spray chamber are identical. Precision of analysis, however, is considerably improved in the latter case. This observation was most pronounced when pyridine solutions of coal derived material were being examined rather than solutions of elemental standards. The relatively high viscosity of the concentrated coal derived solutions (~5-7% w/w) no doubt leads to inconsistency in aspirated sample delivery. Relative standard deviations (RSD) were also improved on going from 10 second to 30 second integration times. Table II compares detection limits and RSD's in pyridine employing direct aspiration (10 second integrations) and pumped delivery (30 second integrations). The RSD's for most elements were less than 10% for the pumped delivery.

The above forced flow procedure was employed to determine trace element content in six additional pyridine soluble SRC's. Each SRC differs from another in either raw coal source, conversion severity (i.e. pressure-temperature), added Na_2CO_3 content or method of residue removal. Table III outlines the various processing parameters and the assigned run number. The measured elemental concentrations are listed in Table IV. Relative standard deviations for each analysis were, in general, less than 10%. This of course was higher for elements near their detection limit. Runs #163B and #166B (Lafayette coal) differ only in the amount of Na_2CO_3 added to the batch. The concentration of each metal for the two runs is remarkably similar. Na_2CO_3 addition is supposed to inhibit corrosion. This operation, no doubt, has been effective since #166B consistently has equal or lower metal concentrations than #163B. The decrease in Si and Fe concentration is most notable. On the other hand, the situation is different for Runs #210 and #220 (Fies coal). Few metals (Si and Sn) show concentration decreases upon addition of 25 lbs Na_2CO_3/ batch. Runs #210 and #220 were made at both higher temperature and pressure, however, metal content comparisons with similar runs (#163B, 166B) at lower temperature did not reveal any significant trends. Runs #198 and #199 employed a more conventional filtration method. Valid comparisons would be #163B vs #198 and #166B vs # 199. In most every case the metal concentration is higher for the filtered method (#199 vs #166B). This was not true for the other pairwise comparison (#198 vs #163B). It is conceiveable that colloidal mineral matter may have escaped the filter process. For certain of the transition metals concentration remained essentially constant as one might suspect if they are truly soluble organically bound species. Runs #217 and #220 enables one again to observe if processing severity influences metal content. For this coal and these conditions there surprisingly is essentially no change in metal content.

TABLE II

TRACE ELEMENT ANALYSIS OF A SOLVENT REFINED COAL[a]
MEASURED IN PYRIDINE AS A FUNCTION OF SAMPLE DELIVERY

Element	Direct Aspiration µg of element / g of SRC	RSD	Pumped Delivery µg of element / g of SRC	RSD
Ag	ND[b]	-	2.7	16.8
Al	325.8	6.8	152.3	1.8
B	121.4	2.5	87.7	14.4
Ba	1.7	43.3	1.5	5.1
Ca	789.9	55.0	601.5	3.0
Cd	ND	-	4.4	19.3
Cr	18.5	15.3	13.3	2.1
Cu	3.0	235.0	15.2	86.9
Fe	38.4	28.3	29.9	4.5
Mg	17.9	3.9	12.6	7.1
Mn	3.1	13.1	1.1	11.9
Mo	8.8	225.6	6.0	5.8
Ni	ND	-	21.7	58.1
Pb	ND	-	59.6	11.4
Si	1074	6.3	737.6	7.3
Sn	171.5	36.2	122.5	4.5
Ti	1217.0	10.3	466.8	3.2
V	13.7	6.4	6.5	7.5
Zn	8.4	32.9	33.8	6.6

[a]Wilsonville, AL, Run 199 (See Table III)

[b]ND = elemental concentration is less than ten times the detection limit as determined in pyridine

TABLE III

PROCESSING CONDITIONS FOR KENTUCKY NO. 9 SOLVENT REFINED COAL

Mine	Run Number	Severity PSI/°F	Na_2CO_3 LB/Batch	Residue Removal
Lafayette	163B	1700/825	0	Critical Solvent Deashing
Lafayette	166B	1700/825	40	Critical Solvent Deashing
Lafayette	198	1700/825	0	Filter
Lafayette	199	1700/825	40	Filter
FIES	210	2100/840	0	Critical Solvent Deashing
FIES	217	2000/785	25	Critical Solvent Deashing
FIES	220	2100/840	25	Critical Solvent Deashing

TABLE IV

TRACE ELEMENT ANALYSIS OF SRC AS A FUNCTION OF SOURCE AND PROCESSING CONDITIONS[a,b]

ELEMENT	RUN 163B	RUN 166B	Run 198	RUN 210	RUN 217	RUN 220	Run X[c]
Ag	ND	ND	ND	ND	ND	6.4	–
Al	120.4	35.9	133.8	28.9	43.3	95.8	102.0
B	47.4	36.3	70.3	45.9	47.9	66.3	–
Ba	0.5	ND	ND	ND	ND	0.7	–
Ca	134.3	133.7	65.7	85.8	133.4	82.8	73
Cd	0.2	ND	0.4	0.5	ND	0.7	–
Cr	1.3	2.2	8.9	0.2	1.3	0.7	2.8
Cu	0.7	7.2	ND	ND	ND	ND	–
Fe	268.2	14.8	167.1	27.9	7.1	27.7	3.9
Mg	10.6	4.8	9.7	1.0	5.6	3.2	–
Mn	5.0	0.2	9.2	0.7	0.3	1.4	16.6
Mo	ND	ND	ND	ND	ND	ND	–
Ni	1.1	ND	ND	8.7	ND	ND	ND
Pb	4.0	ND	2.7	ND	ND	ND	–
Si	270.6	6.7	8.1	11.7	ND	ND	–
Sn	ND	ND	ND	48.0	ND	ND	–
Ti	235.9	136.0	591.0	52.7	97.2	99.3	223.0
V	6.4	5.2	9.4	5.0	4.8	11.7	6.1
Zn	15.2	7.8	10.9	4.2	3.5	1.3	ND

[a] Concentration expressed in μg of element per gram of SRC

[b] ND = Elemental concentration is less than ten times the detection limit as determined in pyridine

[c] Reported in Reference 9; SRC obtained from Tacoma, WA pilot plant; data obtained by NAA

A number of general observations can be made but not readily explained given the limited history available on each SRC. First, Lafayette and Fies mines produce SRC of similar metal content with a few exceptions. Second, those metals which are expected to be most strongly organo-bound (mainly transition metals) do not significantly change concentrations as a function of these processing parameters; whereas, "mineral-related" elements appear to flutuate with processing conditions. It is readily apparent that speciation data is desirable.

A related study has been recently reported(4, 8) on one SRC. While the samples were drawn from a different pilot plant and the mine source is unknown, it is interesting to note that our measured concentrations via ICP-AES and Weiss' concentrations via NAA are comparable. The absence of data in Table IV indicates that the element's concentration was not monitored by Weiss.

Metal Detection In Coal Derived Process Solvent Chromatography. ICP-AES alone will not suggest speciation; however, coupled with chromatography some knowledge of the nature of metal species can be obtained. Size exclusion chromatography (SEC) of a SRC process solvent (92-03-035) with ICP-AES detection has been performed. The solvent was diluted with pyridine (1:1) and injected (200 μl) on the column. Prior to SEC, the process solvent, which had a boiling range defined as initial boiling point (IBP)-800°F, was re-distilled to yield 400-600°F and 600-800°F cuts. Figure 1 shows select metallograms for these three distillates. At the outset, it should be noted that metallic species have generally not been thought to reside in process solvents. Since the 400-600°F and 600-800°F fractions were taken from the IBP-800°F distillate, addition of the former two metallograms should yield the latter. Indeed, if one looks at the Fe 400-600°F metallogram one sees a bimodal distribution with the prominent fraction being of smaller molecular size. The 600-800°F Fe metallogram is likewise bimodal but the predominant fraction is of larger molecular size. The IBP-800°F Fe metallogram exhibits an equal distribution of small and large sized molecules. Similar type results are noted for Cu and Zn although Zn appears to be concentrated in the lower temperature distillate and the higher temperature fraction surprisingly has the smaller size Zn compounds. One explanation for this is that these compounds may be more polar and intermolecular associaton causes the higher distillation temperature. Additional elements (Mg, Ca, Ti, Cr, Mo, Mn, Co, Ni, Cd, Hg and Al) were monitored but not detected except for the trace of Mn found in the IBP-800°F fraction.

These experiments suggest (1) metallic materials are found in process solvents, (2) metals are organically bound since they have survived both distillation and chromatographic separation and (3) a large number of different molecular sized species co-exist. The exact nature of these metal components remains a mystery primarily because (1) the separation behavior of metal complexes via size

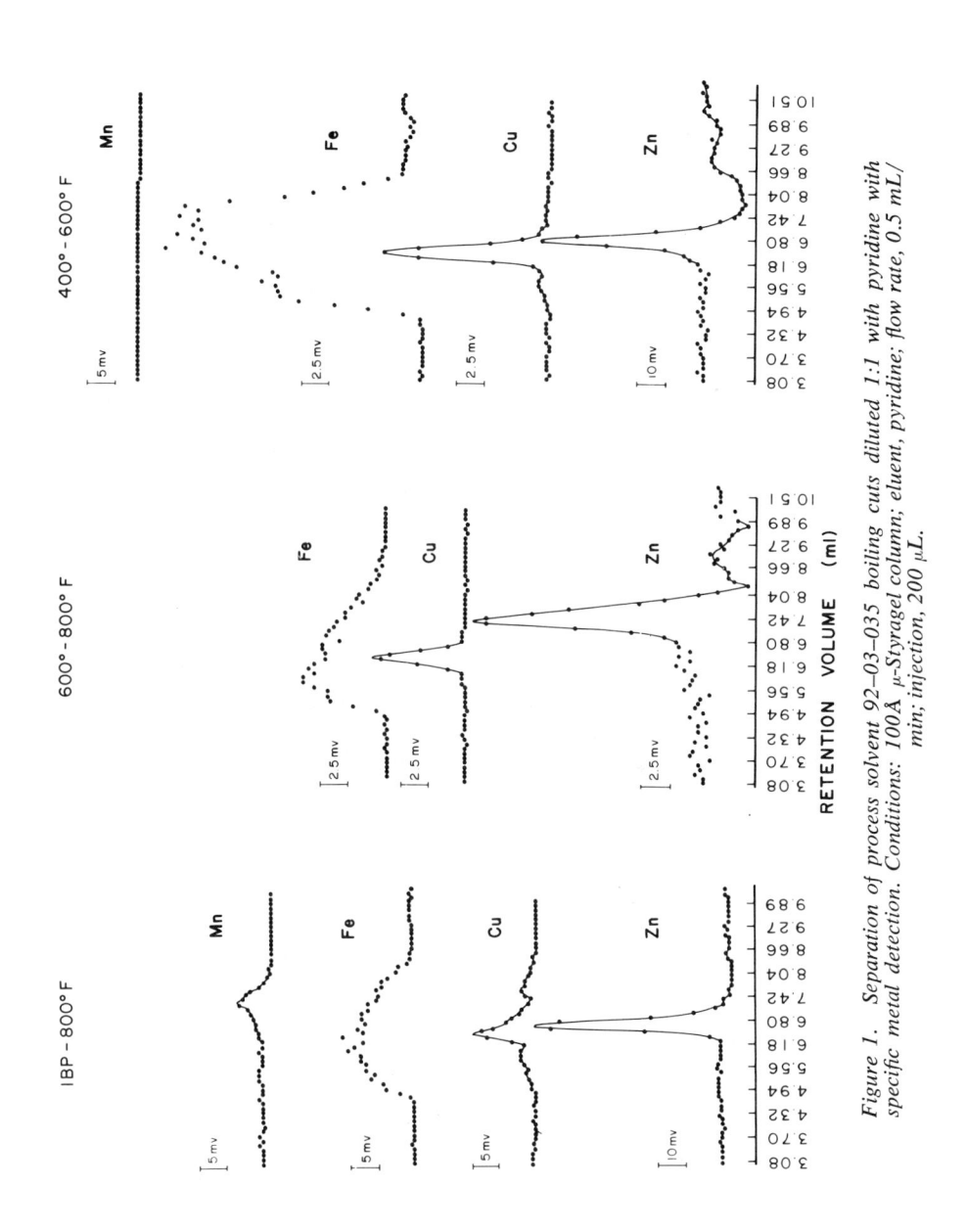

Figure 1. *Separation of process solvent 92-03-035 boiling cuts diluted 1:1 with pyridine with specific metal detection. Conditions: 100Å μ-Styragel column; eluent, pyridine; flow rate, 0.5 mL/min; injection, 200 μL.*

exclusion chromatography is not clearly established, (2) adequate, soluble models are not forthcoming and (3) SEC does not effectively concentrate a particular metal type for isolation and examination by other additional spectroscopic methods. SEC-ICP-AES, in other words, may never yield details as to the exact environment about the metal ion. Multiple peaks may represent different oxidation states or variously intermolecularly associated or solvent associated metal-containing species. Broad peaks may suggest information regarding highly reactive metal species. Clearly a better understanding of the chromatography of metallic species is needed here.

In order to employ more specific, "chemical class" separations and to gain more detailed speciation information, normal phase chromatography of metal compounds with ICP-AES detection has been explored. Normal phase liquid chromatography with ICP detection should provide a better means for eventual speciation of organically bound metals in coal derived products. Its advantage lies in the ability to separate compounds by polarity and allows use of solvents which have a reasonable solubility for coal derived products. Through extensive modeling work and the further development of separation schemes, it should be possible to more fully understand the nature of the organically bound metals present in coal derived products. One drawback to normal phase LC is that retention and irreversible adsorption of labile species such as coordination complexes may occur. The question of whether the solvent system employed is strong enough to remove the more polar materials is always subject to conjecture. Both normal phase and reverse phase gradient elution should solve the latter problem; whereas, the former appears to be most readily approached by using relatively mild interacting reverse phase materials ($-C_{18}$) with aqueous/methanol solvent systems. Only one brief report([14]) has mentioned adsorption LC-ICP-AES in a totally organic phase; therefore, some preliminary modeling has been performed in order to show feasibility of this approach.

Organometallic systems incorporating either Fe or Si served as representative model mixtures. Figure 2 illustrates a simple isocratic separation of a mixture of six organo-iron compounds. The mixture is composed of four organometallic compounds and two coordination compounds (see figure 3). The cyano-amino bonded phase silica (PAC) column used was sufficiently retentive to allow the employment of a good solubilizing solvent ($CHCl_3$) while also obtaining a reasonably good separation. While a separation via polarity is envisioned here, nevertheless the only ionic compound in the mixture, iron clathro chelate, ([15]) was not the most retentive. Interaction of the bonded phase with the free acetyl groups in acetyl ferrocene and diacetyl ferrocene must be comparable to the above in that each of these compounds flanks the ionic chelate in retention time. $Fe_3(CO)_{12}$ and $CpFe(CO)_2I$ are obviously quite non-polar being neutral compounds with highly

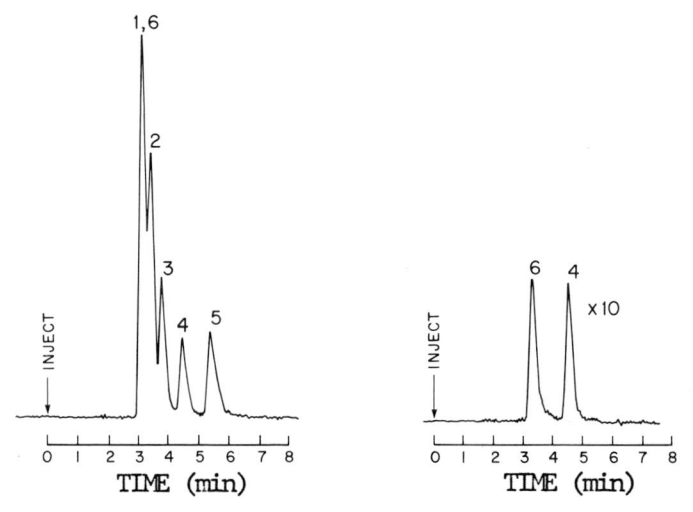

Figure 2. Separation of iron model compounds with Fe and B detection. Conditions: Polar amino cyano column; flow rate, 1 mL/min; eluent, $CHCl_3$/0.5% C_2H_5OH. Key: 1, triirondodecacarbonyl; 2, (Cp)Fe(CO)$_2$I; 3, acetyl ferrocene; 4, iron clathro chelate; 5, diacetyl ferrocene; and 6, $Fe[B(PZ)_4]_2$. Iron compounds are illustrated in Figure 3.

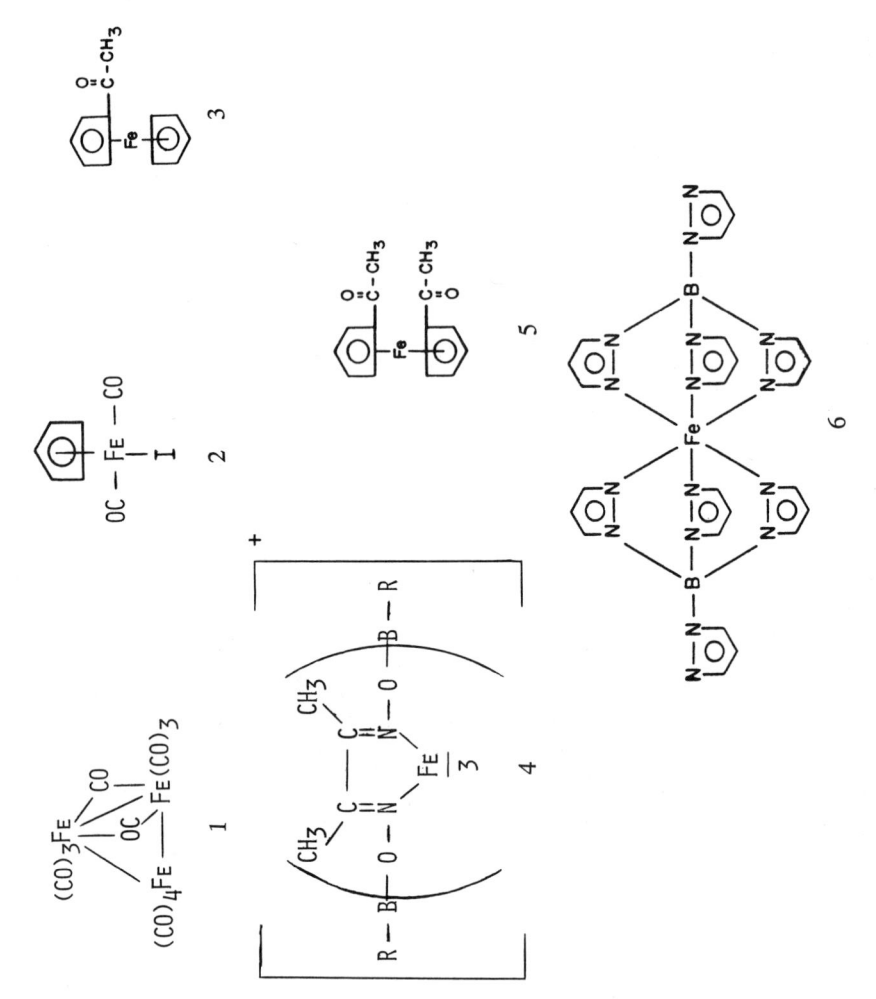

Figure 3. Molecular structures of iron model compounds. Key: 1, triirondodecacarbonyl; 2, (Cp)Fe(CO)₂I; 3, acetyl ferrocene; 4, iron clathro chelate; 5, diacetyl ferrocene; and 6, Fe[B(PZ)₄]₂.

covalent iron-ligand bonding. The higher symmetry of the former no doubt accounts for its earlier elution pattern. The early elution of iron tetrakis(1-pyrazoyl)-borate (16) is somewhat puzzling. The uncoordinated pyrazoyl group apparently has little affinity for the column. The high symmetry of the complex and its zero charge probably operate to lower the affinity also. This separation also demonstrates the multielement capability of ICP-AES detection. The two coordination compounds contain both Fe and B. Upon monitoring both Fe and B emissions, pairs of peaks are observed at the same retention time.

The second mixture studied was composed of several organo-silicon compounds. Their separation was examined as a function of column packing and solvent. Optimized separations are shown in Figure 4 for each column with the corresponding solvent system employed. By using stronger retentive packings (i.e. silica) better solvents for solubility (i.e. heptane, heptane/2% 2-propanol, $CHCl_3$/5% 2-propanol) may be employed while again achieving suitable separations. The general elution pattern for this group of silicon compounds is explained in terms of varying polarity analogous to the similar carbon containing components. An increase in retention time is observed from compound 1, the earliest eluting and hence least polar, through compound 6, the most strongly retained and most polar. The inclusion of heteroatoms accounts for this change in polarity. The separation of compounds 1, 2 and 3 on PAC with non-polar heptane shows the separation of these weakly polar components. The acetylene group is less polar than the ether group which is less polar than the carbonyl group. Compounds 4, 5 and 6 do not elute under these conditions. By changing to a weaker retaining support (CN) and adding a polar modifier in small amounts to heptane (2% 2-propanol), the most polar components are now eluted, however, the early eluting weakly polar components now co-elute. The free hydrogen on the nitrogen of compound 5 makes it more retentive than the diethyl groups on compound 4, while the dihydroxy groups are most retentive. Similarly, by going to the strongest retaining support, silica, and a much stronger mobile phase, an intermediate separation is achieved with resolution of four of the six components. Gradient elution should provide significant improvement and should allow these components to be fully separated. This will be examined in future work.

The preliminary modeling work on a variety of chromatographically stable metal systems established a basis to separate and detect metal components in the process recycle solvent. Preliminary quantitative total metal analysis showed only small amounts of a few metals. For this reason, neat injections of the sample were performed.

Large amounts of chromium, iron and silicon species were observed by normal phase separation with chloroform. Figure 5 shows the SEC of iron species in 400-600°F and 600-800°F boiling fractions. It is interesting to note the large amount of iron

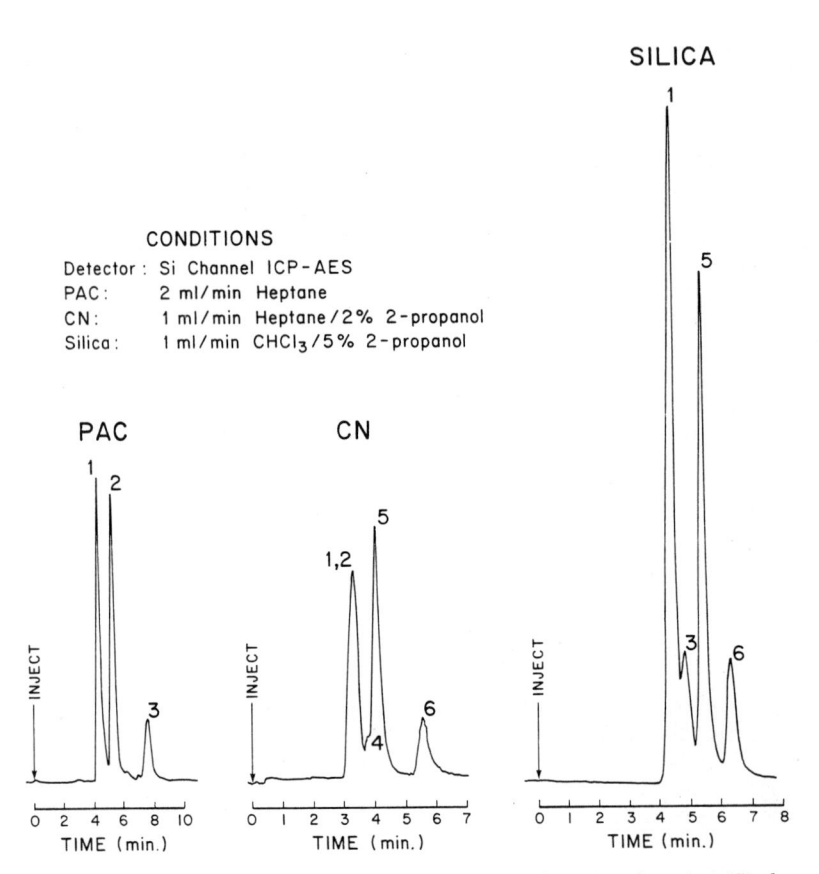

Figure 4. Separation of silicon model compounds with silicon detection (Si channel ICP–AES) as a function of column packing. Key to packing: left, polar amino cyano (PAC); middle, cyano (CN); and right, silica. Key to compounds: 1, $\phi_3 SiC\equiv CSi\phi_3$; 2, $(C_6H_5)_2Si(OC_2H_5)_2$; 3, $Si(OOCCH_3)_4$; 4, $\phi_3Si(NC_2H_5)$; 5, Hexamethyldisilazane; and 6, $\phi_2 Si(OH)_2$.

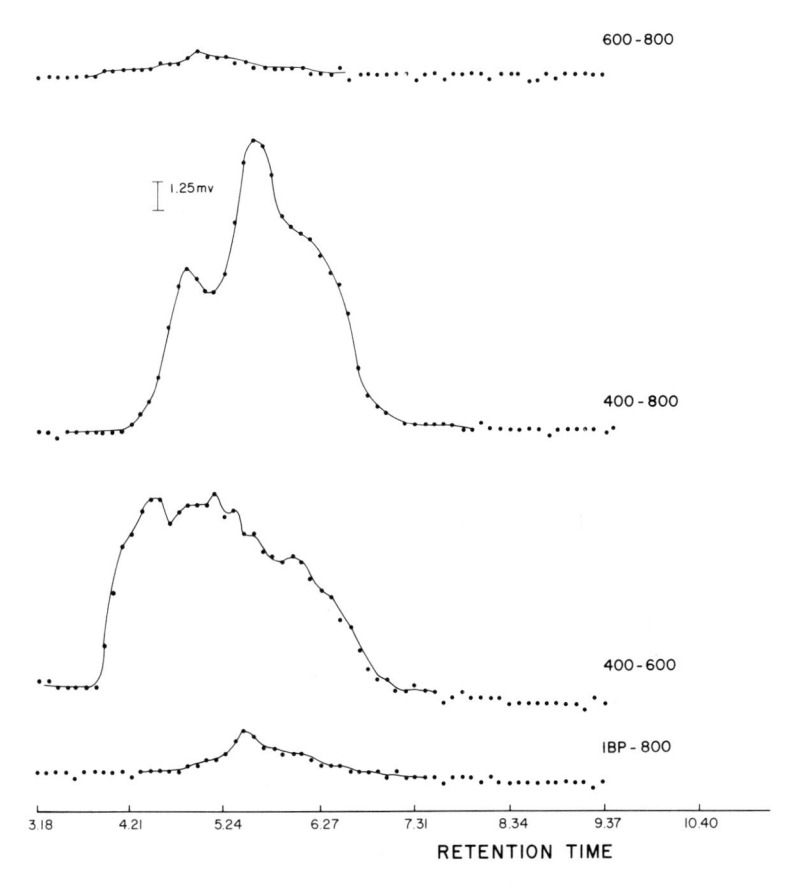

Figure 5. *Separation of process solvent 92–03–035 boiling cuts with Fe detection. Conditions: column, PAC; eluent, CHCl$_3$; flow rate, 1 mL/min; and injection, 200 μL.*

species present in the 400-800°F fraction as compared to the higher boiling fraction. This indicates that most of the iron species are volatile below 600°F. It should also be noted that subsequent analysis showed a very late eluting Fe-containing peak in the 400-800°F fraction.

The same two boiling fractions monitored simultaneously in the chromium channel (Figure 6) showed just the reverse behavior. Chromium appears to be concentrated in the 600-800°F fraction while being absent or below the detection limits in the overall 400-800°F fraction. Similarly, in Figure 7, silicon species are more prevalent in the 600-800°F fraction while appearing to a much smaller degree in the 400-800°F fraction. Other metals are either not present at measurable quantities or they are irreversibly retained on the column. The low concentration of metal species in these samples required overloading of the column with organic constituents and this accounts for the broadness of the observed peaks.

It should be noted that the conditions employed for these separations were the same as those used for the separation of the iron models. The absence of several metals which were observed in SEC of the same process solvent indicate that alternate separations will have to be developed.

Conclusions

Work to present has demonstrated both the existence of organically bound metals in coal derived products as well as the feasibility for direct detection and quantitation in organic solvents by ICP-AES. Before reliable speciation can be accomplished, better chromatographic separations need to be developed. These include preliminary separation into various fractions by polarity followed by subsequent analysis by HPLC. Work to better separate a wide range of polarities via gradient elution is desirable. Also, those materials which are labile enough to react with the normal phase packings need to be examined by reverse phase chromatography both isocratically and via gradient elution. Additional work on modeling of various organically bound metal systems needs to be accomplished especially with those labile systems which up to now have been difficult to separate except by SEC. Work to developed methods of quantitation for chromatographically separated components is necessary. Comparison of total metal content to separated metal content should be made to insure that all metal species have been removed from the column. From this work, the role of organically bound metals in the SRC process, as well as other processes, can be better understood. The ability to track metals throughout a process as well as to identify possible species should greatly aid in determining the effect of specific metals as possible catalysts and/or poisons to the conversion.

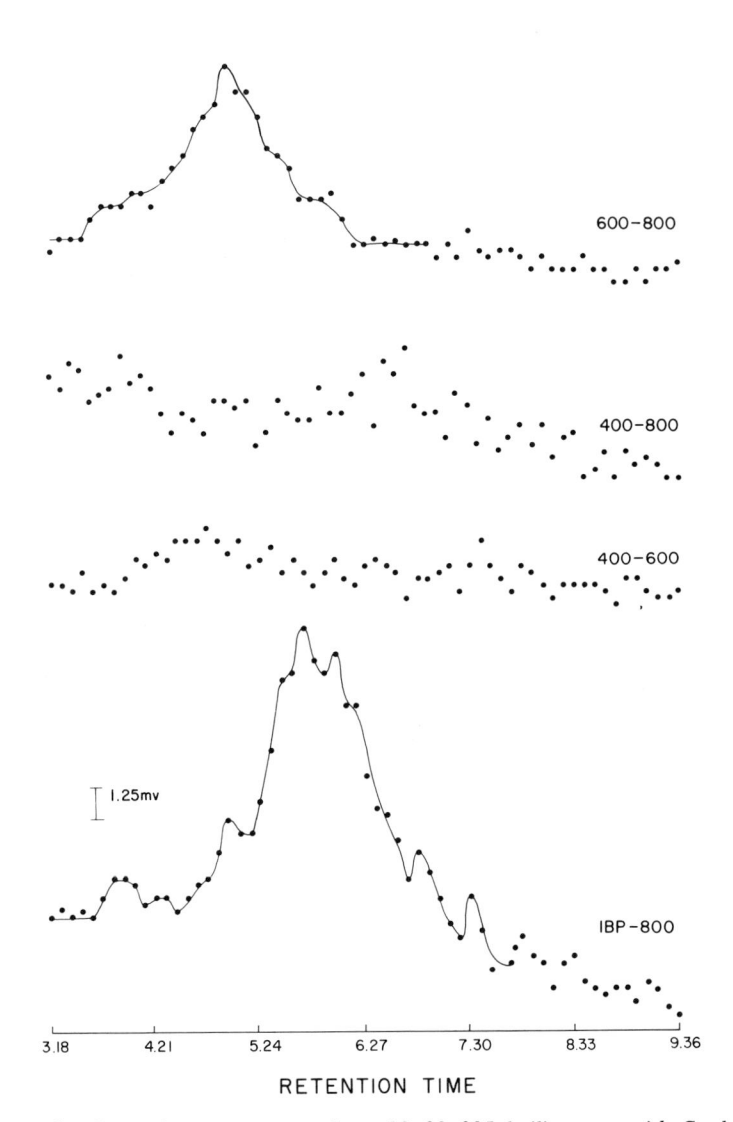

Figure 6. Separation of process solvent 92–03–035 boiling cuts with Cr detection. Conditions are the same as in Figure 5.

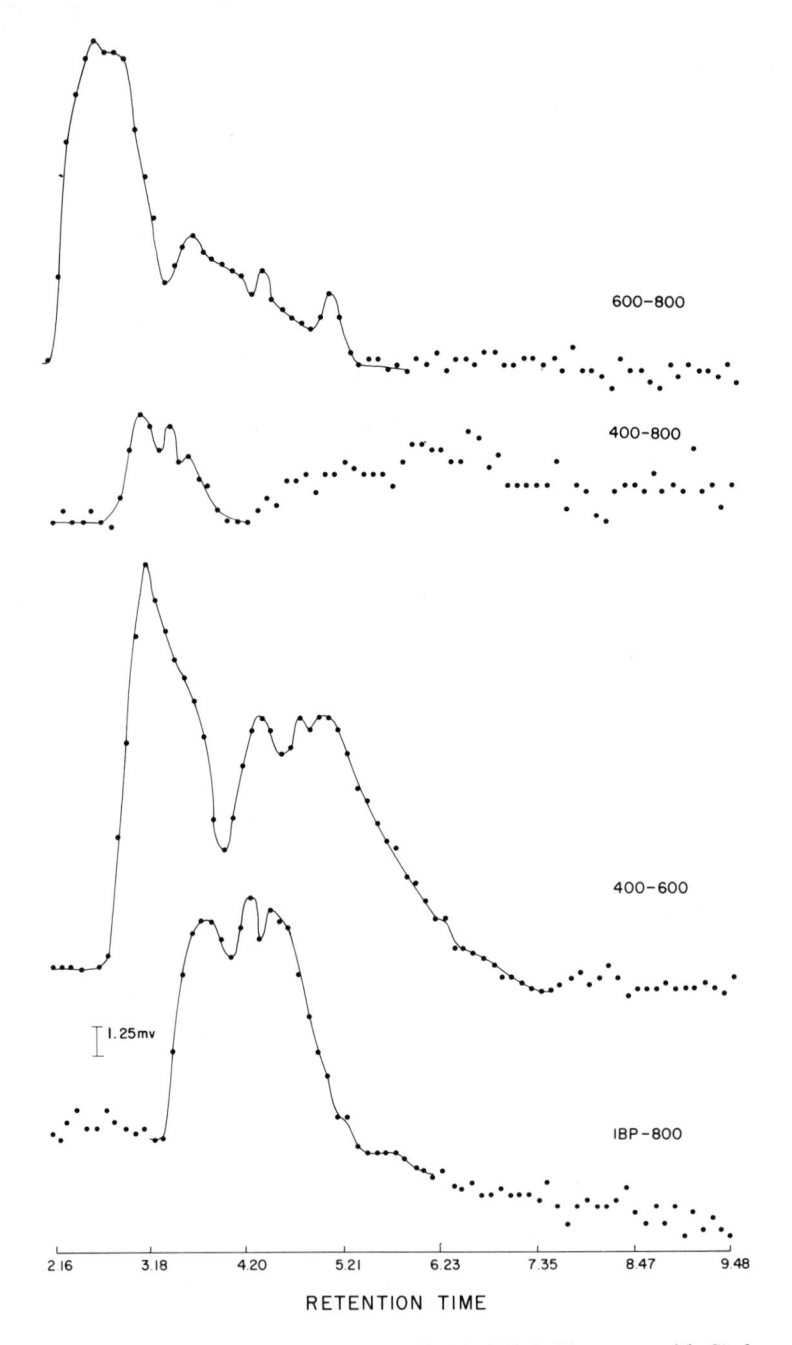

Figure 7. Separation of process solvent 92–03–035 boiling cuts with Si detection. Conditions are the same as in Figure 5.

Acknowledgement - Support by the Commonwealth of Virginia, the Department of Energy under Grant EF-77-01-2696 and the Electric Power Research Institute is gratefully appreciated.

Literature Cited

1. Gorbaty, M. L., F. J. Wright, R. K. Lyon, R. B. Long, R. H. Schlosberg, Z. Baset, R. Liotta, B. G. Silvernagel and D. R. Neskora, Science, 1979, 206, 1029.
2. Lett, R. G., J. W. Adkins, R. R. DeSantis and F. R. Brown, "Trace and Minor Element Analysis of Coal Liquefaction Products", PETC/TR-79/3, August, 1979.
3. Filby, R. H. and S. R. Khalil, "Synthetic Fuel Technology", K. E. Cowser and C. R. Richmond, eds., Ann Arbor, MI, Ann Arbor Science, p. 102 (1980).
4. Weiss, C. S., "The Detection of Trace Element Species in Solvent Refined Coal", Ph.D. Dissertation, Washington State University, 1980.
5. Coleman, W. M., P. Perfetti, H. C. Dorn and L. T. Taylor, Fuel, 1978, 57, 612 and references therein.
6. Hausler, D. W. and L. T. Taylor, Fuel, 1981, 60, 41.
7. Fassel, V. A., C. A. Peterson, F. N. Abercrombie and R. N. Kniseley, Anal. Chem., 1976, 48, 516.
8. Filby, R. H., D. R. Sandstrom, F. W. Lytle, R. B. Greegor, S. R. Khalil, V. Ekambaram, C. S. Weiss and C. A. Grimm, Proceedings DOE/NBS Workshop on "Environmental Speciation and Monitoring Needs for Trace Metal-Containing Substances from Energy Related Projects", F. E. Brinckman and R. H. Fish, eds., NBS Special Publication 618, 1981, p. 21.
9. Maylotte, D. H., J. Wong, R. L. St. Peters, F. W. Lytle and R. B. Greegor, Science, 1981, 214, 554.
10. Bonnett, R. and F. Czechowski, Phil. Trans. R. Soc. Lond. A, 1981, 300, 51.
11. Hausler, D. W. and L. T. Taylor, Anal. Chem., 1981, 53, 1227.
12. Hausler, D. W. and L. T. Taylor, Anal. Chem., 1981, 53, 1221.
13. Taylor, L. T., D. W. Hausler and A. M. Squires, Science, 1981, 213, 644.
14. Winge, R. K. Peterson, V. J. Fassel, V. A. "Inductively Coupled Plasma-Atomic Emission Spectroscopy: Prominent Lines", EPA-600/4-79-017, March, 1979.
15. Gast, C. H., J. C. Kraak, H. K. Poppe and F.J.M.J. Maessen, J. Chromatogr., 1979, 185, 549.
16. Boston, D. R. and N. J. Rose, J. Amer. Chem. Soc., 1968, 90, 6859.
17. Trofimenko, S., J. Amer. Chem. Soc., 1968, 89, 3170.

RECEIVED May 17, 1982

Determination of Chlorine in Organic Combination in the Coal Substance

J. N. CHAKRABARTI

I. S. M., Fuel and Mineral Engineering, Department of Chemistry, Dhanbad 826004, India

Opinions vary regarding the mode of combination of chlorine in coals. Early investigators reported that only part of the chlorine could be removed by extraction with water and suggested that some of it is chemically combined with the coal substance, either as organic chlorine or as adsorbed ions. Later workers, notably Crossley (1952), concluded that virtually the whole of the chlorine is present as alkali chlorides, possibly with small quantities of calcium and magnesium chlorides as well. Edgcombe (1956) found that the alkali in the extracts was equivalent to only part of the total chlorine and concluded that only half of the chlorine could be present as alkali chlorides. He held the view that the volatile chlorine represented that part of the total chlorine which was not present as alkali chlorides and that it existed in the coal as chloride ions held on the coal substance by an ion-exchange linkage.

The nature of chlorine in coal is of major interest in its industrial applications. For example, combustion of high-chlorine coal in boilers results in serious fouling by the formation of alkali-bonded deposits. A correct measure of the chlorine as inorganic chlorides is required in the case of coals that are to be carbonized, since at the temperature of carbonization these salts may react with the refractory walls, forming a glaze. This glazed coating reduces the effective life of a coke oven or of a gas retort. This is precisely the reason why the presence of an appreciable quantity of chlorine in a sample of coal is ordinarily considered detrimental to the brickwork of coke ovens and gas retorts. However, high chlorine contents in coals may not always be prejudicial to the stability of the refractories, as an appreciable percentage of it may be in organic conbination. In the latter case, there may be either volatilization or decomposition or both. Halogenated organic compounds have hardly any effect on refractories. If there is any liberation of hydrogen chloride gas

0097-6156/82/0205-0185$06.00/0
© 1982 American Chemical Society

or simple chlorine gas due to decomposition, the refractories may be simply bleached. It is thus evident that a correct estimation of chlorine in organic combination is of great significance.

In his investigations on Indian coals, the author proved for the first time the existence of chlorine in organic combination in the coal substance. It was found that in all the ten coals examined, the proportion of chlorine in organic combination was appreciable and in some cases its proportion rose to 30-36 percent of the total chlorine.

It has been suggested that on an average 50% of the total chlorine present in coals of East Midlands Coalfield in the United Kingdom (which may have total chlorine contents as high as 1%) must be assumed to be associated with organic matter, probably as hydrochlorides of pyridine bases. (Given and Yarzab, 1978.)

Work on chlorine contents of coals from Illinois Basin has shown that all the chlorine present could not be accounted for in minerals present as halite (NaCl) or sylvite (KCl) and therefore a substantial proportion is associated with the organic matter.

Outline of the Method

The chlorine in organic combination in coal is allowed to combine with hydrogen in the presence of nickel catalyst in a silica combustion tube. The hydrogen chloride gas so formed is combined with ammonia circulating through the reaction tube, forming ammonium chloride as a sublimate, which is dissolved out by washing with water. Any unreacted hydrogen chloride is arrested by hot barium carbonate kept at the exit end. The chloride formed is then estimated volumetrically.

$$H^+ + Cl^- \rightarrow HCl \qquad (i)$$

$$NH_3 + HCl \rightarrow NH_4Cl \qquad (ii)$$

$$2\ HCl + BaCO_3 \rightarrow BaCl_2 + H_2O + CO_2 \qquad (iii)$$

Apparatus and Reagents

A. Apparatus.

    (a) A silica tube 55-60 cm long and 15 cm in diameter
    (b) Two silica or porcelain boats, one for coal and the other for holding barium carobnate
    (c) Three Meker burners
    (d) Kipp's apparatus for generating hydrogen
    (e) Conical flask provided with delivery tube and containing aqueous ammonia
    (f) Vacuum desiccator

B. Reagents

  (a) Granualted Zinc
  (b) Dilute sulfuric acid
  (c) Solution of lead acetate
  (d) Solution of potassium hydroxide
  (e) Aqueous ammonia
  (f) Barium carbonate (pure)
  (g) Acetic acid
  (h) Concentrated nitric acid
  (i) Silver nitrate solution, 0.05 N
  (j) Nitrobenzene
  (k) Standard potassium thiocyanate solution, 0.05 N
  (l) Ferric alum indicator, saturated solution in water, sufficient nitric acid being added to remove the brown color, if any
  (m) 2,4,6-Trichlorophenol, A.R.
  (n) Sodium chloride, A.R.
  (o) Sugar charcoal, specially prepared.

Procedure and Standardization

A. Procedure

The silica tube is mounted horizontally by two clamps, the vertical distance being adjusted so as to utilize fully the heating effect of three Meker burners. One end of the combustion tube is provided with a cork and delivery tube and connected to a constant supply of pure hydrogen (e.g. Kipp's apparatus). The other end of the combustion tube is kept open via a constricted tube (see Fig. 1).

B. Standardization

First, an artificial mixture of pure sodium chloride, sugar charcoal and 2,4,6-trichlorophenol is put in the boat. This is introduced into the tube near the end connected to the source of hydrogen and is placed at a distance of 8 cm (see Fig. 1). A nickel foil rolled in the form of a spiral is placed at a distance of 20 cm from the same end. Another boat containing pure barium carbonate is placed just near the open end of the tube. At the outset, a steady stream of pure hydrogen, saturated with ammonia, is maintained through the tube. For this purpose, the gas is bubbled successively through water, solutions of lead acetate, potassium hydroxide, and aqueous ammonia (not shown in Fig. 1). The burner immediately below the nickel foil is started first. When the exterior surface of the tube is red hot (temperature 700-800°C) heating of the boat containing barium carbonate is started. The temperature of the boat containing the mixture

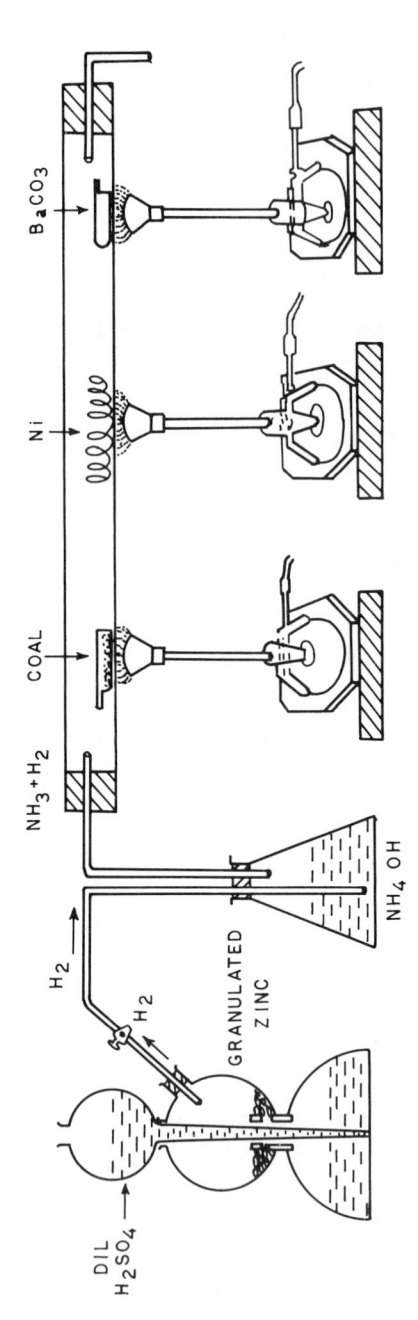

Figure 1. Experimental apparatus for determination of chlorine in organic combination in the coal substance.

of 2,4,6-trichlorophenol, sodium chloride and sugar charcoal is raised gradually and the full temperature effect is allowed only after 15 minutes. Hydrogen combines with chlorine linked to organic compounds in the presence of nickel. This then combines with ammonia, forming ammonium chloride, and is deposited on the inner walls of the tube between the nickel catalyst and the open end. The least traces of unreacted hydrogen chloride are retained by hot barium carbonate. The reaction requires a period of 45 minutes. After the reaction is over, the tube is allowed to cool. This is then washed with water to dissolve the deposited ammonium chloride, after withdrawing the boats and the spiral. Barium carbonate is also washed into the solution containing ammonium chloride. This is then rendered acidic by acetic acid, and boiled to expel hydrocyanic acid, if any. Chlorine is estimated volumetrically by Volhard's method.

This method has been found to be very suitable for obtaining a true measure of the chlorine linked to organic compounds. The inorganic chlorides, however, are not affected by this treatment. This is verified at the very outset by taking a mixture of pure sodium chloride, sugar charcoal and 2,4,6-trichlorophenol in the boat and estimating the chlorine liberated. As the organic compound, i.e. 2,4,5-trichlorophenol, is likely to contain some moisture, it is dried in a vacuum desiccator prior to being weighed. The results (Table I) show that the organic chlorine alone is liberated in this way. The marginal differences between the observed and calculated values of chlorine are probably due to the fact that the substance was not 100% pure. However, the differences are within the limits of experimental error.

Having thus standardized the method for estimating organic chlorine, coal samples crushed to pass through a 240 mes BS, or 6 mesh (64 µm) 15 sieve, are put in the boat and their organic chlorine contents determined as described for the standardization.

For accurate determinations of chlorine, it is advisable to carry out a blank determination using sugar charcoal instead of coal and to determine the chlorine collected, if any, as ammonium chloride and also in the barium carbonate.

TABLE I

Standardization with a Pure Organic Substance

Experiment No.	Substance	Chlorine % Found	Chlorine % Calculated
1	2,4,6-trichlorophenol	53.49	53.93
2	Do	53.56	53.93
3	Do (0.144g) + NaCl(0.7917g)	54.10	53.93
4	Do(0.1509g)+NaCl(1.023g)	54.09	53.93

The results of experiments with ten coals from different localities are shown in Table II. For a comparative study, the total chlorine content of each coal sample was also determined by the Eschka method.

TABLE II

Chlorine Contents of Indian Coals

Sample No.	Organic Chlorine on d.a.f. Coal (%)	Total Chlorine on d.a.f. Coal (%)	Ratio of Total/Organic Chlorine	Organic Chlorine as % of Total Chlorine
1	0.097	0.395	4.07:1	24.50
2	0.124	0.495	4.0:1	25.05
3	0.127	0.418	3.29:1	30.38
4	0.043	0.426	9.9:1	10.09
5	0.032	0.322	10.06:1	9.93
6	0.124	0.538	4.33:1	23.04
7	0.045	0.551	12.2:1	8.16
8	0.082	0.340	4.1:1	24.10
9	0.068	0.537	7.9:1	12.66
10	0.127	0.350	2.75:1	36.20

Literature Cited

1. Given, P.H.; Harzab, R.F. In "Methods for Coal and Coal Products", Anal. Vol. II, 1978, 19-20.

RECEIVED July 14, 1982

Electron Probe Microanalysis
A Means of Direct Determination of Organic Sulfur in Coal

ROBERT RAYMOND, JR.

Los Alamos National Laboratory, Earth and Space Sciences Division, Los Alamos, NM 87545

An analytical method of measuring organic sulfur directly by use of the electron probe microanalyzer (EPM) avoids the uncertainty of calculating organic sulfur by difference. The EPM enables rapid, nondestructive determination of the organic sulfur content of individual macerals. Thus, total organic sulfur content of a coal may be computed by a mean modal analysis of the macerals. Twenty-nine coals, collected from 13 states within the contiguous USA, and ranging in rank, age, and organic sulfur content (0.2 to 5.3 wt% dmmf) have been analyzed. When plotting organic sulfur content of the coals vs organic sulfur content of respective vitrinite components, the best linear fit of the data has a correlation coefficient of 0.99, a slope of 0.98, and a y-intercept of -0.03. Empirically, organic sulfur content of a coal (dmmf) essentially equals organic sulfur content of its vitrinite. Note that analyzed samples contained as little as 41.9 wt% vitrinite macerals (dmmf).

For EPM organic sulfur analysis representative samples (-20 to -100 mesh) are potted in epoxy pellets, polished, and carbon coated. Vitrinite grains are identified during analysis by shape and texture, with resulting organic sulfur contents equivalent to those determined when using oil-immersion, reflectance techniques for vitrinite identification. According to t-statistics, analyzing 15 vitrinite grains achieves a maximum variability of less than 0.20 wt% in the coals that contain less than 2.00 wt% organic sulfur. Instrument time to analyze 15 grains, and therefore to determine the organic sulfur content of a sample, is less than 10 minutes. To test the EPM method,

0097-6156/82/0205-0191$06.00/0
© 1982 American Chemical Society

we chose coals containing either no inorganic sulfur (organic sulfur = total sulfur), or for which ASTM organic sulfur values were corrected for unextracted iron (which takes into account any unextracted pyrite). These 13 samples contained up to 2.09 wt% organic sulfur and variations between EPM and ASTM values were less than 0.16 wt%.

As our need for efficient coal utilization increases, inability to interpret the occurrence and the amount of sulfur in coal becomes increasingly serious. The effect that sulfur has on coal utilization, and on the environment is staggering. We can theorize that sulfur form and content vary within a coal as a direct result of the geochemistry of precursor peatforming environments, source areas surrounding the peats, and diagenetic conditions occurring later during coalification. Still, we have little knowledge about the emplacement of sulfur in peats, and of the transitions it goes through during coal maturation. Largely because of this lack of knowledge is our basic inability to anticipate and control the effects of sulfur during coal utilization.

The American Society for Testing and Materials (ASTM) Standard Test Method D2492-68 for sulfur analysis in coal specifies the analytical determination of values for total, pyritic and sulfate sulfur. Organic sulfur is calculated by subtracting pyritic and sulfate sulfur from the total. The procedures are aimed at providing rapid, inexpensive, and reproducible data for coal utilization. The pyritic, sulfate, and organic sulfur contents reported by the processes adequately reflect the amount of sulfur that can be removed by sizing, specific gravity separation, and hindered settling techniques. But any error in total, pyritic or sulfate sulfur determination will show up as an error in organic sulfur determination. Reasons for error in pyritic sulfur determinations are reported by Edwards et al. (1) and Greer (2).

The electron probe microanalyzer (EPM) uses a finely focused electron beam that impinges on a polished sample, generally at 15-20 keV, producing x rays characteristic of the elements present in the sample. In general, EPM results are accurate and reproducible to ±2% of the amount present for most major elements. Relative accuracy decreases with decreasing elemental concentrations. For elemental concentrations of about 1 wt%, relative accuracy should be within ±5%. The EPM revolutionized petrology and geochemistry, but the paucity of published papers reporting its use in coal and peat studies shows that EPM has been neglected in coal research. Electron microanalysis of coal generally has been limited to the scanning electron microscope (SEM) with energy dispersive x-ray (EDX) capability. The unique capabilities of the EPM can provide information that is inaccessible through other methods. Examples are x-ray chemical shift determinations, light element detection, and resolution of x-ray lines that have EDX interferences by wavelength dispersive x-ray (WDX) techniques. Modern combination EPM-SEM instruments equipped with both EDX and WDX capabilities are especially powerful tools for coal analysis.

EPM has important advantages over conventional methods of analysis for organic sulfur in coal: analysis by EPM is done directly, thus avoiding problems associated with calculating organic sulfur content by difference; organic sulfur contents of individual macerals can be measured in situ in a sample; and organic sulfur analysis with the EPM is non-destructive and rapid. The following shows how the EPM may be used for quantitative analysis of organic sulfur in coal.

Background

Initially Raymond and Gooley (3) calculated the organic sulfur content of a coal with the EPM using a mean modal analysis technique. For this work well-documented samples from the Pennsylvania State University Coal Bank were analyzed. The samples had previously been ground to -20 mesh size. Upon receipt, the samples were split, potted in epoxy, and made into polished 2.5 cm rounds, approximately 120 μm thick. The rounds were mounted on 27x46 mm glass slides, and photomicrograph mosaics of portions of the samples were prepared at approximatley 400X magnification. Mosaics are necessary since the samples must be carbon-coated for conduction during electron bombardment, and once the samples are carbon coated, reflectance levels are obscured and identification of all the various macerals is extremely difficult. Coal macerals present in the mosaics were identified by oil-immersion, reflected light techniques. Where possible, 15 examples of each maceral type within the section were located. In most cases less than 15 pieces of one or more particular macerals could be found, and so fewer than 15 were therefore included in the analysis. Our EPM has a 400X magnification optical system, and using morphology of the various coal grains in conjunction with the mosaics, the points of EPM analysis could be exactly located.

Standardization datum for EPM analysis is determined as an average of seven background and matrix corrected intensities on a recently developed petroleum coke standard (4). After each coal analysis the computer prints x-ray intensities and wt% sulfur as determined by comparison with the x-ray intensity of the petroleum coke standard. X-ray intensity of iron is continually monitored during analysis and any intensities in excess of background levels are assumed to result from contamination by pyrite, so those results are rejected. At the same time, an energy dispersive system output is monitored for any elements that might suggest the presence of other sulfides or sulfate minerals. Therefore our reported sulfur contents are most likely organic.

Another possibility is elemental sulfur, but the homogeneous distribution of sulfur throughout each maceral is indicative of organic rather than elemental sulfur.

The results of a sample analysis include a number of organic sulfur measurements for all macerals present in the sample. An example of the analysis of a high volatile A bituminous (hvAb) coal can be seen in Table I. The mean organic sulfur content (S_o) for each maceral-type determined by EPM was multiplied by the wt% of that maceral in the dry coal sample. The Pennsylvania State University Coal Section determines the wt% of individual macerals in the coal by multiplying maceral density by volume % in the coal. Volume % is determined by counting 1000 points, a method that the Pennsylvania State University Coal Section has been able to show has a reproducibility of 2-3%. The sulfur contents contributed to the total organic sulfur content of the coal by the individual macerals is also shown in Table I. The summation of the contributions of the individual macerals gives the total organic sulfur concentration (dry) as determined by EPM.

To measure the validity of the EPM technique coals were chosen in which sulfate sulfur as determined by ASTM methods equaled zero and in which pyritic sulfur was minimal as determined by ASTM methods and as observed by optical microscopy. Since inorganic sulfur contents are small, any discrepancies between EPM and ASTM organic sulfur contents due to inaccurate pyrite or sulfate analysis also should be small. As can be seen in Table II the EPM analyses very closely approach those of the ASTM.

The comprehensive EPM method discussed above is extremely time consuming. The method requires point counting the sample to determine the wt% of the various macerals. An oil-immersion photomosaic has to be prepared for identification of analytical sites once the sample has been placed in the EPM. Finally, greater than 60 EPM analyses must be performed to determine the organic sulfur content of a single coal sample. Data derived from analysis of numerous coals using the comprehensive method,

Table I: Comprehensive EPM method of organic sulfur analysis of L. Elkhorn, KY, hvAb coal (after _3_).

MACERAL	wt% S_o	MACERAL wt%	wt% S_o / MACERAL
Vitrinite	0.61	52.8	0.32
Pseudovitrinite	0.56	16.4	0.09
Fusinite	0.27	6.2	0.02
Semifusinite	0.44	6.2	0.03
Sporinite	0.64	6.0	0.04
Micrinite	0.59	2.8	0.02
Macrinite	0.51	0.7	(0.004)

Total S_o (dry) = 0.52 wt%

Table II: ASTM/EPM comparative study
(after (3); all contents as dry wt%)

COAL	RANK	ASTM PYRITIC S	ASTM ORGANIC S	EPM ORGANIC S
U. Elkhorn #3	hvAb	0.01	0.61	0.63
Ohio #5	subC	0.01	0.92	0.94
L. Elkhorn	hvAb	0.03	0.52	0.52
Hazard #7	hvAb	0.03	0.51	0.58
U. Sunnyside	hvAb	0.06	0.59	0.66
Blind Canyon	hvAb	0.13	0.33	0.41
Dietz #3	subC	0.13	0.15	0.18

though, provided us with a better and more confident EPM approach.

The comprehensive method was performed on 29 coals that represented 27 seams from 13 states in the contiguous USA. Rank of the coals ranged from subbituminous C to low volatile; total sulfur contents ranged from 0.28 to 8.60 wt% (dry)(Table III). The organic sulfur contents of the coals determined by the comprehensive EPM method are plotted vs the organic sulfur content of respective vitrinite components (Figure 1). The best linear fit of the data has a correlation coefficient of 0.99, a y-intercept of -0.03, and a slope of 0.98. Empirically, the organic sulfur content of a coal essentially equals the organic sulfur content of its vitrinite.

Raymond (5) showed that a general relationship exists in most coals with respect to organic sulfur contents of the macerals: sporinite, resinite $\geq$ micrinite, vitrinite > psuedovitrinite $\geq$ semifusinite $\geq$ macrinite > fusinite (Table IV). How then can the vitrinite organic sulfur content be representative of all macerals present in a coal sample? In most of the 29 coals discussed above (as is the case in most coals) the vitrinite macerals dominate (Table III and V). Thus the vitrinite organic sulfur content has a major influence on the organic sulfur content of the coal. But what of the coals containing as little as 41.9 wt% vitrinite macerals? Table VI contains the wt% of the various macerals found in two coals and the organic sulfur contents of those macerals. As can be seen in Table VI, the vitrinite organic sulfur contents approximate the organic sulfur contents determined from the weighted mean of the other macerals present. Therefore, as well as commonly being the most dominant maceral, the vitrinite contains an organic sulfur content approximately equivalent to the mean of all the macerals.

Two factors make it advantageous to measure the organic sulfur content of vitrinite in order to find the organic sulfur content of a coal. The most obvious is that the number of EPM analyses will be fewer. For each of the coal samples listed in

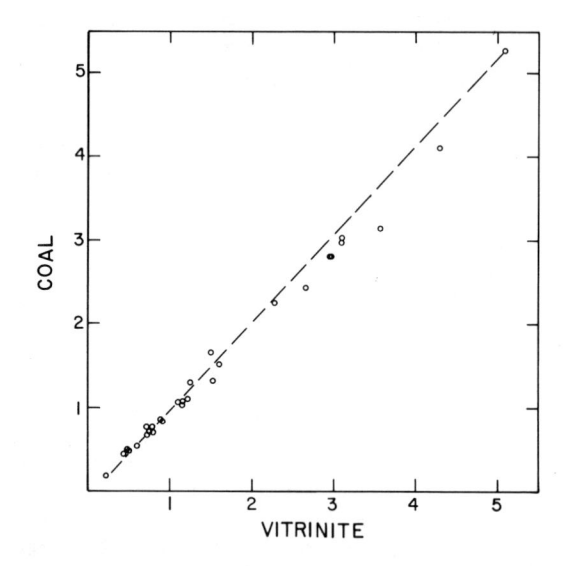

Figure 1. Organic sulfur content of coal vs. organic sulfur content of respective vitrinite components for coals listed in Table III (dmmf wt %). All measurements by EPM.

Table III: Various characteristics of the 29 coal samples on which the comprehensive EPM method was performed (Total S on dry basis; all other values on dmmf basis), V = vitrinite macerals, I = inertinite macerals, L = liptinite macerals, S_o Mod. = organic sulfur value determined by maceral mean modal analysis, S_o Vit. = organic sulfur value determined by vitrinite alone.

Rank	Tot.S	Seam Name	State	%V	%I	%L	S_o Mod.	S_o Vit.
lvb	2.07	L. Kittanning	PA	76.0	24.0	0.0	0.46	0.48
"	3.29	U. Hartshorne	OK	92.2	7.8	0.0	0.84	0.89
"	6.01	U. Kittanning	PA	76.4	23.5	0.1	0.74	0.78
mvb	0.82	Sewell	W.VA	66.9	28.7	4.4	0.65	0.77
"	1.62	U. Freeport	PA	87.6	9.1	3.3	0.48	0.52
hvAb	0.46	Blind Canyon	UT	88.5	9.8	1.7	0.43	0.44
"	0.54	Hazard #7	KY	70.9	20.1	9.0	0.74	0.77
"	0.55	L. Elkhorn	KY	75.9	17.5	6.6	0.55	0.61
"	0.62	U. Elkhorn	KY	41.9	38.9	19.2	0.67	0.73
"	0.65	U. Sunnyside	UT	89.8	8.5	1.7	0.71	0.75
"	1.96	Clintwood	VA	64.5	31.2	4.3	0.81	0.88
"	2.34	Americana	ALA	68.7	24.9	5.4	1.32	1.52
"	2.62	Pittsburgh	PA	83.6	13.7	2.7	1.08	1.19
"	2.93	Pittsburgh	W.VA	75.1	20.9	4.0	2.41	2.65
"	3.73	L. Kittanning	PA	85.0	11.8	3.2	1.07	1.14
"	5.64	L. Clarion	PA	83.9	11.0	5.1	1.28	1.23
"	6.55	Clarion	PA	72.6	16.9	10.5	1.64	1.50

Continued on next page.

TABLE III (continued)

Rank	Tot.S	Seam Name	State	%V	%I	%L	S_o Mod.	S_o Vit.
hvBb	3.39	Ohio #4	OH	75.0	17.7	7.3	2.80	2.93
"	3.60	Illinois #2	IL	78.3	18.4	3.3	3.02	3.09
"	4.55	Kentucky #14	KY	83.4	12.0	4.6	1.16	1.13
"	5.85	Bevier	MO	69.6	28.7	1.7	2.97	3.08
"	7.27	Tebo	MO	75.3	19.1	5.6	3.13	3.55
hvCb	2.59	Illinois #6	IL	84.7	10.9	4.4	2.26	2.28
"	3.86	Illinois #5	IL	89.2	10.1	0.7	2.80	2.96
"	6.14	L. Cherokee	IA	74.3	22.3	3.4	5.29	5.10
"	8.60	unknown	IA	89.4	9.3	1.3	4.08	4.31
subB	0.28	Dietz #3	WY	83.8	15.6	0.6	0.19	0.21
subC	0.93	Ohio #5	OH	84.6	10.0	5.4	1.04	1.16
"	1.78	Wildcat	TX	88.8	10.0	1.2	1.50	1.59

Table IV: General relationship between macerals and organic sulfur contents for high and low sulfur coals (S_o wt% on dmmf basis) S = sporinite, R = resinite, Mi = micrinite, V = vitrinite, Pv = pseudovitrinite, Ma = macrinite, Sf = semifusinite, F = fusinite (after 5).

Rank	Coal	S	R	Mi	V	Pv	Ma	Sf	F
lvb	L. Kittanning	–	–	0.57	0.48	–	0.39	0.36	0.30
hvAb	U. Sunnyside	0.56	0.53	0.69	0.75	0.53	0.43	0.47	0.39
hvAb	Blind Canyon	0.70	0.53	0.48	0.44	0.40	0.30	0.30	0.26
hvAb	Americana	1.23	–	1.07	1.52	1.30	0.83	0.71	0.57
hvBb	Tebo	4.07	4.99	3.87	3.55	2.81	–	1.03	0.75
subC	Ohio #5	–	0.87	1.28	1.16	0.35	0.82	0.32	0.20

Table V: Analysis of maceral constituents for 29 coals
(maceral wt% on a dmmf basis)

	$\bar{x}$ wt%	RANGE
Vitrinite	78.5	41.9-92.2
Inertinite	17.3	7.8-38.9
Liptinite	3.7	0.0-19.2

Table VII, Raymond et al. (6) analyzed up to 400 vitrinite grains for organic sulfur content both with and without the aid of photomosaics. Using a t-statistic approach they calculated the number of analyses (n) for each run necessary to give a desired maximum variability of 10%, at the 95% confidence level, from the true mean as defined by 100 analyses. As can be seen in Table VII, in no case was it necessary to analyze more than 14 vitrinite areas. The second advantage to analyzing only vitrinite is that Raymond et al. (6) were able to achieve essentially identical results both with and without the use of photomosaics. Using texture and morphology to identify areas of vitrinite after the sample had been placed in the EPM was as successful as identifying the vitrinite using oil-immersion microscopy prior to analysis. It should be noted that only two of the four researches were familiar with the EPM. This, combined with the variety of coals chosen, supports the unbiased nature of the test even though the sample set consisted of only four coals. Thus the need for photomosaics is eliminated.

The Rapid EPM Method

EPM analysis for organic sulfur content can be performed easily on -20 to -100 mesh coal samples. Samples need only be mounted in epoxy and polished similar to how coal samples are commonly prepared for petrographic examination. 15 areas within non-contiguous vitrinite grains are analyzed with the EPM. Without the need to produce a photomosaic, the organic sulfur content of vitrinite, and therefore of a coal, may be determined in less than 10 minutes.

To test the EPM method coals were analyzed for which the ASTM organic sulfur values were corrected for unextracted iron. As discussed by Suhr and Given (7) such a correction would take into account the effect of any pyrite that remained unextracted following the ASTM Standard Method D2492-68. As can be seen from the data in Table VIII, the EPM organic sulfur contents are very close to those of the corrected ASTM values.

Discussion

The chemistry of a coal may vary greatly within a seam, both from the top to the bottom and transversely throughout it. These variations are real, and they represent different conditions that

Table VI: Relationship between organic sulfur contents of vitrinite, remaining macerals, and whole coal (all wt% on dmmf basis) V = vitrinite, Pv = pseudovitrinite, F = fusinite, Sf = semifusinite, Ma = macrinite, Mi = micrinite, S= sporinite, R = resinite.

U. Elkhorn hvAb Coal

	V	Pv	F	Sf	Ma	Mi	S	R
wt% of sample	38.9	2.2	7.1	5.3	13.2	13.3	17.6	2.4
wt% S_o	0.73	0.45	0.30	0.38	0.64	0.60	0.94	1.03

S_o Vit. = 0.73 S_o (wt'd $\bar{x}$) remaining macerals = 0.67 S_o coal = 0.69

Ohio #4 hvBb Coal

	V	Pv	F	Sf	Ma	Mi	S
wt% of sample	72.0	3.0	3.9	5.7	0.3	7.8	7.3
wt% S_o	2.93	2.56	0.73	1.51	0.92	2.90	3.89

S_o Vit. = 2.93 S_o (wt'd $\bar{x}$) remaining macerals = 2.50 S_o coal = 2.80

Table VII: Number of analyses (n) necessary to give a maximum desired variability when analyzing for organic sulfur with EPM (after 6).

COAL	RANK	SULFUR wt% (dry)	n
Tebo	hvBb	7.27	13-14
Ohio #5	subC	0.93	10-12
U. Sunnyside	hvAb	0.65	5-6
L. Kittanning	low vol	2.07	7-9

Table VIII: Coals containing pyrite - EPM S_o vs documented ASTM S_o (ASTM data after (7); all contents as dry wt%).

L. KITTANNING hvAb/hvBb COAL

SAMPLE	ASTM S_o (diff.)	ASTM S_o (corr.)	EPM S_o
1273	1.54	1.49	1.50
1276	2.12	2.07	2.09
1277	2.09	2.04	2.08
1279	0.55	0.44	0.60
1282	0.61	0.53	0.57
1299	1.28	1.19	1.09

occur during coal maturation processes. Using the EPM method, the potential exists to achieve very rapid, multiple organic sulfur analyses, which in turn will allow for rapid, detailed measurements of variations in organic sulfur content occurring across coal seams. Indeed, Raymond (8) and Raymond and Davies (9) have been able to apply the EPM method to the L. Kittanning coal seam and have been able to show how organic sulfur values for that seam correlate with organic sulfur values for recent peats deposited under similar conditions in Florida Bay. The EPM offers the opportunity to investigate the occurrence of organic sulfur in coal, on a maceral or whole coal basis, that heretofore was shrouded in the heterogeneity of the sample and the complexity of the analysis.

Acknowledgments

I would like to thank the Pennsylvania State University Coal Section for the samples they provided for EPM analysis, and for the constituent maceral data and ASTM analyses on those samples.

I would especially like to thank Dr. Peter Given for providing me with splits of the L. Kittanning samples listed in Table VIII. Over the past four years Tom Gregory, Roland Hagan, and Dave Mann, Los Alamos National Laboratory, have provided the technical assistance in sample preparation and analysis necessary to keep the development of the EPM organic sulfur technique on line. Ron Gooley, Los Alamos National Laboratory, and Tom Davies, Exxon Production Research, provided the scientific expertise to question various aspects of the EPM procedure, and by so doing, helped to make the procedure that much stronger. The general assistance of Tom Davies, Roland Hagan, and Alan Allwardt, UC Santa Cruz, in the statistical analyses is truly appreciated. This work was performed at the Los Alamos National Laboratory and supported by the Department of Energy under contract W-7405-ENG-36.

Literature Cited

1. Edwards, A. H., Jones, J. M., Newcombe, W. Fuel 1964, 43, 55-62.
2. Greer, R. T. in "Scanning Electron Microscopy/1977/I," O. Johari (ed.), ITT Research Institute, Chicago, 1977, 79-93.
3. Raymond, R., Jr.; Gooley, R. in "Scanning Electron Microscopy /1978/I," O. Johari (ed.), SEM Inc., AMF O'Hare, IL, 1978, 93-107.
4. Harris, L. A., Raymond, R., Jr.; Gooley, R. in "Microbeam Analysis," D. B. Wittry (ed.), San Francisco Press, San Francisco, 1980, 147-148.
5. Raymond, R., Jr. in "Microbeam Analysis," E. E. Newbury (ed.), San Francisco Press, San Francisco, 1979, 105-110.
6. Raymond, R., Jr.; Davies, T. D.; Hagan, R. C. in "Microbeam Analysis," D. B. Wittry (ed.), San Francisco Press, San Francisco, 1980, 149-150.
7. Suhr, N.; Given, P. H. Fuel 1981, 60, 541-542.
8. Raymond, R., Jr. Compte Rendu of the IX Inter. Cong. of Carb. Strat. and Geol., Urbana, IL (in press).
9. Raymond, R., Jr.; Davies, T. D. GSA Abst. with Progs., 1979, V. 11, no. 7, p. 501.

RECEIVED May 17, 1982

10

Chemical Fractionation and Analysis of Organic Compounds in Process Streams of Low Btu Gasifier Effluents

R. L. HANSON, R. E. ROYER, J. M. BENSON, R. L. CARPENTER, G. J. NEWTON, and R. F. HENDERSON

Lovelace Biomedical and Environmental Research Institute, Inhalation Toxicology Research Institute, Albuquerque, NM 87185

> Low Btu gas requires cleanup before it can be burned in a turbine. Several cleanup devices have been used on the output stream of the stirred bed gasifier at the Morgantown Energy Technology Center. Both vapor phase and particulate phase organic compounds were sampled at several locations in the cleanup process. Vapor phase samples were collected using Tenax-GC polymer adsorbent and by condensation traps. The particulate phase organic materials were collected on glass fiber filters placed before the Tenax adsorbent, by condensation traps and by a multicyclone train. Methods of chemical analysis included: the determination of mass of total dichloromethane extractable materials and gas chromatographable organic compounds, fractionation using gel permeation chromatography and silica column chromatography and analysis by gas chromatography/mass spectrometry to identify compounds. We found that the cleanup system preferentially removed higher molecular weight organic compounds from the process stream. Greater quantities of organic compounds were found associated with the particulate phase than the vapor phase in cooled and diluted samples of the process stream. Gas chromatographic/mass spectrometric analysis indicated the presence of aliphatic compounds and many polynuclear aromatic compounds containing oxygen, nitrogen and sulfur heteroatoms.

Coal gasification is being developed as an alternative method of energy production. Low Btu gasification of coal is being evaluated for use in combined-cycle electrical generation in which the low Btu coal gas is combusted in a gas turbine and

0097-6156/82/0205-0205$06.00/0
© 1982 American Chemical Society

the hot exhaust gases sent through a steam recovery system. Low Btu gas can also be upgraded to higher Btu gas by methanation processes.

Currently available gasifiers have two major restrictions: (A) they are only compatible with noncaking coals and hence are unable to use the large eastern USA coal reserves; and (B) except for Lurgi gasifiers they are limited to low operating pressures which result in limited throughput. A research and development program at the Morgantown Energy Technology Center (METC) is under way to overcome these restrictions. Program goals at METC include development of gas cleanup systems to allow the use of combined-cycle low Btu gas-fired turbines.

The Lovelace Inhalation Toxicology Research Institute (ITRI) has been involved with METC in a cooperative program to address basic human health risks associated with low Btu coal gasification. A portion of this research involves chemical characterization of organic compounds in gaseous, liquid and solid process and waste streams.

Complex organic mixtures require chemical fractionation prior to chemical analyses if minor components are to be identified (1). Chemical fractionation methods used in this study were gel chromatography, acid-base neutral solvent partitioning and chromatography on silica gel columns. Identification of compounds was done by gas chromatography/mass spectrometry (GC/MS). Proton nuclear magnetic resonance (PMR) spectroscopy was used to characterize fractions and subfractions which were not amenable to GC/MS analysis.

Experimental

  The METC Low Btu Coal Gasifier. The METC coal gasifier has been described previously (2, 3). Briefly, it is a pressurized version of the McDowell-Wellman (Wellman-Galusha) atmospheric pressure stirred-bed coal gasifier (Figure 1) and differs from commercial fixed-bed producers in its smaller size (1.1 m ID) and various provisions for stirring the bed. Cleanup devices consisted of a cyclone for removing large particles, a humidifier, tar trap and Venturi scrubber for removal of tars and oils and a side stream for experimental gas cleanup with the cleaned gas burned in a flare. The experimental cleanup devices were an electrostatic precipitator, a direct cooler, Holmes-Stretford desulfurizer, alkali scrubber and indirect cooler.

  Sample Collection. Producer gas was diluted with cool dry nitrogen by a radially injected diluter which resulted in aerosol formation (2). For collection of particles and vapor phase materials the diluted producer gas was passed through a glass fiber filter and then through a Tenax-GC® polymer adsorbent trap (4) at a flow rate of about 12 L/min. Table I lists the types of samples collected and the positions at which they were collected. In order to obtain larger amounts of condensates, a series sampling train consisting of four modified Greenberg-Smith impingers

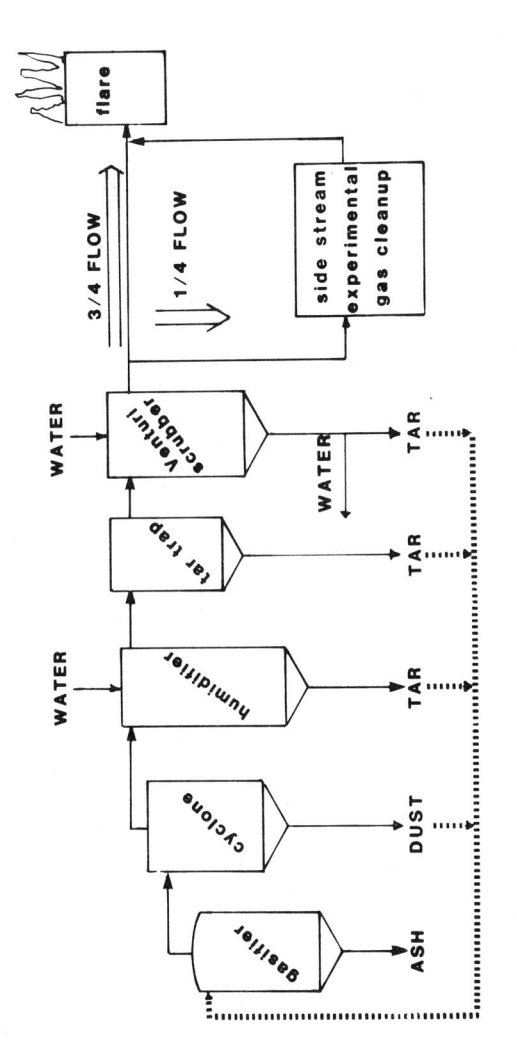

Figure 1. Schematic diagram of the METC low Btu gasifier and cleanup system.

was used to sample undiluted gas after the cyclone. Modifications consisted of increasing collecting surfaces (glass beads or steel wool) inside the main vessels, eliminating the jet and collector and adjusting sample flow to 20 L/min. Condensers were placed in temperature-controlled baths with Number 1 at ambient (0° to 10°C), Number 2 in an ice water bath (approximately 0°C) and Numbers 3 and 4 in dry ice acetone (-77°C). Tar was obtained from the humidifier tar trap and with the outlet water from the Venturi scrubber decanter. Ash was obtained from the gasifier and cyclone.

Table I

Samples Collected at Gasifier

Position	Type of Samples
Base of Gasifier	Ash
Base of Cyclone	Ash
After Cyclone	Filter, Tenax Adsorbent, Condenser
Base of Humidifier	Tar
Base of Tar Trap	Tar
After Tar Trap	Filter, Tenax Adsorbent
Venturi Scrubber Decanter	Tar, Water
After Venturi Scrubber	Filter, Tenax Adsorbent
After Side Stream Cleanup	Filter, Tenax Adsorbent

Sample Extractions. Both filter particulate samples and Tenax adsorbent samples were extracted with organic solvents to remove the adsorbed organic material. Pentane was used for Soxhlet extraction of Tenax samples whereas dichloromethane was used for ultrasonic extraction of particulate filter samples (5). Extracts were concentrated by rotary evaporation. For gravimetric analyses of some of the extracts, the concentrate from the rotary evaporator was transferred to a small vial and taken to dryness under a stream of nitrogen.

Venturi scrubber outlet water and the condenser samples were solvent partitioned with dichloromethane. The dichloromethane was removed by rotary evaporation followed by final drying under a stream of nitrogen.

Sample Fractionations. Organic materials were first fractionated using a column (1 m by 1.5 cm I.D.) filled with Sephadex LH-20. The column is first eluted with 240 mL of tetrahydrofuran which elutes the least polar and the higher molecular weight compounds first followed by the more polar ones. A final, very polar, fraction is eluted with 300 mL of methanol. Several standards were run to characterize the elution of classes of organic compounds from the column as shown in Figure 2.

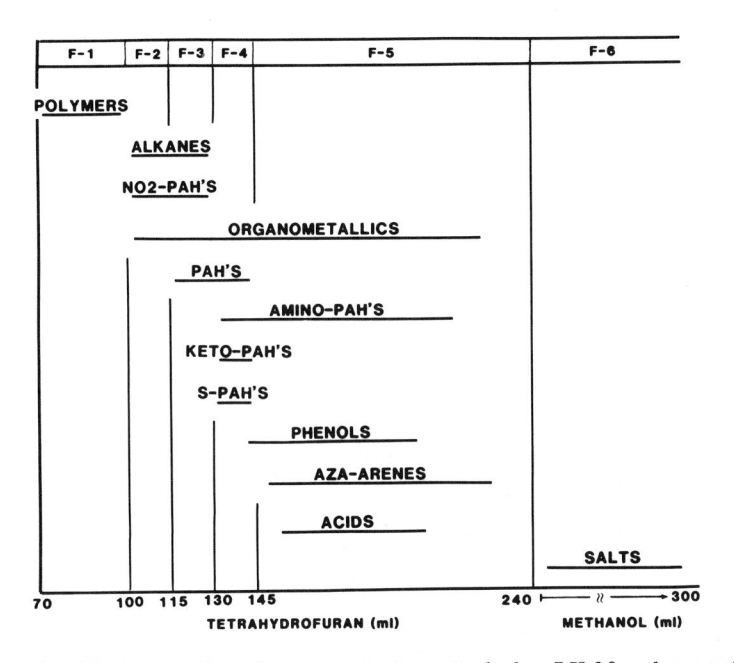

Figure 2. Elution profiles of compounds from Sephadex LH-20 column using tetrahydrofuran (THF) as eluant. The lines for each compound indicate the volume of THF in which the compounds eluted. Elution of compounds was monitored with an ISCO model UA-5 absorbance monitor at 280 nm.

Fractions 3 and 4 from the Sephadex column, which contained primarily nonpolar polycyclic aromatic compounds, were combined and then subfractionated on a column of silica gel. The silica (Sigma Chemical Co., 60-200 mesh type I chromatographic grade) gel was activated at 110°C for 1 hr and placed in the column (30 cm x 2 cm I.D.) as a slurry in toluene. Combined Fractions 3 and 4 (200 mg) were placed on the column in 5 mL of toluene and eluted sequentially with 115 mL of toluene (subfraction A), 110 mL of propanol (Subfraction B) and 300 mL of methanol (Subfraction C). When Fractions 3 and 4 were subfractionated separately 1:1 toluene:n-propanol (V:V) was used in place of propanol. For a filter sample Fractions 1 and 2 were also separated on the silica gel column.

Because of its more polar nature, Fraction 5 from the Sephadex column was subfractionated by aqueous/organic partitioning. A 100 mg sample of Fraction 5 was dissolved in 20 ml of dichloromethane and extracted twice with equal volumes of 1N H_3PO_4 and twice with equal volumes of 5% sodium carbonate. The organic phase (neutrals/phenols) was dried over sodium sulfate, reduced by rotary evaporation and the remaining solvent was removed under a stream of dry nitrogen at 50°C. The aqueous H_3PO_4 (organic bases) and Na_2CO_3 (organic acids) extracts were washed with hexane and then neutralized with dilute sodium hydroxide or HCl solution, respectively. The bases and acids were back extracted with dichloromethane and treated in the same manner as the neutrals/phenols. The aqueous phase remaining from extraction of the bases was allowed to evaporate over a period of several days in a fume hood and the dry salts were extracted with dichloromethane to remove any organic material (water soluble compounds).

Chemical Analysis. The quantity of gas chromatographable organic compounds were determined as described previously (4, 5). For GC/MS analysis, a Finnigan model 4023 gas chromatograph/mass spectrometer and INCOS data system were used to obtain 70 ev electron impact mass spectra of gas chromatographable compounds. A 25 m OV-101 quartz capillary column was used. Carrier gas was helium with a flow velocity of 30 cm/sec. The temperature program was 50°C for two minutes and then 5° per minute to 270°. The final temperature was held for 15 minutes. Samples were injected (splitless injection) in 1 μL of dichloromethane at a concentration of 2 μg per μL.

The PMR spectra were taken at 80 MHz with a Varian FT-80A NMR spectrometer. Samples (0.5 mL) were prepared in deuterochloroform at concentrations of 10-25% by weight. Tetramethylsilane (TMS) was used as an internal reference. Chemical shifts (δ) are given in parts per million from TMS. Five hundred pulses were taken to obtain very high signal to noise ratios.

Results and Discussion

Process Stream Samples. The concentrations of organic extractable material in the process stream are listed in Table II.

These results show that the cleanup system removed the particulate phase organic compounds. The reconstructed ion chromatogram from GC/MS analysis of the samples collected after the cyclone are shown in Figure 3. The GC/MS analyses of samples indicated that compounds with molecular weights from about 100 to 400 daltons were present in samples collected after the cyclone whereas the cleanest gas after the direct cooler had fluoranthene (MW=202) as the largest organic compound. The gas chromatographable percentage for various gasifier sample extracts and Sephadex LH-20 fractions of a condenser sample are given in Table III.

Table II

Concentrations of Organic Material in Gasifier Process Stream

Sampling Position	n	Vapor Phase (Tenax Adsorbent) g/m^3	Particle Phase (Tenax Prefilter) g/m^3
Between Cyclone and Humidifier	(1)	0.48	12.5
Between Tar Trap and Venturi Scrubber	(2)	0.15 - 0.64	0.83 - 3.0
After Venturi Scrubber	(2)	0.45 - 1.17	1.34 - 2.12
After Side Stream Cleanup	(2)	0.12 - 0.90	0.0014 - 0.0018

Table III

Gas Chromatographable Percentage for Extracts and Fractions of Gasifier Samples

Sample	Gas Chromatographable %
Tenax Filter Extracts After Cyclone	36
Tenax Filter Extracts After Tar Trap	54
Tenax Filter Extracts Cleaned Gas	100
Tar Trap Tar Extracts	39
Condenser Sample Obtained After Cyclone	31
Sephadex LH-20 Fractions of Condenser Material Obtained After Cyclone	
Fraction 1	34
Fraction 2	27
Fraction 3	100
Fraction 4	28
Fraction 5	58
Fraction 6	20

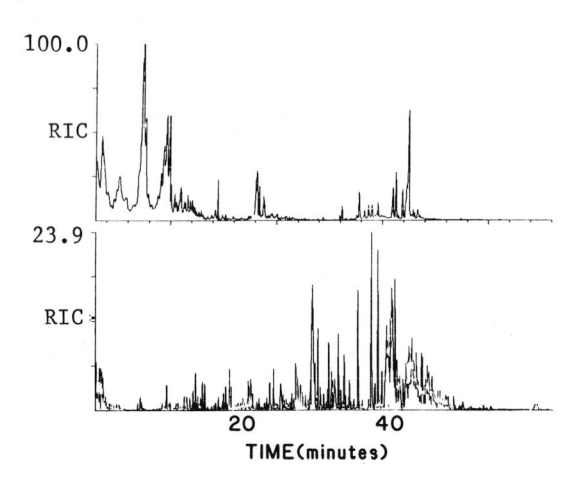

Figure 3. Reconstructed ion chromatograms from GC/MS analysis of vapor phase organic extractable from Tenax (upper trace) and dichloromethane extractables from glass fiber Tenax filter (lower trace). Samples of the raw gas collected after the cyclone. A 25 m OV–101 quartz capillary column was used in a Finnigan model 4023 gas chromatograph mass spectrometer. Temperature program was 2 min at 50°C, 5°C/min to 270°C with 15 min at 270°C.

Since higher molecular weight organic compounds were found after the cyclone than further down the process stream the condenser samples and a filter sample from after the cyclone were studied further. Because all the gas chromatograms of condenser samples indicated the presence of components with a wide range of volatility, a portion of the dichloromethane sample collected at 0°C was fractionated by the Sephadex LH-20 column. The mass distribution of this condenser sample and its subfractions are given in Table IV.

A filter sample collected after the gasifier cyclone was extracted with CH_2Cl_2 and it was found that 96.8 percent of the mass on the filter was soluble. A portion of this extract was fractionated on the Sephadex LH-20 column. The mass distribution of the Sephadex LH-20 fractions and silica gel subfractionation of Fractions 1 and 2 are given in Table V.

The differences in mass distribution for the Sephadex LH-20 fractions listed in Table IV and Table V are due to different modes of sample collection. The condenser sample contains condensable material that is present in the vapor phase and is not collected by the filter even though the process stream for the filter sample was diluted with nitrogen.

The three silica gel column subfractions of Fraction 1 were analyzed by GC/MS. The toluene subfraction (A), which had the majority of the mass, produced the reconstructed ion chromatogram shown in Figure 4. This chromatogram and the mass spectra of the peaks indicated the presence of a homologous series of aliphatic compounds as expected from Sephadex LH-20 Fraction 1. These aliphatic compounds are gas chromatographable and do not account for the nonvolatile components in the gasifier tar samples. Most likely, higher molecular weight polynuclear aromatic hydrocarbons eluted in Sephadex LH-20 Fraction 4 and the more polar compounds eluted in Fractions 5 and 6 of many of the tar samples contribute to the non-gas chromatographable components.

Cleanup Device Samples. The percent dichloromethane extractables from solid and liquid samples collected by the gasifier cleanup devices are given in Table VI. A portion of the dichloromethane extractables from the Venturi scrubber decanter outlet water was fractionated on the Sephadex LH-20 column. The organic compounds in this sample eluted from the column primarily (approximately 80%) in Fraction 5. A portion of Fraction 5 was subfractionated into acidic, basic and neutral components. This resulted in 15 percent of the mass recovered in the acidic subfraction, 1.5 percent in the basic subfraction, 11 percent in the neutral subfraction and the remaining 72.5 percent being water soluble. Many nitrogen heterocycles were also found in the basic subfraction by GC/MS analysis.

The tar phase from the Venturi scrubber decanter outlet and the tar from the tar trap were fractionated on the Sephadex LH-20 column. A much broader distribution of the organic compounds

Table IV

Mass Distribution of 0°C Condenser Sample

Sephadex LH-20 Fractions and Subfractions	% Mass in Fraction	
Fraction 1	14	
Fraction 2	6	
Fraction 3	14	
a. Toluene[a]		11
b. Toluene:n-propanol		3
c. Methanol		0.1
Fraction 4	6	
a. Toluene[a]		4
b. Toluene:n-propanol		2
c. Methanol		0.2
Fraction 5	30	
a. Acids[b]		4.5
b. Bases		0.3
c. Neutrals		3.3
d. Amphoterics and Water Soluble Compounds		22
Fraction 6	30	
TOTAL	100	

[a] Subfractions obtained by elution from a silica gel column using toluene, toluene:n-propanol (1:1; V:V) and methanol.
[b] Subfractions obtained by acid, base solvent partitioning with dichloromethane.

Table V

Mass Distribution of Dichloromethane Extract
of a Raw Gas Filter Sample

Sephadex LH-20 Fractions and Subfractions	% Mass in Fraction	
Fraction 1	11	
a. Toluene		6.9
b. Toluene:n-propanol		3.4
c. Methanol		0.7
Fraction 2	19	
a. Toluene		8.9
b. Toluene:n-propanol		9.7
c. Methanol		0.4
Fraction 3	31	
Fraction 4	10	
Fraction 5	23	
Fraction 6	6	
TOTAL	100	

Table VI

Dichloromethane Extractables From Gasifier
Cleanup Device Samples

Effluent	% Extractable
Bottom Ash	0.01
Cyclone Ash	0.30
Tar Trap Tar	(100)
Scrubber Water	0.17
Scrubber Tar	(100)

from the tars was found than occurred for the dichloromethane extractable organics from the outlet scrubber water (6-9). Mass distributions in the LH-20 fractions of the Venturi scrubber decanter outlet tar and of the tar trap tar are given in Table VII. Mass distributions for the tars from these two cleanup devices are similar, indicating that these tars may be similar in composition. A portion of Fraction 5 from the tar trap tar was solvent partitioned into acids, bases, neutrals and water solubles. Most of mass appeared in the neutral subfraction (67%) with 9%

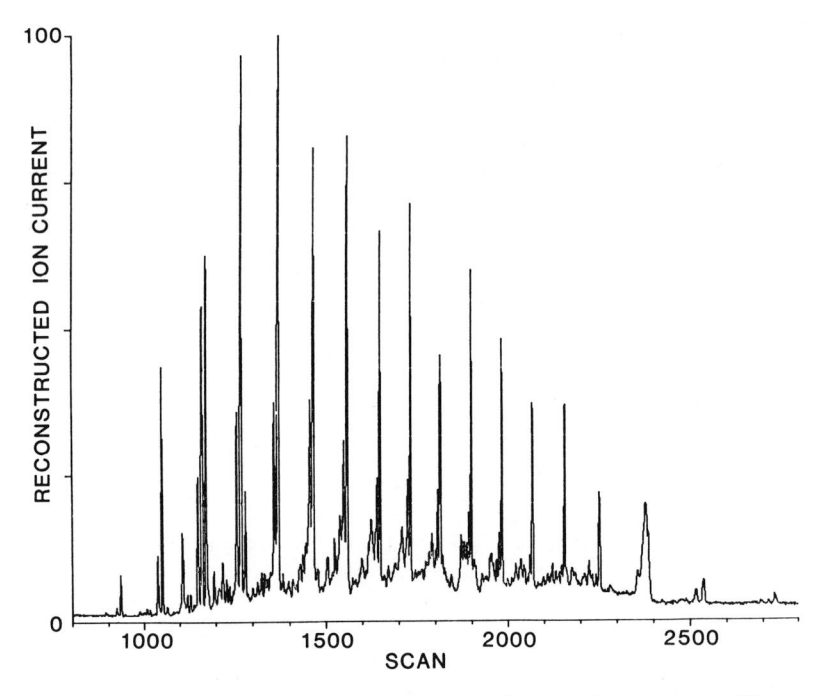

Figure 4. Reconstructed ion chromatogram of toluene subfraction from silica gel column of Sephadex LH-20 fraction 1 of filter sample collected after the gasifier cyclone. The major peaks are the aliphatic hydrocarbons in this sample.

acids, 4% bases and 20% water soluble. This neutral subfraction was found to have carbazole, methylcarbazoles, phenanthridone and phenols present when analyzed by GC/MS. Two of these phenols were identified as 2-hydroxyfluorene and 9-phenanthrol.

Table VII

Mass Distribution of Sephadex LH-20 Fractions of Tar Samples from Venturi Scrubber Decanter Outlet and the Tar Trap

Sephadex LH-20 Fraction (#)	Venturi Scrubber Decanter Outlet (Weight Percent)	Tar Trap (Weight Percent)
1	7	11
2	12	15
3	32	29
4	22	18
5	20	21
6	7	6

Analysis of the basic subfraction of Fraction 5 by GC/MS indicated that it contained a series of aza-arenes and methylated aza-arenes ranging in size from methylquinoline to azapyrene as indicated in Table VIII.

Fractions 3 and 4 from the tar trap tar, which primarily contained polycyclic aromatic hydrocarbons and their alkylated derivatives, were combined and subfractionated on a silica gel column. In sequential elution, toluene eluted 46 percent of the mass (subfraction A), propanol eluted 48 percent (subfraction B), methanol eluted 3 percent (subfraction C) and 3 percent was recovered by washing the silica gel with tetrahydrofuran (subfraction D). The toluene subfraction was found by GC/MS analysis to consist of a series of PAH, both parent compounds and numerous methylated derivatives. In addition, the heterocyclic compound dibenzothiophene was present as a major peak. Tentative identifications are listed in Table IX. Compounds whose identifications were confirmed by GC retention time comparisons with standards are noted. Although the identities of many of the individual PAHs have not been confirmed, it is clear that most of the subfraction is PAH in nature.

For the tar trap tar Sephadex LH-20 Fractions 1 and 2 and silica gel subfractions B and C of combined Fractions 3 and 4 were too nonvolatile for GC/MS analysis so they were analyzed by PMR. The spectra are shown in Figures 5a, 5b, 5c, and 5d respectively. The various regions of the spectra were assigned by using a system based on that of Bartle and Jones [10]. Signals for various chemical shift ranges were assigned as shown in Table X. Integration over the chemical shift ranges listed gave the

Table VIII

Compounds Tentatively Identified by GC/MS in the Basic Subfraction of Sephadex LH-20 Fraction 5 of Tar Trap Tar

	M.W.	Number of Isomers Found
Methylquinoline	143	3
2,7-Dimethylquinoline[a]	157	
Dimethylquinolines	157	4
Triomethylquinolines	171	7
Benzo(h)quinoline[a]	179	
Acridine[a]	179	
Phenanthridine[a]	179	
Benzo(f)quinoline	179	
Methylbenzoquinolines	193	4
Methylacridine	193	1
Methylphenanthridines	193	2
Dimethylbenzoquinolines	207	6
Azapyrene	203	

[a]Had same retention times as standards.

Table IX

Compounds Tentatively Identified by GC/MS in Toluene Subfraction of Sephadex LH-20 Fraction 3 and 4 of Tar Trap Tar.

	M.W.	Number of Isomers Found
Dimethylnaphthalene	156	4
Acenaphthene	154	
Trimethylnaphthalene	170	3
Fluorene[a]	166	
Methylfluorene	180	3
Dibenzothophene[a]	184	
Phenanthrene[a]	178	
Anthracene[a]	178	
Methyldibenzothiophene	198	
Methylphenanthrene	192	4
Dimethylphenanthrene	206	4
Fluoranthene[a]	202	
Pyrene[a]	202	
Phenylnaphthalene	204	
Methylpyrene	216	1
Trimethylphenanthrene	220	2
Benzo[a]fluorene	216	
Benzo[a]pyrene[a]	252	

[a]Had same retention times as standards.

Table X

The Mole Percent of Different Types of Hydrogen Atoms in Non-Gas Chromatographable Fractions and Subfractions of Coal Gasifier Tar as Determined by PMR

| δ range[a] | Assignment | Mole Percent of Hydrogen ||||
		Fraction 1	Fraction 2	Subfraction B	Subfraction C
0.5-1.0	CH_3 γ or further from aromatic ring	11	8	1	-
1.0-2.0	CH_2 β or further from aromatic ring or CH_3 β to aromatic ring	49	19	24	4
2.0-3.3	CH, CH_2 or CH_3 α to aromatic ring	11	27	30	34
5.5-9.0	aromatic and phenolic hydrogen	29	46	44	50

[a] In parts per million from tetramethylsilane.
[b] See reference 5.

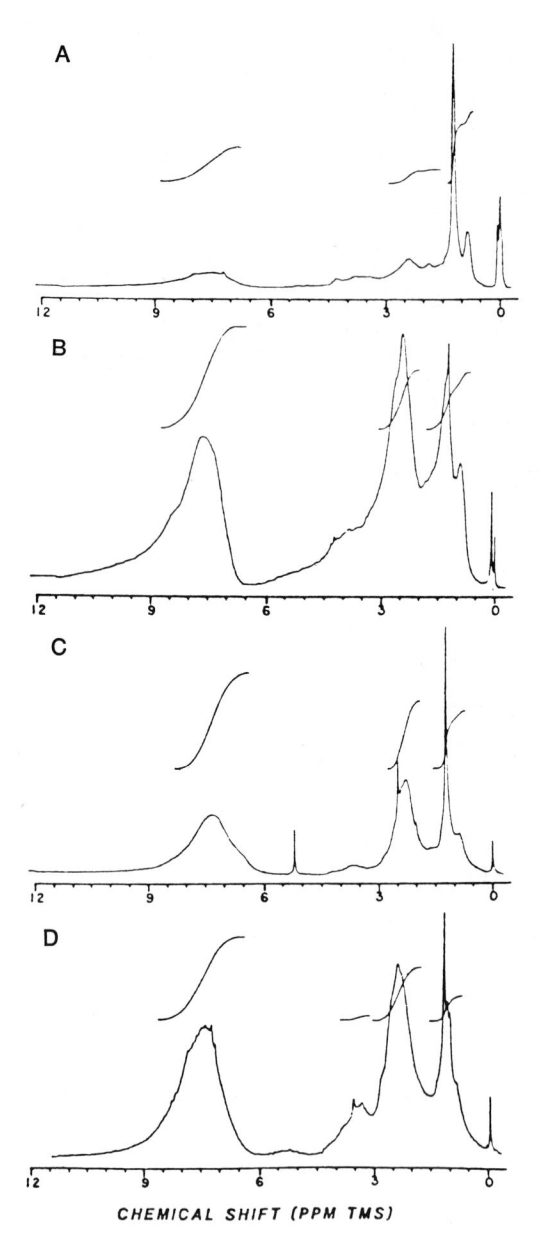

Figure 5. PMR spectra of tar trap tar sample. Key: A, Sephadex LH-20 fraction 1; B, Sephadex LH-20 fraction 2; C, propanol subfraction of Sephadex LH-20 fractions 3 and 4; and D, methanol subfraction of Sephadex LH-20 fractions 3 and 4.

values for mole percent of the different types of hydrogen in Fractions 1 and 2 and subfractions B and C as shown.

The PMR spectra showed that material in Fraction 1 had more aliphatic character than that in Fraction 2, which indicates that some class separation occurs on Sephadex LH-20 even for materials which are excluded from the gel pores. The increase in mole percent of aromatic hydrogen in Fraction 2 over Fraction 1 was accompanied by a larger percentage of the aliphatic hydrogen being α to an aromatic ring, as would be expected. Material eluting in subfraction B (from Sephadex LH-20 Fractions 3 and 4) appeared to be similar to Fraction 2 by PMR, except for the smaller peak ascribed to methyl groups β or further from an aromatic ring. It is possible that these materials were separated on the Sephadex LH-20 column due to sieving action of the gel and that the material in subfraction B was lower in molecular weight than that in Fraction 2. In subfraction C, nearly all the hydrogens were aromatic or on carbon α to an aromatic ring. In addition, the signal in the range δ 3-4 is significant, probably indicating the presence of hydrogens on carbon atoms substituted with O or N atoms (11). The source of the peak very near TMS in Fractions 1 and 2 is not known. The peak at δ 5.3 in the spectrum of subfraction B is residual CH_2Cl_2 solvent.

Many similar types of gasifier process effluents and coal tar have been studied from a variety of coal gasifiers. The composition of organic extracts of scrubber water particulate matter from a high pressure, entrained flow gasifier has been reported (12). Glass capillary GC/MS was used to separate 140 components in coal tar (13). Several liquid phases for coating of capillary columns were evaluated in this research. Coal tar from a coking operation was analyzed with glass capillary GC/MS (14). The coal tar was fractionated to obtain fractions containing acids and phenols, bases, hydrocarbons, neutral polar and aromatic components before the GC/MS analyses. Coal liquefaction distillate cuts from the SRC-II process have been studied using microbial mutagenesis assays together with solvent fractionation and GC/MS analysis (15). This report indicated that hydrotreating reduced the mutagenicity of samples. Much of this loss was attributed to the conversion of three and four-ring primary and aromatic amines by the hydrotreatment. Work has not yet revealed the presence of primary aromatic amines in the mutagenic samples, fractions and subfractions of gasifier materials (16).

Conclusions

Results indicated the complex nature of organic materials associated with low Btu coal gasification. Several classes of organic compounds were found, including neutral sulfur and nitrogen heterocycles, phenols, bases including aza arenes and polycyclic aromatic hydrocarbons. Some organic compounds were not gas chromatographable, so various liquid chromatographic fractionation procedures and pH partitioning were used to fractionate and characterize these compounds. The PMR spectoscopy also provided useful information about the structure of the non-gas chromatographable organic compounds.

The following conclusions can be drawn concerning the gasifier cleanup system. The concentration of particle associated organic compounds is reduced. The cleanup system removes higher molecular weight organic compound which are primarily aromatic and alkyl substituted aromatic compounds. Diluting and cooling the low Btu producer gas, as would happen in fugitive releases, produces a respirable aerosol. It is mainly the heavier polynuclear aromatic hydrocarbons that are transferred from the vapor to the particle phase by diluting and cooling the producer gas.

Acknowledgements

The authors acknowledge the excellent cooperation and assistance received from the staff of the Morgantown Energy Technology Center. The authors thank D. Horinek, S. Sinclair, and S. Crain for their technical assistance. We thank our colleagues on the ITRI sampling team for obtaining samples. The review and support of Drs. C. H. Hobbs and R. O. McClellan is also acknowledged.

This research was performed under U. S. Department of Energy Contract Number DE-AC04-76EV01013.

Literature Cited

1. Novotny, M.; Strand, J. W.; Smith, S. L.; Wiesler, D.; Schwende, F. J., Fuel 60, 213, 1981.
2. Newton, G. J.; Carpenter, R. L.; Yeh, H. C.; Weissman, S. H.; Hanson, R. L.; and Hobbs, C. H., "Sampling of Process Streams for Physical and Chemical Characterization of Respirable Aerosols," in Proceedings of ORNL CONF-780903, 25-28, September 1978, Potential Health and Environmental Effects of Synthetic Fossil Fuel Technologies, NTIS, Springfield, VA, 1979, pp 78-94.
3. Wilson, M. W.; and Bisset, L. A., "Development and Operation of METC 42-inch Gas Producer During 1964-1977," METC/RI-78/12, October 1978.
4. Hanson, R. L.; Royer, R. E.; Carpenter, R. L.; and Newton, G. J., "Characterization of Potential Organic Emissions from a Low Btu Gasifier for Coal Conversion," in Polynuclear Aromatic Hydrocarbons, Ann Arbor Science Publishers, Inc., Ann Arbor, MI, 1979, pp 3-19.
5. Hanson, R. L.; Carpenter, R. L.; Newton, G. J.; Rothenberg, S. J., J. Environ. Sci. Health, Part A, 1979, 14, 223.

6. Royer, R. E.; Mitchell, C. E.; Hanson, R. L. Submitted to Environ. Sci. Technol., 1981.
7. Hill, J. O.; Royer, R. E.; Giere, M. S.; Mitchell, C. E. Submitted to Environ. Sci. Technol., 1981.
8. Benson, J. M.; Hill, J. O.; Mitchell, C. E.; Newton, G. J.; Carpenter, R. L. In Press Arch. Environ. Contam. Toxicol., 1982.
9. Benson, J. M.; Mitchell, C. E.; Royer, R. E.; Clark, C. R.; Carpenter, R. L.; Newton, G. J. Submitted to Arch. Environ. Contam. Toxicol., 1981.
10. Bartle, K. D.; Jones, D. W. In "Analytical Methods for Coal and Coal Products, Vol. II"; Karr, C. Jr., Ed.; Academic Press, New York, 1978; pp 103-160.
11. Aczel, T.; Williams R. B.; Brown, R. A.; Pancirov, R. J. In "Analytical Methods for Coal and Coal Products, Vol. I", Karr, C. Jr., Ed.; Academic Press, New York, 1978; pp 499-540.
12. Hansen, L. D.; Phillips, L. R.; Mangelson, N. F.; Lee, M. L., Fuel 59, 323, 1980.
13. Borwitzky, H.; Schomburg, G., J. Chromatogr. 170, 99, 1979.
14. Novotny, M.; Strand, J. W.; Smith, S. L.; Wiesler, D.; Schwende, F. J., Fuel 60, 213, 1981.
15. Wilson, B. W.; Petersen, M. R.; Pelroy, R. A.; Cresto, J. T., Fuel 60, 289, 1981.
16. Benson, J. M.; Hill, J. O.; Royer, R. E.; Hanson, R. L.; Mitchell, C. E.; Newton, G. J., "Toxicological Characterization of the Process Stream and Effluents of a Low Btu Coal Gasifer," in Proceedings of the 20th Hanford Life Sciences Symposium, Richland, Washington, 1980, in press.

RECEIVED May 17, 1982

11

Solvent Analysis of Coal-Derived Products Using Pressure Filtration

BRUCE R. UTZ, NAND K. NARAIN, HERBERT R. APPELL, and BERNARD D. BLAUSTEIN

U.S. Department of Energy, Pittsburgh Energy Technology Center, Pittsburgh, PA 15236

A room-temperature, pressure-filtration method for solvent characterization of coal-derived products is described. Tetrahydrofuran solubility and cyclohexane solubility of coal-derived products were determined. A comparison of the method was made with a room-temperature, reduced-pressure Soxhlet-extraction method, and demonstrated that the pressure-filtration method can solubilize as much coal-derived product as the special type of Soxhlet extraction. Pressure filtration was also shown to be precise, and gave standard deviations as low as S = 0.10% when using tetrahydrofuran and S = 0.26% when using cyclohexane. Values of S can be slightly larger depending on the vehicle used in the coal-vehicle slurry. A major advantage of pressure filtration is that solvent filtration of a sample can be completed within 15 minutes, whereas a Soxhlet extraction can take many hours or even days. With high-conversion coal-oil slurry, reaction products, the solvent classification can be completed in 1 1/2 to 2 hours.

Solvent solubility is widely used to classify coal-derived products (1,2). The most popular methods are based on some form of Soxhlet extraction. Methods involving Soxhlet extraction are normally time consuming; thus complete solvent solubility classification is laborious. An alternate method, used initially by Bertolacini et al. (3) and modified for our use, employs pressure filtration. Pressure filtrations carried out at room temperature have been used to classify a number of coal-derived products obtained under a variety of liquefaction conditions.

The purpose of this paper is to present data on pressure filtrations as a method to classify coal-derived products, and to show the advantages of using pressure filtrations in place of methods employing Soxhlet extractions. The types of samples that require special care in handling and interpretation, and

This chapter not subject to U.S. copyright.
Published, 1982, American Chemical Society.

the precision of the pressure-filtration method are also discussed. No attempt is made to compare pressure filtration with any "standard" Soxhlet method used by various laboratories. The comparisons between room-temperature pressure filtrations and room-temperature continuous extractions were made only to determine relative extraction efficiencies.

Experimental

All experiments were carried out using the pressure-filtration apparatus (Figure 1) and the Soxhlet-extraction apparatus (Figure 2). The solvents used for this study were tetrahydrofuran (THF) and cyclohexane. The vehicles used were a filtered Koppers creosote oil (b.p. 250-410°C) and SRC Heavy Distillate recycle oil (b.p. 275-540°C). The coals (Table I) used were Kentucky 9/14, Colonial Mine (-250 to +325 mesh); and Illinois #6, Burning Star Mine (-100 mesh). Both coals were vacuum dried.

Pressure Filtration Apparatus and Procedure. The two most important components of the pressure-filtration apparatus are reservoir R, a Whitey 1000-mL stainless steel sample cylinder, and holder F, a 100-mL Millipore pressure filter/holder. A weighing pan containing a Millipore Teflon filter and a Teflon O-ring was weighed. Most filtrations were done with a 10-micron teflon filter, which was adequate for removal of insoluble solids from the suspension. The lower collar #5 of the Millipore filtration apparatus F was unscrewed, the Teflon filter and O-ring were fitted into place, and collar #5 was put back on. A representative sample weighing 5-7 grams was removed from the reaction slurry and added to a tared Pyrex 30-mL or 90-mL weighing bottle with a ground glass lid. The sample was then diluted with solvent and mixed by stirring with a glass rod. Approximately 30 mL of fresh solvent was added to apparatus F to dilute the solution/suspension during addition. The sample was then added by washing out the weighing bottle, and the filtration holder was filled to a total volume of approximately 100 mL.

Lid #4 and quickconnect #3 were attached. Quickconnect #2 was disconnected and 400 mL of solvent was added to the stainless steel sample cylinder (R). Quick-connect #2 was reconnected and two-way valve #1 was turned to pressurize the system to approximately 40 psig with nitrogen. The flow rate, approximately 35-45 mL/min, was controlled at the valve, allowing a fine stream of filtrate to form. Once the extraction/filtration was complete (13-15 min), nitrogen was used to blow the residue dry for 1-2 minutes. The valve used to pressurize the apparatus was turned off and the

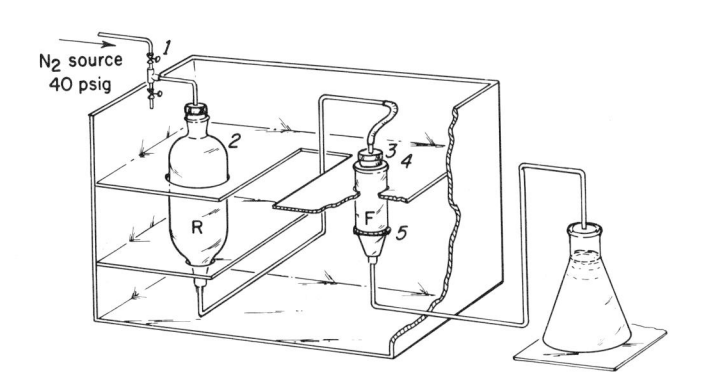

Figure 1. Pressure filtration apparatus. See text for discussion.

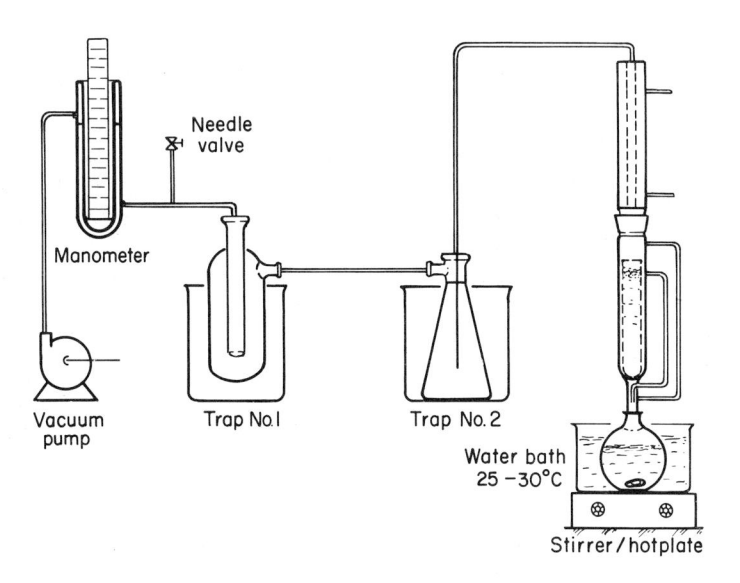

Figure 2. Reduced-pressure, room temperature Soxhlet-extraction apparatus.

Table I. Analyses of Coal

	Kentucky 9/14, Colonial Mine	Illinois #6, Burning Star Mine
Proximate Analysis (Wt. %, moisture free)		
Volatile Matter	39.0	39.7
Fixed Carbon	51.7	48.9
Ash	9.3	11.3
Ultimate Analysis (Wt. %, moisture free)		
Hydrogen	4.9	4.7
Carbon	72.1	69.8
Nitrogen	1.5	1.2
Sulfur	3.4	3.1
Oxygen (ind.)	8.8	9.9
Ash	9.3	11.3

the solvent may be larger than those obtained from a room-temperature extraction, because the solvent may be able to dissolve more of the sample at higher temperatures. Since pressure filtrations were conducted at room temperature, a meaningful comparison of data could best be made if Soxhlet extractions were also conducted at room temperature. This required Soxhlet extraction at reduced pressures (Figure 2). A Sargent-Welch "Ser Vac" pump, used with a simple manometer and Hoke "Milli-Mite" needle valve, was used to maintain pressures of 9-12 mm Hg. Trap #1 was cooled by a Neslab "U Cool" immersion cooler at -5°C in an ethylene glycol bath. Trap #2 was cooled at -30°C with a Neslab temperature-controlled Cryo Cool CC-80 II in an ethylene glycol bath. The condenser, using ethylene glycol, was at 5°C for cyclohexane extractions and at -10° to -20°C for THF extractions using a Lauda K-4/R refrigerated circulator. Trap #1 was a simple vapor trap, but trap #2 was a filtering flask to accommodate a larger volume of escaping vapor due to the volatile nature of solvents. The extraction solvent was stirred and warmed in a 1 or 2 liter round bottomed flask using a Corning Combo hot plate/stirrer. The water bath was kept at 25°-30°C.

Soxhlet extractions were usually carried out on reaction slurries weighing 100 to 150 grams to allow sufficient material for product isolation, characterization, and mass balances. The 100-150 g sample was diluted in 600-700 mL of solvent, the suspension stirred for 15 minutes and then filtered through a predried cellulose thimble. The filtrate was then pressure-filtered through a 5-micron filter to remove any small particles that may have passed through the thimble. The residue was then extracted until the filtrate in the Soxhlet extractor was nearly colorless (2-5 days). The thimble, with residue, was dried in a vacuum oven at 100°C for at least 4 hours, or until there was no further weight loss. The filtrate was again pressure-filtered through a 5-micron filter, and the weight of filtered solid was added to the residue weight after drying. As in pressure filtration, the conversion to soluble material was based on the residue weight and the weight of the coal in the sample.

<u>Measurements of Precision</u>. Experiments to measure precision were done on a coal-oil slurry treated at 440°C using either Koppers creosote oil or SRC Heavy Distillate recycle solvent as a vehicle. The coal-oil slurries were reacted in a 1-liter continuously stirred tank reactor (CSTR) using 25 wt % of Illinois #6 Coal (Burning Star Mine), 2000 psig hydrogen, and a slurry feed rate of 524 mL/hr with Koppers creosote oil and 414 mL/hr with SRC HD recycle solvent. Experiments were also done on a coal-oil slurry treated at 250°C for 15 minutes using Koppers creosote oil as a vehicle. Reacted coal-oil

depressurization valve, on two-way valve #1, was opened. Lower collar #5 was unscrewed; the Teflon filter with residue and Teflon O-ring were taken out, put into their appropriate weighing pan, and dried under a vacuum at 100°C for 1 hour and then immediately weighed.

Modified Pressure Filtration Procedure. A modified procedure to determine cyclohexane solubility of reaction slurries having high to moderate solids content required the use of some THF. A 5-7 g reaction slurry sample was diluted with approximately 25 mL of THF; the mixture was stirred and transferred to the filtration holder containing 30 mL of THF, and the normal filtration procedure was carried out. The residue was removed, dried, and weighed.

The THF filtrate, which also contains cyclohexane soluble vehicle, was concentrated by roto-evaporation to remove the THF, 5 mL of THF was added back to the slurry, and this THF slurry was dripped into 150 mL of stirred cyclohexane. The flask that contained the concentrated THF solution was sonicated with 10-30 mL of cyclohexane in order to clean it thoroughly. Since the filter/holder has a capacity of 100 mL, only two-thirds of the 150-mL cyclohexane suspension was added. Approximately half of the cyclohexane suspension in the filter/holder was pressure-filtered to allow addition of the remaining suspension. Five hundred mL of cyclohexane was then added to reservoir R, and the sample was pressure-filtered. The residue was then removed, vacuum-dried, and weighed. Since the THF-insoluble material was also cyclohexane-insoluble, combining the weight of the THF-insoluble residue and the insoluble residue obtained from the cyclohexane filtration allows the calculation of a cyclohexane solubility value. The results from the modified procedure on one sample will thus give both THF and cyclohexane solubility values. The modified pressure-filtration procedure takes 2 to 2 1/2 hours.

The following equation was used to calculate conversion of coal to soluble products:

$$\frac{\text{Wt. of coal in sample - residue wt.}}{\text{Wt. of maf coal in sample}} \times 100 = \% \text{ solubles} \qquad (3)$$

It should be understood that the ash value does not represent the actual amount of mineral matter in coal and the mineral matter in the residue is probably a reacted form of mineral matter. These slight differences do not significantly interfere with the "% solubles" calculation.

Reduced-Pressure, Room-Temperature Soxhlet Extraction: Apparatus and Procedure. Solubility values obtained from a typical Soxhlet extraction conducted near the boiling point of slurries were heated to 60°C in order to improve mixing, and 5-7 g aliquots were added to glass weighing bottles having ground glass lids. Individual samples were then analyzed by the pressure filtration method.

Results and Discussion

Three different types of material were used to study the pressure-filtration method. Those were 1) Kentucky 9/14 coal, 2) slurries from low-temperature reactions (low conversion to soluble products, high solids content), and 3) slurries from high-temperature reactions (high conversion to soluble products, low solids content). These particular types were chosen because the first represents a material that is difficult to extract; the second, a material that was difficult to extract with some solvents and more easily extracted with others; and the third, a material that can be extracted with most solvents.

Extraction of Coal: A Study of Variables. Approximately 10% of Kentucky 9/14 coal is soluble in THF, using a reduced-pressure Soxhlet extraction. When extracting coal, the length of time that the solvent is in contact with the coal is very important (4). The solvent must be in contact with the coal particle long enough to cause swelling (if possible), penetrate the deepest pores, and then extract and remove those soluble components from within the coal matrix (5,6). With Soxhlet extractions the contact time is essentially unlimited, and the sample is extracted for as long as necessary. Contact time in pressure filtrations can be controlled by decreasing the flow rate or by using more solvent. Table II demonstrates the variations in THF solubility that occur when extracting Kentucky 9/14 coal by pressure filtration.

The results obtained by varying the filtration time demonstrate that by increasing contact time, the solvent has a greater opportunity to interact with the coal sample and extract the soluble components. As can be seen in Table II, greater amounts of solvent which lead to longer filtration times, and smaller sample sizes, increase the amount of material extracted from coal. In filtrations where the filtration times were 13-15 minutes, we were able to obtain an extraction value approaching the 10% value obtained by room-temperature, reduced-pressure Soxhlet extraction.

Results for Slurries Reacted at Low Temperatures. Reduced-pressure Soxhlet extraction of the low-temperature-reaction sample, using THF, continued for approximately 4 days. On the other hand, pressure filtration of duplicate low-temperature samples took only 13-15 minutes.

Table II. Pressure Filtration of Kentucky 9/14 Coal

Variable	Weight coal* (g)	Amount of solvent (mL)	Filtration time (min)	THF solubles (%)
Sample Size	1.837	500	1	3.4
	2.060	500	1	3.4
	1.001	500	1	3.8
	1.001	500	1	3.5
	0.501	500	1	6.4
	0.502	500	1	6.2
	0.937	1000	13-14	7.0
	0.931	1000	12-14	7.3
	0.500	1000	14-15	8.6
	0.506	1000	14-15	8.3
Filtration time	1.001	500	1	3.8
	1.001	500	1	3.5
	1.002	500	4-5	5.5
	1.008	500	4-5	5.8
	1.002	500	7-8	6.2
	1.005	500	7-8	5.8
	0.937	1000	13-14	7.0
	0.931	1000	13-14	7.3
	1.029	1000	14-15**	8.3

* -250 + 325 mesh
** Coal sonicated in 50 mL THF for 10 minutes before filtration

The THF solubility values (Table III, SCT 65 & 69) demonstrate that the extractions using the pressure-filtration method are thorough, and the solubility values are comparable to the continuous Soxhlet-extraction method. When the same procedure was performed using cyclohexane, a large difference was seen between low-temperature Soxhlet-extraction values and the pressure-filtration values.

The negative numbers for solubility values shown in Table III show that the residue weight after reaction is greater than the initial weight of coal in the sample. (See % solubles equation in experimental section.) This apparent anomaly can be accounted for by some of the vehicle being incorporated into the coal matrix during reaction at 250°C (7-10). Experiments have been conducted that show that during the solubilization of coal at low temperatures some of the nitrogen rich components of the vehicle creosote oil are chemically incorporated within the coal matrix, and some of the vehicle may be physically entrapped or difficult to extract because of diffusional limitations (11). Creosote oil, is completely soluble in cyclohexane, and its extraction from the partially disrupted coal matrix should occur readily, unless physical or chemical incorporation has taken place. Thus the negative values for "% solubles" are a result of the incorporation of vehicle into the residue. The typical filtration times using cyclohexane in pressure filtration may not allow for a thorough extraction of high-solids content, coal-oil slurries and may require a modified procedure as discussed below.

Results for Slurries Reacted at High Temperatures. Coal-oil slurries reacted at high temperatures, where most materials are in solution and the coal matrix has been degraded, does not necessitate the solvent swelling of solid coal particles. The results (Table III, SCT-83) compare the effectiveness of both solvents using pressure filtration and Soxhlet extraction. The pressure-filtration results for both cyclohexane and THF appear to be comparable to those using the Soxhlet-extraction procedure.

Modified Cyclohexane Pressure Filtration. Cyclohexane is not an efficient solvent when extracting material from low-temperature reaction slurries that include significant amounts of unreacted coal. Since this solvent does not have the ability to swell the coal matrix, it must gradually force its way into the pores of the reacted coal matrix and slowly extract the cyclohexane soluble material. This requires a long contact time and pressure filtration does not satisfy this requirement. Soxhlet extraction with cyclohexane required 3-4 days for the filtrate to become essentially colorless. When using cyclohexane as a solvent, the disparity between the

Table III. Extraction of Slurries Reacted at High & Low Temperatures

Run # SCT-	Coal	Wt. Ratio Coal:Oil	Reaction Temp. °C	Time min.	THF Extraction Soxhlet	THF Extraction Pressure Filtration	Cyclohexane Extraction Soxhlet	Cyclohexane Extraction Pressure Filtration
65	KY 9/14	1:5	250	15	18.7	18.8	-5.3	-30
69	KY 9/14	1:5	250	15	18.1	17.9	-4.3	-32
83	IL #6	1:3	450	40	90.2	88.8	28.5	34.3

% Solubles

Soxhlet and the pressure-filtration extraction values obtained with coal-derived products from low-temperature reactions necessitated the development of an alternate pressure-filtration procedure. If cyclohexane cannot swell and penetrate the unreacted coal in the reacted slurry adequately, then THF can be used to swell and penetrate the unreacted coal. All the THF soluble material (which includes the cyclohexane soluble vehicle) can then be efficiently extracted and separated from the unreacted coal matrix.

The results using the modified procedure are shown in Table IV. The pressure-filtration values in SCT-87 compare well with the Soxhlet-extraction values from SCT 65 and 69. In other experiments (SCT 91 and 95), part of the reacted slurry was used to perform solvent classification by pressure filtration and the remainder for solvent classification by Soxhlet extraction. Agreement of cyclohexane solubility values within 3-6% is a considerable improvement over the difference of approximately 25% obtained with the standard pressure-filtration procedure, but further improvements in the technique are being sought. The solubility values for the two methods compare fairly well, considering that these "high solids content" slurries are difficult to extract with cyclohexane.

Precision of Pressure Filtration. Pressure filtration can solubilize as much material from coal-derived products as a room-temperature Soxhlet extraction, but it also must be precise in order to be a viable method (1,2). Measurements of precision are shown in Table V and were made on samples obtained both from high-temperature reactions using either creosote oil or SRC HD recycle oil as a vehicle, and from low-temperature reactions using creosote oil as a vehicle. The standard deviations and relative standard deviations from the results for the high-temperature slurry products demonstrate that the pressure-filtration method has the necessary precision to be a viable method. Values for samples from low-temperature reactions show that THF filtrations provide the necessary precision, but the precision for the cyclohexane filtrations should be improved.

In summary, the pressure-filtration technique is a useful and precise analytical tool in performing the solvent classification of certain coal-derived products. The method will not only give meaningful results comparable to continuous, room-temperature Soxhlet extractions but, most importantly, save many hours in the laboratory. We have found it rapid enough to monitor continuous operations. With coal-oil slurry reaction mixtures obtained from high-temperature coal-liquefaction reactions, the solvent classification can be completed in 1 1/2 to 2 hours.

Table IV. Comparisons Using the Modified Cyclohexane Pressure Filtration for Slurries Reacted at Low Temperatures

Run # SCT	Modified cyclohexane pressure filtrations (% solubles)	Cyclohexane Soxhlet extractions (% solubles)
65		-5.3
69		-4.3
87*	-6.8	
91	-5.7	-8.9
95	-9.8	-4.1

* Value based on 5 replicates.

Table V. Standard Deviations and Relative Standard Deviations for Pressure Filtrations of Coal-Oil Slurries

	Vehicle	Standard Deviation (%)*/ Relative Standard Deviation THF	Cyclohexane
High Temperature Reactions	Creosote oil	0.10/0.005	0.26/0.004
	SRC HD recycle	0.12/0.005	0.61/0.007
Low Temperature Reactions	Creosote oil	0.29/0.003	1.1**/0.012

* 5-7 Replicates for each set.
** Only 4 replicates used.

Acknowledgement

We thank Dr. Richard Tischer for his assistance and expertise. We also wish to thank Mr. Gary Steigel for supplying us with samples from the high-temperature continuous reactor.

Reference in this report to any specific commercial product, process, or service is to facilitate understanding and does not necessarily imply its endorsement or favoring by the United States Department of Energy.

Literature Cited

1. Mima, M. J.; Schultz, H.; McKinstry, W. E. "Methods for the Determination of Benzene Insolubles, Asphaltenes, and Oils in Coal-Derived Liquids", PERC/RI 76/6, September 1976.
2. Schultz, H.; Mima, M. J. "Comparisons of Methods for the Determination of Asphaltenes, Oils, and Insolubles. Part 1 - A Coal-Derived Liquid." PETC.TR-80/3, May 1980.
3. Bertolacini, R. J.; Gutberlet, L. G; Kim, D. K.; Robinson, K. K. "Catalyst Development for Coal Liquefaction", EPRI Report AF-574, November 1977.
4. Dryden, I. G. C. Fuel 1951, 30, 145-158.
5. Kiebler, M. W. "Chemistry of Coal Utilization", Wiley and Sons; New York, 1945; 715-760.
6. Marzec, A.; Juzwa, M.; Betlej, K.; Sobkowiak, M. Fuel Proc. Technol. 1979, 2, 35-44.
7. Grens, E. A. II; Hershkowitz F.; Holten, R. R.; Shinn, J. H.; Vermeulen T. Ind. Eng. Chem., Process. Des. Dev. 1980, 19, 396-401.
8. Wright, C. H.; Schmalzer, D. K. "Coal Liquefaction Preheater Studies", U. S. Department of Energy Project Reviews Meeting, Section 4, March 1979.
9. Bickel, T. C.; Stohl, F. V.; Thomas, M. G. "Coal Liquefaction Process Research", Sandia National Laboratories, Quarterly Report, April - June 1980.
10. Longanbach, J. R.; Droege, J. W.; Chauhan, S. P. "Short Residence Time Coal Liquefaction", EPRI Report AF 780, Battelle Columbus Laboratories, June 1978.
11. Utz, B. R.; Narain, N. K.; Appell, H. R. U.S. Department of Energy, Pittsburgh Energy Technology Center, 1980, private communications.

RECEIVED May 17, 1982

12

Scanning Electron Miscroscope-Based Automated Image Analysis (SEM–AIA) and Mössbauer Spectroscopy

Quantitative Characterization of Coal Minerals

F. E. HUGGINS, G. P. HUFFMAN, and R. J. LEE

U.S. Steel Research, Monroeville, PA 15146

Many coal-utilization problems derive from the mineral matter in coal and require for their solution quantitative determinations of this component. For such determinations, we use automated image analysis (AIA) in the scanning electron microscope (SEM-AIA) and ^{57}Fe Mössbauer spectroscopy. SEM-AIA involves measurement of the areas of typically 1024 particles in a polished coal section and classification of each particle into one of 29 categories on the basis of its energy-dispersive x-ray spectrum. From such data, the relative weight percentages of the different minerals are derived. Mössbauer spectroscopy is used for classifying certain iron-bearing categories that could arise from two or more distinct minerals with identical energy-dispersive spectra. Because these techniques are used directly on the coal, provide information on mineral particle size, and are more sensitive to minor constituents than other techniques, they compare well with other techniques such as Fourier transform infrared spectroscopy and x-ray diffraction commonly used for quantitative coal mineralogies.

There is currently much interest in the mineral matter included in coal and the behavior and properties of such matter during coal-utilization technologies. For the most part, the mineral matter has deleterious effects during utilization, and considerable attention must often be given to minimizing problems arising from this inorganic, noncombustible component in coal. On the positive side, however, the mineral matter appears to have desirable catalytic properties in certain coal-conversion technologies and can be used by geologists in exploration to delineate coal-quality trends within a coal deposit. Although some mineralogical information can be inferred from standard chemical and other analyses of the coal and the ash remaining after controlled combustion, a full

0097-6156/82/0205-0239$06.00/0
© 1982 American Chemical Society

understanding of mineral-related phenomena in coal utilization requires a detailed knowledge of the mineralogy of the coal.

A number of techniques have been used to determine mineralogies of coal. As discussed in a recent review (1), the most common techniques are x-ray diffraction, infrared spectroscopy, optical microscopy, and electron microscopy. X-ray diffraction and infrared spectroscopy can be considered "bulk" methods because they are generally best performed on the mineral-matter concentrate obtained by removal of the macerals by low-temperature ashing (2). The microscope methods can be considered "particulate" methods because mineral grains in the coal are sized and classified individually. These microscope methods are usually used without separation of minerals and macerals because the coal macerals can serve as a background matrix to separate mineral particles and to provide contrast for the dimensional measurement of the particle.

Although particulate methods are potentially more informative than bulk methods because of the extra information obtained on particle size, their application in coal mineralogy has been relatively limited. The main reason for this has been the lack of automation, as such microscopic measurements are time-consuming and tedious if sufficient data have to be generated manually. Recently, however, the application of microcomputers and image-analysis systems to microscopy has progressed to the extent that most of the operation, data collection, and analysis can now be done automatically. A number of reports of such automated microscope techniques used in coal research have appeared recently in the literature (3-8).

In this paper we describe the use of a scanning electron microscope (SEM) equipped with automatic-image-analysis (AIA) capability for the determination of coal mineralogies. Because one of the main limitations of the SEM-AIA technique is its inability to distinguish the various iron-bearing minerals, ^{57}Fe Mössbauer spectroscopy is used to supplement data from SEM-AIA with respect to iron-bearing minerals. The combination of these two techniques usually provides a detailed, quantitative characterization of the minerals in coal.

Experimental Methods

Sample Preparation. Our sample-preparation procedures for SEM-AIA on coal have been modified from those described earlier (8). Approximately 7 g of -60 mesh coal is mixed into a slurry with about 2 g of an epoxy plastic and pressed in a stainless-steel die to form a cylindrical sample pellet 1.6 cm in diameter and 3 to 4 cm in length. A slice, approximately 1 cm thick, is cut from the pellet with a diamond wafering saw at an angle of 54.7° to the axis of the cylinder. This angle of cut has been found (9) to minimize bias that might arise from possible preferred orientation of clay minerals due to compression of the coal/epoxy mixture. The resulting oval-shaped pellet is placed face down in a one-inch bakelite ring form and surrounded by more epoxy to form a cylindrical pellet that fits conveniently into an automatic polishing machine. The coal/epoxy pellet is then ground, parallel to the cut face, with grids impregnated with cubic boron nitride and finally polished with 3 μm and 1 μm diamond pastes.

The use of cubic boron nitride and diamond paste for grinding and polishing minimizes contamination by materials that can be mistaken for minerals in the SEM analysis. The sample pellet is then cut parallel to the polished face to reduce its height to about 0.6 cm, mounted on an aluminum stub, and carbon-coated for use in the SEM.

Sample-preparation procedures for Mössbauer spectroscopy are relatively simple. Depending on the iron content of the coal, 0.3 to 1.0 g of very finely crushed (-400 mesh) coal is placed in a plexiglass compression holder of 1.3-cm-diameter. The filled holder is then placed in a metallic shield with a 1.3 cm diameter opening that enables the sample to be exposed to the γ-ray beam from the Mössbauer source. Identical procedures with a different metallic shield arrangement inside a vacuum shroud are used for cryogenic experiments.

SEM Automated Image Analysis. SEM-AIA has been used in our laboratory for a broad range of applications, from characterization of air-pollution samples to measurement of inclusions in steel. In this paper, only its application to minerals in coal will be described, as more general descriptions of the range and potential of SEM-AIA have been given elsewhere (10,11).

SEM-AIA of coal uses two distinct signals emitted from the sample in response to irradiation by the 20-kV electron beam. One of these signals is the back-scattered-electron (BSE) radiation, which provides excellent contrast between mineral particles and the background matrix of coal macerals and epoxy (Figure 1); it is used to locate the mineral particles and measure their area. A four-channel BSE detection system is used to collect this radiation because it increases the contrast and signal-to-noise ratio and minimizes geometric variation in the signal. The other signal used is the characteristic x-ray spectrum arising from the elements in the mineral particle. These x-rays are collected by means of an energy-dispersive x-ray spectrometer and are used to type the mineral particle.

The SEM-AIA system consists of an SEM (ETEC Autoscan) interfaced by means of two minicomputers to a beam-control unit (LeMont Scientific) and an energy-dispersive x-ray analyzer unit (Tracor Northern). A diagram of the system is shown schematically in Figure 2. The beam-control unit consists of a minicomputer-controlled digital scan generator and an analog comparator. The digital scan generator moves the electron beam in a specific pattern through the field of view to locate particles and to measure their area.

To locate particles, the beam is stepped across the sample in a coarse grid pattern, with typically 300 x 300 grid points covering the field of view. At each point, the intensity of the BSE radiation is sampled, and the minicomputer decides whether the beam is on or off a mineral particle. The intensity level of the BSE radiation that discriminates between mineral and background is set manually at the beginning of each field of view using the analog comparator. Once a particle is located, the grid density is increased to 2048 x 2048 grid points to increase the precision of the area measurement of the particle. The size and shape of the particle are determined from the contiguous group of points above the analog comparator level (11). The beam-control unit then positions the electron beam at the center of the particle and control of the operation is transferred to the minicomputer in the energy-dispersive x-ray unit.

The energy-dispersive x-ray spectrum of the particle is collected for two to three seconds in a multichannel analyzer, and the percentages of x-rays from 11 principal elements in coal ash (Na, Mg, Al, Si, P, S, Cl, K, Ca, Ti, Fe) are stored in the minicomputer memory along with dimensional and locational data about the particle transferred from the beam-control minicomputer while the x-rays are being recorded. The x-ray spectrum is classified by means of a sorting routine into one of 29 compositional categories. Control is then returned to the beam-control minicomputer and the next particle is sought by the beam-control unit.

The analysis is terminated once 1023 particles have been located, measured, and classified. However, to get a good representation of particles across a broad size range, a limit is put on the number of particles measured in various size ranges (Table I). The area of the sample scanned to collect the required number of particles in each size range is recorded. The ratio (M_j) of the area scanned in the top size range to that in the jth size range is used as a weighting factor in the calculation of the weight percentages of the different minerals. The formula for this calculation is based on the area of the particles of a given type, multiplied by the density of the mineral and the factor, M_j:

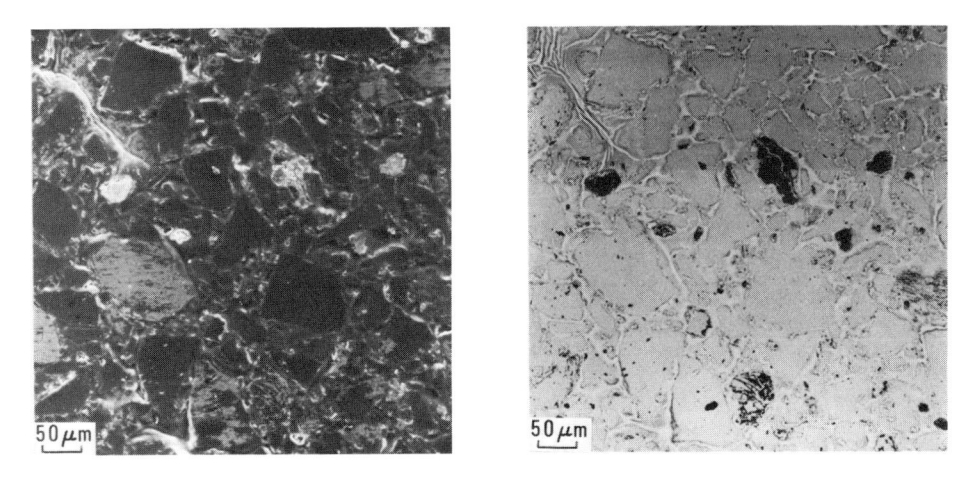

12. HUGGINS ET AL. *SEM–AIA and Mössbauer Spectroscopy* 243

Figure 1. Secondary electron image (left) and inverted backscattered electron (BSE, right) image of a polished sample of a bituminous coal (D seam, Colorado). The minerals are black and the maceral epoxy background is light gray in the inverted BSE image.

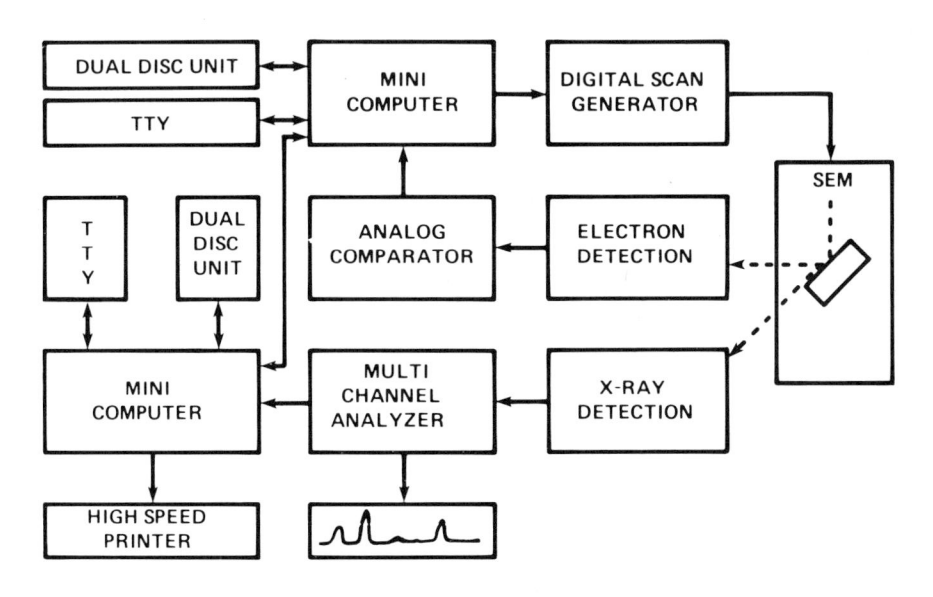

Figure 2. Schematic diagram of the SEM–AIA system.

Table I. Size-Range Classes in SEM-AIA of Coal

Class	Lower Limit (μm)	Upper Limit (μm)	Maximum Number of Particles
1	10.0	50.0	300
2	5.0	10.0	200
3	2.5	5.0	175
4	1.0	2.5	175
5	0.2	1.0	173

Table II. SEM-AIA Mineral Categories

Major*	Accessory**	Mixed[+]
Quartz	Montmorillonite	Pyrite/Jarosite
Kaolinite	Chlorite	Silicate/Sulfur
Illite	Gypsum	Silicate/Pyrite
Pyrite	Jarosite	Mixed sulfate/sulfide
Fe-rich	Fe Sulfate	Mixed Carbonate
Calcite	Halite	Mixed Chloride
	Sylvite	
Mixed Silicates	Dolomite	Trace element
	Ankerite	
	Rutile	Unknown
	Apatite	
	Sulfur	2 spare

* Usually each >3 wt% of mineral matter and sum of major categories >90 wt %.

** Usually each <3 wt % of mineral matter and sum of accessory categories <5 wt %.

[+] Sum of mixed categories usually <10 wt %.

$$W_i = 100 \, \rho_i \sum_{j=1}^{5} M_j \sum_{k=1}^{N_{ij}} A_{ijk} \Bigg/ \sum_{i=1}^{29} \rho_i \sum_{j=1}^{5} M_j \sum_{k=1}^{N_{ij}} A_{ijk}$$

where W_i is the weight percentage of the ith mineral.
ρ_i is the density of the ith mineral.
N_{ij} is the number of particles of type i in the jth size range.
A_{ijk} is the area of the kth particle of type i in the jth size range.

This procedure and calculation yield results of much better reproducibility than the method followed earlier (6) of simply measuring the first 1000 particles encountered. The derived results on weight percentages and on the size distribution of the different minerals are stored on floppy diskettes along with the original data on the individual particles.

As a result of experiments done to test the accuracy and precision of SEM-AIA for coal minerals (8), it was found that pyrite is overestimated by about one-third because its BSE contrast is much higher than that of all other common coal minerals that have very similar BSE contrasts. The weight percentages from SEM-AIA are adjusted for this overestimation and all data reported in this paper are so adjusted.

The 29 compositional categories used in the sorting routine consist of 18 individual minerals, 7 mixed categories, a trace-element category (most intense peak in the energy-dispersive spectrum is not one of the 11 elements listed above) and an unknown category. There are also two spare categories available for special use; for example, in certain lignites from Montana, barite ($BaSO_4$) was a significant (~5 wt%) accessory mineral and was included in the analysis by simply reprogramming the sorting routine. The categories are listed in Table II; they and the sorting routine have been described in detail elsewhere (6,8).

Mössbauer Spectroscopy. ^{57}Fe Mössbauer spectroscopy of coals and coal derivatives is yielding much new information about the iron minerals in coal and their behavior during coal utilization. However, in this paper, only its use to supplement SEM-AIA data will be detailed and the interested reader is referred to published reports (12-17) describing the theory and practice of Mössbauer spectroscopy and its application to a broad range of coals and coal products.

Our Mössbauer techniques, when used simply to supplement SEM-AIA of a coal, are quite standard. All measurements are conducted in transmission mode, with the coal sample usually at room temperature. Spectra are not obtained routinely with the sample at cryogenic temperatures unless the coal is weathered or unusual in some other respect. In most coals the different minerals each contribute a quadrupole doublet, consisting of two similar peaks, to the spectrum (Figure 3, top). However, if the mineral is magnetically ordered, a six-peak pattern will be observed (Figure 3, bottom). Different Mössbauer parameters are derived from the peak positions in these two-peak or six-peak patterns and the iron-bearing phase can then be identified from the values of these parameters.

The different iron-bearing minerals that we have observed in Mössbauer spectra of various U. S. coals are listed in Table III, along with their Mössbauer parameters. From examination of this table, it is quite obvious that there are a number of different minerals that could give rise in SEM-AIA to Fe-rich x-ray spectra (siderite, iron metal, hematite, magnetite, goethite, other oxyhydroxides) or to Fe-S x-ray spectra (pyrite, marcasite, szomolnokite, melanterite, coquimbite, other iron sulfates). Thus, discrimination among the possibilities for these SEM categories is necessary and can be achieved by Mössbauer spectroscopy, although it should be noted that pyrite and marcasite are sufficiently different from sulfates in terms of the Fe:S ratio that separate categories are incorporated in the SEM-AIA sorting routine. The Mössbauer spectrum is also quite sensitive to minor degrees of weathering ([18]) and provides a means by which oxidation can be readily recognized. Oxidation usually gives rise to a wider variety of minerals in the coal than would normally be found. As a result, some modification of the SEM-AIA operation may be needed.

Discussion

Over the last two years we have determined the mineral matter in about 100 coal samples from some 30 U. S. coal seams by using the combination of SEM-AIA and Mössbauer spectroscopy. These coals have varied across a wide rank spectrum, from Montana lignites to Narragansett Basin anthracites. In this section the relative merits and limitations of applying these techniques will be illustrated and discussed.

Characterization of Minerals in U. S. Coals.
Analyses are shown in Table IV for nine coals of different rank. The Mössbauer data are expressed as percentages of the total iron contained in specific minerals, whereas the SEM data are expressed as weight percentages of the mineral matter in the coal. These coals were also chosen to illustrate the

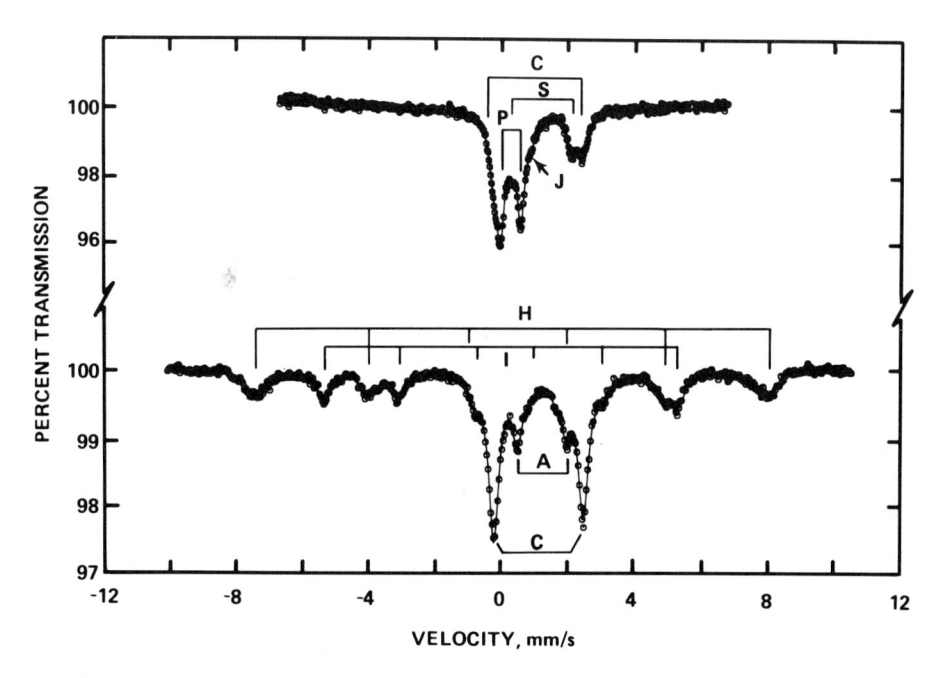

Figure 3. Room temperature Mössbauer spectra of coal samples from the Pratt seam in Alabama (top) and from an anthracite deposit in Rhode Island (bottom). In the top spectrum, peaks indicated arise from iron in the common coal minerals pyrite (P), clays (C), siderite (S), and jarosite (J). In the bottom spectrum, peaks indicated arise from iron in clays (C), and in the rare coal minerals, ankerite (A), iron metal (I), and ferric oxide (H).

Table III. List of Iron-bearing Minerals Recognized in Room-Temperature Mössbauer Spectra of U. S. Coals

Mineral	Isomer Shift (mm/s)	Quadruple Splitting (mm/s)	Magnetic Splitting (kG)	Comments on Occurrence
Pyrite, FeS_2	0.31	0.61	—	Very common, in most coals.
Marcasite, FeS_2	0.29	0.51	—	Very rarely observed; subordinate to pyrite.
Clay (illite, chlorite)	1.12	2.65	—	Common — especially in low sulfur
Siderite, $FeCO_3$	1.23	1.80	—	Common — and higher rank coals
Clay (montmorillonite)	1.12	2.79	—	Rare
Jarosite, $(K,Na)Fe_3(SO_4)_2(OH)_6$	0.38	1.15*	—	Fairly common, from FeS_2 oxidation.
Szomolnokite, $FeSO_4 \cdot H_2O$	1.26	2.71	—	Fairly common, from FeS_2 oxidation.
Melanterite, $FeSO_4 \cdot 7H_2O$	1.27	3.18	—	Rare, restricted to "wet" coals.
Coquimbite, $Fe_2(SO_4)_3 \cdot 9H_2O$	**	**	—	Rare
Ankerite, $Ca(Fe,Mg)(CO_3)_2$	1.23	1.50	—	Rare; New England anthracites only examples of ankerite dominant over siderite.

Table III (Continued)

Mineral	Mössbauer Parameters			Comments on Occurrence
	Isomer Shift (mm/s)	Quadruple Splitting (mm/s)	Magnetic Splitting (kG)	
Fe metal, Fe	0.00	0.00	330	Very Rare; New England anthracites only - from reduction of siderite?
Goethite, α-FeOOH	0.37	Var+	0 - 390+	Common in weathered coals, from pyrite oxidation.
Lepidocrocite, γ-FeOOH	0.37	0.65	—	Rare; in weathered coals.
Magnetite Fe$_3$O$_4$	0.31 0.65	0.00 0.00	495 465	Rare; oxidation product or processing contaminant.
Hematite, α-Fe$_2$O$_3$	0.38	-0.10	515	Very rare; dehydration of FeOOH

* Varies somewhat (1.05 - 1.25 mm/s) depending on composition.
** Complex spectrum
+ Varies depending on magnetic properties. Commonly superparamagnetic in coals.

Table IV

Mineralogical Analyses of Selected U. S. Coals

Coal (Seam, County, State, Rank)

	Pust Dawson, MT Lignite	Rosebud Rosebud, MT Sbb	Herrin No. 6 ----, IL HVcb	D Seam Gunnison, CO. HVbb	Gholson Shelby, AL HVab	Pratt Jefferson, AL MVb	Pocahontas No. 3 Wyoming, WV LVb	No. 8 Seam ----, PA Anthracite	"Anthracite" Bristol, MA Anthracite
Mössbauer									
% Fe in pyrite	100	65	93	15	16	45	19	78	16
Siderite				78	79	17	20	12	9
Clay					5	27	58	7	41
Jarosite			tr			11	4	3	
Other*		16	4	7					31-a
Other		19-m	tr-m						4-i
SEM-AIA, wt%									
Quartz	16	24	15	8	3	7	10	9	32
Kaolinite	38	40	5	29	1	20	26	5	tr
Illite	tr	1	5	1	1	32	13	29	24
Chlorite	–	–	tr	tr	2	1	7	tr	8
Montmorillonite	–	1	1	15	–	tr	1	tr	–
Mixed silicate	2	6	25	17	7	10	20	21	10
Pyrite	4	6	27	1	5	11	2	23	3
Fe-rich	–	1	tr	4	37	4	3	2	2
Calcite	7+	8	6	6	28	4	4	1	6
Ankerite	1	–	–	tr	7	tr	tr	tr	9
Gypsum	–	1	tr	–	–	–	tr	–	1
Jarosite	–	tr	1	tr	tr	tr	tr	1	–
Fe-sulfate	–	2	–	–	tr	1	1	–	–
Rutile	tr	–	–	tr	tr	1	tr	–	tr
Apatite	–	1	–	1	–	–	1	–	–
Barite	8	2	–	2	–	–	1	1	tr
Sil/sul	tr	1	7	–	2	1	1	tr	tr
Sil/pyr	–	–	2	–	tr	1	1	5	1
Others**	4	1	4	2	4	3	1	4	5
Unknown	20+	4	2	14	5	4	10		

tr - < 0.5 wt % for SEM-AIA; tr - <2% of total Fe for Mössbauer.
* Minerals indicated are: m-melanterite, a-ankerite, i-iron metal.
** Other mixed categories, for the most part.
+ Mostly Ca-rich macerals (see text).

heterogeneous nature of U. S. coals in terms of the variation of individual minerals. Even minerals traditionally considered as major components of coal mineralogy (kaolinite, illite, pyrite) can be negligible in certain coals.

The Mössbauer data indicate that the Fe-rich category can be equated to siderite except for the Massachusetts anthracite, in which iron metal also significantly contributes to this category. Except possibly for the two lowest rank coals, these nine coals do not show appreciable effects of weathering and no oxides or oxyhydroxides were observed in any Mössbauer spectrum. The presence of jarosite in the majority of these samples, however, indicates that some oxidation of pyrite has been initiated. The presence of significant melanterite in the Rosebud sample suggests that such oxidation has progressed furthest in this sample.

The SEM-AIA categories listed in Table IV consist of 15 individual mineral categories. This number of minerals is about double that determined by standard XRD or IR techniques, which are generally limited to the determination of no more than seven major minerals (1). The SEM-AIA technique, then, is more sensitive to minor mineral components than these more traditional techniques. Also, it is considerably more flexible in that a novel mineral can be introduced into the analysis very easily and without the need for calibration data, unlike these other techniques.

As can be seen in Table IV, one of the major SEM-AIA categories in most analyses is the mixed silicate category. This category arises from two main sources: either physically associated silicates, such as quartz-clay combinations derived from partings, or mixed-layer clays of variable compositions outside the compositional limits of the individual clay categories (kaolinite, illite, montmorillonite, chlorite). For example, about 30 percent of the particles in the mixed-silicate category of the Massachusetts anthracite had compositions consistent with an illite-chlorite mixed-layer clay. The weight percentages in this category can be further subdivided as needed.

It must be noted here that XRD and IR techniques are mineralogically specific, whereas SEM-AIA is only compositionally specific and the correctness of the SEM-AIA mineral analysis depends on the uniqueness of the individual mineral compositions. For example, SEM-AIA cannot discriminate between pyrite and marcasite, or among the TiO_2 polymorphs, or between halloysite and kaolinite [both $Al_2Si_2O_5(OH)_4$]. However, except for the Fe-rich and Fe-S categories, which are resolved by Mössbauer spectroscopy, all other duplications of mineral compositions involve minerals that are uncommon in coal. In general, the compositions of the common coal minerals are unique and a sufficient means by which the minerals can be identified.

This lack of mineralogical specificity in SEM-AIA, however, can be used to advantage when examining mineral-derived matter in cokes, chars, and other products of coal conversion technologies. As the SEM-AIA technique is insensitive to crystal structure, it makes no difference whether the minerals during conversion undergo phase transformations, become amorphous (e.g., dehydration of clays), or even melt, as long as the compositional fingerprints remain unaltered. Therefore, the SEM-AIA technique can be a most useful means to correlate directly minerals in the coal with the mineral-derived phases in the conversion product. For example, we have been able to demonstrate (unpublished work) that during carbonization, the minerals principally only undergo decomposition reactions involving the loss of volatile molecules, such as H_2O, H_2S, and CO_2. There is no evidence for extensive reaction among the different minerals, despite exposure to temperatures above 1000°C for periods of hours during carbonization, because the weight percentages of the different SEM-AIA compositional classes are not greatly different between the coals and the cokes derived from them.

Characterization of Inorganic Elements in Macerals. As noted above, one of the advantages of SEM-AIA is the fact that the mineral matter is analyzed without separation from the macerals. Occasionally, however, the presence of diffuse, organically bonded inorganic elements in the macerals will complicate the analysis. This complication is most serious for lignites, the macerals of which often contain significant calcium and other inorganic elements. As shown in Figure 4, the simple two-phase BSE image (minerals and maceral/epoxy background) in high-rank coals is replaced by a three-phase BSE image (minerals, Ca-rich macerals, and epoxy) in lignites.

If the particle/background discrimination level in the analog comparator is set high enough to exclude Ca-rich macerals, then an appreciable fraction of the inorganic matter in lignites is excluded from the analysis. If, however, the level is set to include the Ca-rich macerals, the SEM-AIA procedure finds the macerals and very little else because of their large total area compared with that of discrete minerals. Of these two procedures, we prefer the first so that the discrete minerals are emphasized in the analysis. A mass-balance calculation comparing the discrete mineral analysis with the chemical analysis of the ash will then give an approximate value for the fraction of inorganic elements in combination with the macerals. However, as the inorganic-

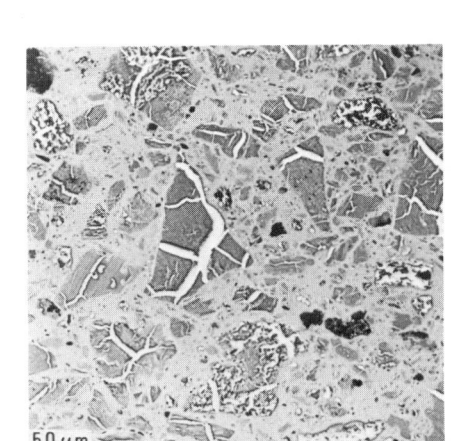

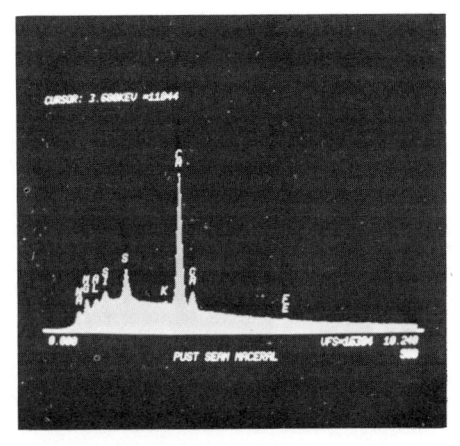

Figure 4. Inverted backscattered electron image of a polished sample of Montana Pust seam lignite (left) and energy-dispersive x-ray spectrum of an individual maceral (right). Note, in comparison to Figure 1, the enhanced contrast of the macerals due to the presence of inorganic elements.

element concentrations in the macerals vary, some inorganic-rich maceral areas are also measured, and therefore such analysis of lignites can only be regarded as semi-quantitative. This difficulty should ultimately be resolved by modification of the discrimination-level definition from the present overall average value to one sensitive to local variation.

For the lignite analysis shown in Table IV, most of the measured inorganic-rich maceral areas were assigned to the unknown category, although those that were particularly rich in calcium were assigned to the calcite category. In this latter situation, the very low total intensity of x-rays, a parameter recorded in the analysis, serves to distinguish these Ca-rich areas from calcite. Indeed, very little calcite was present in this lignite.

The same problem of Ca-rich macerals has also been noted in SEM-AIA of subbituminous coals and highly weathered bituminous coals (unpublished data), but to a much less serious degree. Similar complications arise when an organic element, usually sulfur or chlorine, contributes a strong background signal to the x-ray spectrum. For small particles, especially, this extra x-ray peak can be sufficiently intense to cause the particle to be assigned to a mixed or unknown category. In Table IV, for example, a strong organic sulfur background is responsible for the relatively high percentage of particles in the silicate/sulfur category for the Herrin No. 6 coal, and a strong chlorine background is mostly responsible for the large number of unknown particles in D-seam coal from Colorado.

Although the presence of Ca-rich macerals in lignites and the occasionally strong organic-element background in other coals complicate SEM-AIA, the information obtained about these components is, in itself, very useful. Scanning electron microscopy may be one of the few techniques that enable these diffuse components to be investigated in any detail.

Particle-size Analysis. Another advantage that SEM-AIA offers over bulk methods is the information that can be obtained relating to the distribution of particle size. Indeed, not only are bulk methods uninformative about particle size, but considerable precautions must be taken to ensure that particle-size effects do not adversely affect the determinations of the minerals (1,8).

An excellent application of SEM-AIA particle-size data was discovered in an investigation of pyrite oxidation as a result of a coal weathering (18). Table V summarizes mineralogical data obtained by SEM-AIA on a suite of four samples of Middle Kittanning coal. These samples were obtained from a strip mine

Table V. Mineralogical Data on Samples of
Middle Kittanning Seam by SEM-AIA

Mineral	Weight Percent of Mineral Matter			
	Unweathered	Highwall	Intermediate	Outcrop
Quartz	3	11	13	7
Kaolinite	14	17	17	18
Illite	18	24	8	5
Mixed Silicate	9	15	11	7
Pyrite (FeS_2)	50	24	20	9
Goethite (FeOOH)	1	4	27	51
Calcite	3	1	1	-
Others	1	1	2	2
Unknown	1	2	2	2
$\frac{FeOOH}{FeOOH + FeS_2}$:	0.02	0.14	0.57	0.85
by Mössbauer	0.00	0.16	0.53	0.78

between the highwall of the mine and an outcrop. The weathered nature of the coal increases from left to right in Table V. Mössbauer data confirmed the identity of the iron-rich category as goethite (α-FeOOH) and excellent agreement was found between SEM-AIA and Mössbauer techniques for the goethite/pyrite ratio, as shown in Table V.

Data on the distribution of the average diameter of goethite particles over the size ranges listed in Table I are shown in Figure 5. These data show that the slightly oxidized highwall sample has the highest percentage of goethite particles in the smallest size range. The goethite distribution profiles become less strongly biased towards small particle size from the highwall to the outcrop. The distribution profile for goethite in the outcrop sample, is in fact, quite similar to that for pyrite in the unoxidized coal.

These data confirm that particle size is an important factor controlling the conversion of pyrite to goethite in coal undergoing atmospheric weathering. Such data on particle size could not be easily obtained by other methods.

Conclusions

In this paper, we have described how minerals in coal can be quantitatively determined by the combination of SEM-AIA and Mössbauer spectroscopy. These techniques can be applied to all coals, from lignite to anthracite, with little or no modification. Although such applications are not discussed in any detail here, these techniques can also be applied to cokes, chars, solvent-refined coals, and other products of coal conversion.

Mineral characterization by SEM-AIA and Mössbauer spectroscopy is perhaps less specific than by FTIR or XRD. However, this disadvantage is compensated by the fact that more information is obtained by SEM-AIA. Such added information includes data on particle-size parameters and semiquantitative determinations of inorganic constituents of macerals, as well as data for a larger number of minerals in the routine analysis. In addition, as the minerals do not have to be separated from macerals for the analysis, SEM-AIA should be amenable for studies of mineral associations and possibly for mineral-maceral associations.

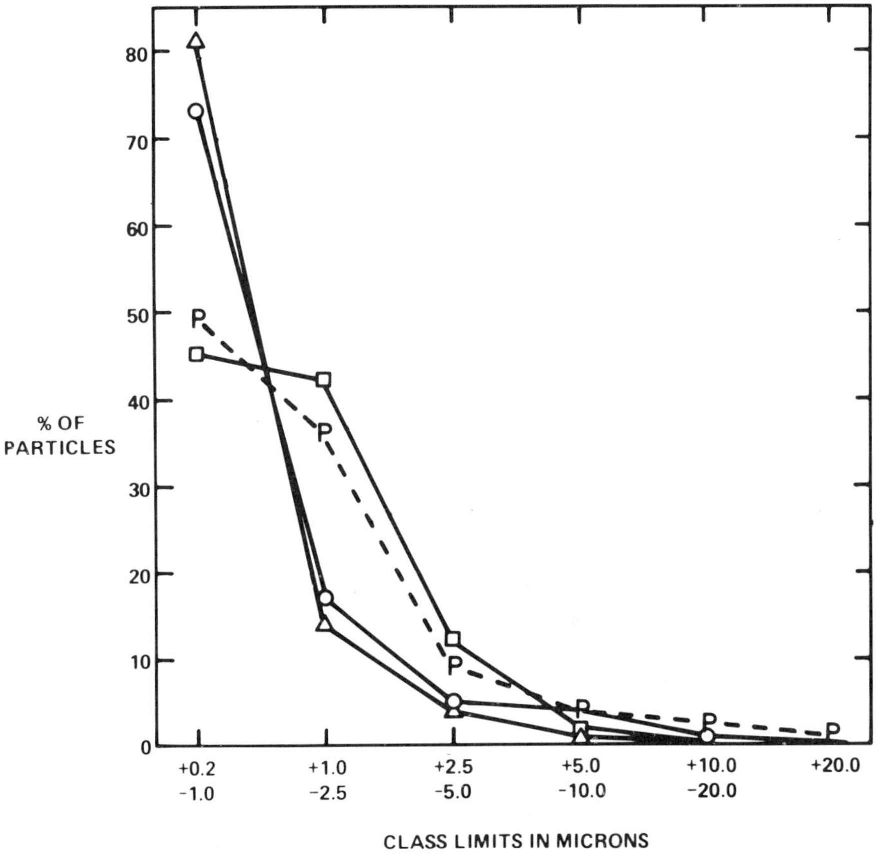

Figure 5. Distribution of particle-size parameter (average diameter of cross-sectional area) for geothite (α-FeOOH) in samples of Middle Kittanning coal. Key: △, high wall sample; ○, middle sample; □, outcrop sample; and P, original pyrite sample. See Table I for definition of class limits.

The material in this paper is intended for general information only. Any use of this material in relation to any specific application should be based on independent examination and verification of its unrestricted availability for such use, and a determination of suitability for the application by professionally qualified personnel. No license under any United States Steel Corporation patents or other proprietary interest is implied by the publication of this paper. Those making use of or relying upon the material assume all risks and liability arising from such use or reliance.

Literature Cited

1. R. G. Jenkins and P. L. Walker, Jr. "Analytical Methods for Coal and Coal Products", Vol. II; Academic Press: New York, 1978, 265-292.
2. H. J. Gluskoter, Fuel 1965, 44, 285-291.
3. L. A. Harris, T. Rose, L. DeRoos, J. Greene, Econ. Geol., 1976, 71, 695-697.
4. P. L. Walker, W. Spackman, P. H. Given, E. W. White, A. Davis, R. G. Jenkins, EPRI Report AF-417, 1977.
5. P. L. Walker, W. Spackman, P. H. Given, A. Davis, R. G. Jenkins, P. C. Painter, EPRI Report AF-832, 1978.
6. R. J. Lee, F. E. Huggins, G. P. Huffman, SEM/1978/I, 561-568.
7. A. Moza, L. G. Austin, G. G. Johnson, Jr., SEM/1979/I, 473-476.
8. F. E. Huggins, D. A. Kosmack, G. P. Huffman, R. J. Lee, SEM/1980/I, 531-540.
9. T. Erickson, R. Wappling, J. Physique, 1978, 37(C6), 719-723.
10. R. J. Lee, R. M. Fisher, NBS Special Publication 553, 1980, 63-83.
11. R. J. Lee, J. F. Kelly, SEM/1980/I, 303-310.
12. G. P. Huffman, F. E. Huggins, Fuel, 1978, 57, 592-604.
13. F. E. Huggins, G. P. Huffman, "Analytical Methods for Coal and Coal Products", Vol. III; Academic Press: New York, 1979, 371-423.
14. I. S. Jacobs, L. M. Levinson, H. R. Hart, Jr., J. Appl. Phys., 1978, 49, 1775-1780.
15. P. A. Montano, "Mössbauer Spectroscopy and its Chemical Applications"; American Chemical Society: Washington, D.C., 1981, 135-175.
16. G. P. Huffman, F. E. Huggins, G. R. Dunmyre, Fuel, 1981, 60, 585-597.
17. C. C. Hinkley, G. V. Smith, H. Twardowska, M. Saporoschenko, R. H. Shiley, R. A. Griffen, Fuel, 1980, 59, 161-165.
18. F. E. Huggins, G. P. Huffman, D. A. Kosmack, D. E. Lowenhaupt, Coal Geol., 1980, 1, 75-81.

RECEIVED May 17, 1982

13
Analytical Instruments in the Coal Preparation Industry
Current Status and Development Needs

LEON N. KLATT

Oak Ridge National Laboratory, Analytical Chemistry Division, Oak Ridge, TN 37830

The expanded use of domestic coal supplies as a substitute for imported petroleum will require significant expansion and modernization of the coal preparation industry. The Oak Ridge National Laboratory has assembled a multidisciplined team to study the needs of this industry and to undertake a development program with the goal of improving the efficiency and practice of coal beneficiation. Automated on-line analytical instruments are key elements in accomplishing this goal. On-line ash monitors are commercially available and have been successfully used in coal preparation plants world wide. On-line sulfur meters, based upon prompt neutron activation analysis have been developed. These systems can be modified to determine ash and calorific value. A moisture monitor, based upon microwave attenuation can be added to the sulfur meter. Nuclear density gauges are the only on-line process control instruments used in coal preparation plants. World wide development activities related to on-line analytical instruments are summarized, and development needs for product quality and process control instruments are discussed.

Coal preparation has been an important segment of the coal industry in the United States for approximately one hundred years. Its initial use was in the production of high quality coal used in the manufacture of coke. By 1965 approximately 95% of metallurgical grade coal was processed through a coal preparation plant. The use of cleaned coal as a fuel for electric power generation and as an industrial boiler fuel is relatively recent. Studies conducted for the U. S. Department of Energy (1) and the U. S. Environmental Protection Agency (2) concluded that coal preparation combined with fuel gas desulfurization is the least

0097-6156/82/0205-0259$06.25/0
© 1982 American Chemical Society

costly method of meeting New Source Performance Standards for sulfur dioxide emissions. The use of cleaned coal as a boiler fuel should improve boiler efficiencies; however, the comparative data required to quantify the impact that cleaned coal has on the performance and reliability of power plants are not available (3).

Coal preparation can be defined as a process or series of processes which reduces the mineral and pyritic sulfur content of run-of-the-mine coal. Although chemical and physical methods can be used to achieve the beneficiation, past and current industrial practices employ only physical methods, and are based upon the differences in specific gravity between the coal, mineral matter, and pyritic sulfur. A simplified block diagram of a coal preparation plant is shown in Figure 1. The run-of-the-mine coal is sized classified by a wet screening operation. The coarse coal is cleaned in jigs or heavy media baths. The smaller sized material is further subdivided into intermediate and fine fractions. The intermediate material is processed via a wide variety of unit operations, ranging from heavy media cyclones to wet concentrating tables. The fine fraction is usually processed via froth flotation. Figure 2 summarizes the quantity of clean coal produced since 1940 and compares this with the total U.S. coal production.

Different coal deposits differ widely not only in maceral, sulfur and mineral content but also differ widely in other properties, e.g., grindability, volatility and BTU content. Significant variations in the properties of coal can also occur within a deposit and within an individual seam. As a result, each coal deposit presents to the coal preparation engineer and plant operator a unique set of factors that must be considered in the design and operation of a preparation plant.

The fundamental information required by the coal preparation engineer and plant operator is summarized in coal washability curves. Figure 3 is an example of a washability curve. This data is obtained by subjecting the coal sample to a series of specific gravity fractionations and analyzing each float and sink subfraction for ash content. Other analyses, such as sulfur and BTU content, are often conducted, depending on the end use of the coal. Washability data for different preselected particle size fractions of the original coal sample are also obtained. This latter information is particularly important to the coal preparation engineer because different processing techniques and equipment are required to treat different size material.

Analytical Instrumentation

Laboratory Instrumentation. Successful operation of a modern coal preparation plant requires the availability of a well equipped and well staffed coal analysis laboratory. This laboratory must be capable of performing tests to characterize

13. KLATT *Analytical Instruments in the Coal Industry* 261

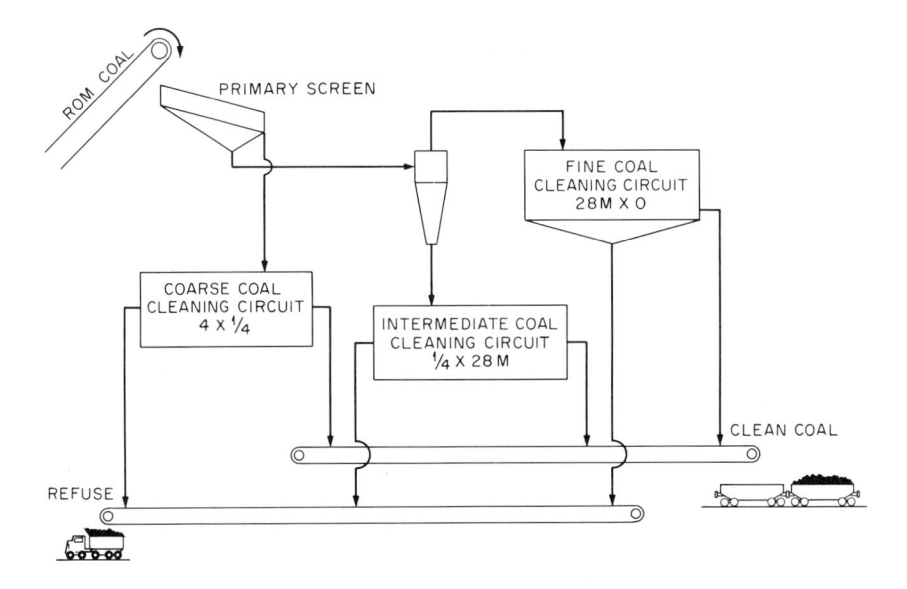

Figure 1. Simplified diagram of a modern coal preparation plant.

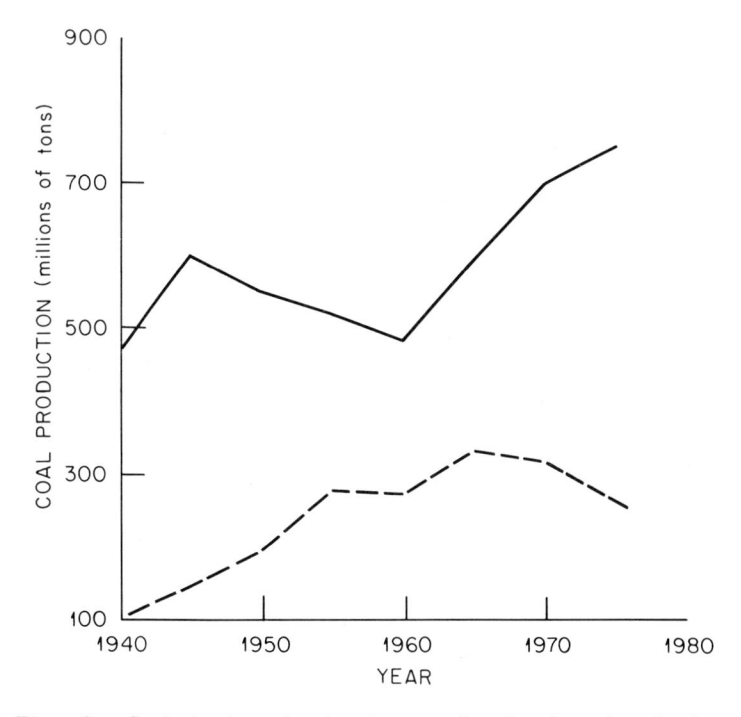

Figure 2. Coal mine (———) and coal preparation plant (– – –) production.

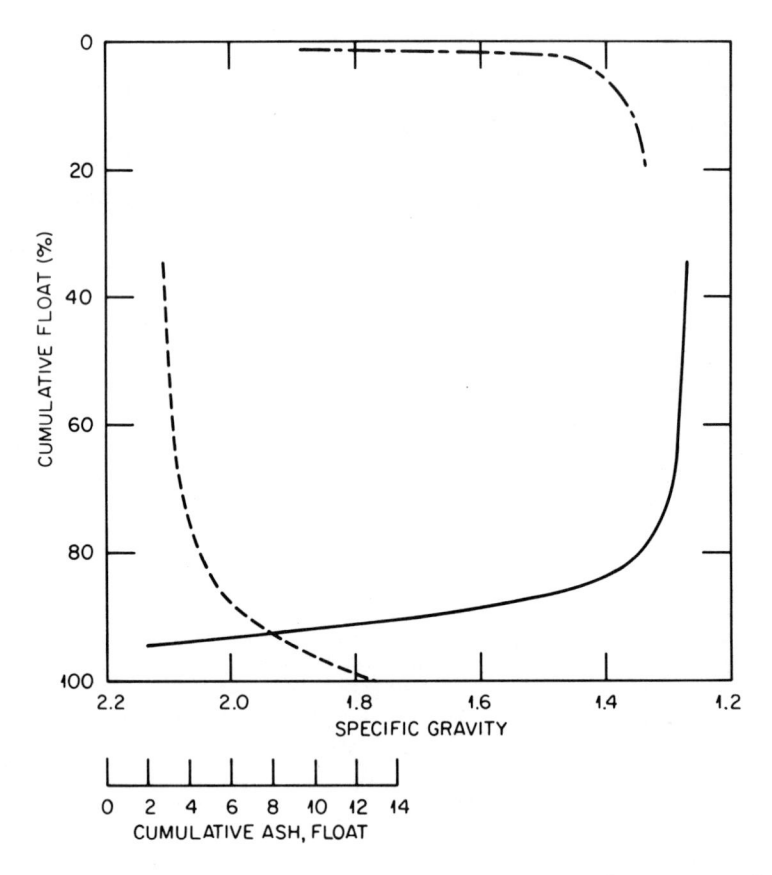

Figure 3. Coal washability curves. Key: – – –, cumulative ash; — – — –, ± 0.10 specific gravity distribution; and ———, yield.

the feed coal, to verify product quality, and to provide process control information. The analyses commonly performed are described in standard test procedures published by the American Society for Testing and Materials (4).

On-Line Instrumentation-General Considerations. Rapid response time and the ability to sample large quantities of material constitute the principal design criteria for on-line instruments in the coal preparation industry. Other important factors include cost, reliability, and maintainability; however these are common to on-line instrumentation in any industry. Rapid response is required because the residence time of material in the process equipment varies from a few seconds to a few minutes; average residence time in the plant is typically less than ten minutes. Furthermore, significant variations in the composition of the run-of-the-mine coal fed to the plant can occur in a step functional manner. The heterogenous nature of coal and the quantity of material processed through a typical circuit within a preparation plant requires a simple means of obtaining information representative of the process stream. Non-invasive sampling methods are preferred because material handling problems are minimized.

Because of the above requirements, most of the laboratory techniques used to analyze coal cannot be transferred to on-line instruments. Methods which use high energy photons, gamma or x-ray, can meet the above design criteria, and consequently, all the systems developed or under development for the coal industry, and which yield elemental composition information employ methodology which involves the measurement of gamma or x-rays.

On-line instrumentation in coal preparation plants can be classified into two broad categories, instruments that monitor a process or material parameter and instruments that monitor process equipment status. Liquid level and pressure gauges which monitor sump levels and pump outlet pressure are examples of instruments in the latter classification and will not be discussed in this report.

On-Line Instrumentation Used in the Coal Preparation Industry. Ash monitors, nuclear density gauges, and pH monitors are the only on-line instruments currently used to measure process or material parameters in U.S. coal preparation plants.

Two on-line ash monitors are commercially available, the Sortex Ash Monitor (Gunson's Sortex Ltd., London, England) and the KHD Analyzer (KHD Industrieanlogen, Humbolt, Weday, Federal Republic of Germany). Both instruments use the backscatter of high energy photons to determine the ash content of the sample; the Sortex unit employs ^{238}Pu, which emits a 17.6 KeV x-ray, while the KHD unit uses ^{241}Am, which emits a 60 KeV gamma ray. The Sortex Ash Monitor has been installed in eight U.S. coal preparation plants; thirty additional units have been installed worldwide.

The first installation and use of an ash monitor in the U.S. was a Sortex unit at the Moss No. 3 Preparation Plant (Clinchfield Coal Company, Division of Pittston Coal Group, Lebanon, Virginia). The Moss No. 3 plant produces three blends of metallurgical coal and a steam coal. The Sortex Ash monitor is used to monitor the metallurgical coal product, and its output is used by the plant superintendent to select the blending ratios of run-of-the mine coal being cleaned at the plant.

When a material is exposed to x-rays or gamma rays, a portion of the radiation is absorbed and a portion is scattered. The absorption process follows an exponential relationship

$$I = I_o \exp[-\mu_m \rho l] \qquad (1)$$

where I_o is the incident x-ray intensity, I the intensity after traversing an absorber thickness l, μ_m is the mass absorption coefficient, and ρ is the density of the material. The mass absorption coefficient increases approximately as the 3.5 power of the atomic number and decreases with increasing photon energy. The total scattering process is composed of elastic and inelastic components. The probability for elastic scattering of a photon increases with the square of the atomic number and decreases with increasing photon energy. The probability for inelastic scattering is proportional to the atomic number and also decreases with increasing photon energy. Consequently, for a given incident photon energy, an increase in ash forming elements in the coal decreases the total number of photons backscattered, and a measurement of the backscatter intensity provides a measure of the total ash forming elemental content of a coal sample.

A schematic drawing of the sensor system employed in the Sortex Ash Monitor is shown in Figure 4. The x-rays emitted by the ^{238}Pu source irradiate the coal sample and penetrate up to 38 mm. This radiation is absorbed and scattered; the fraction backscattered is counted with a gas proportional counter. The aluminum filter is used to compensate for iron fluorescent x-rays, which are excited by the incident x-rays. The filter preferentially absorbs most of the iron fluorescent x-rays (∼6 KeV), and its thickness is chosen based upon the iron content of the coal samples. The aluminum filter also compensates for sulfur variations in the coal. This occurs because a major fraction of sulfur is present as pyrite and the decrease in x-ray backscatter intensity due to sulfur is partially offset by an increase in iron fluorescence. Organic sulfur is generally constant, and it is corrected for in the calibration of the instrument (5).

Figure 5 contains the flow diagram for the installation of the Sortex Ash Monitor at the Moss No. 3 coal preparation plant. The clean coal product is sampled, crushed to a top size of 5 mm, and presented to the ash monitor. The discharged sample is

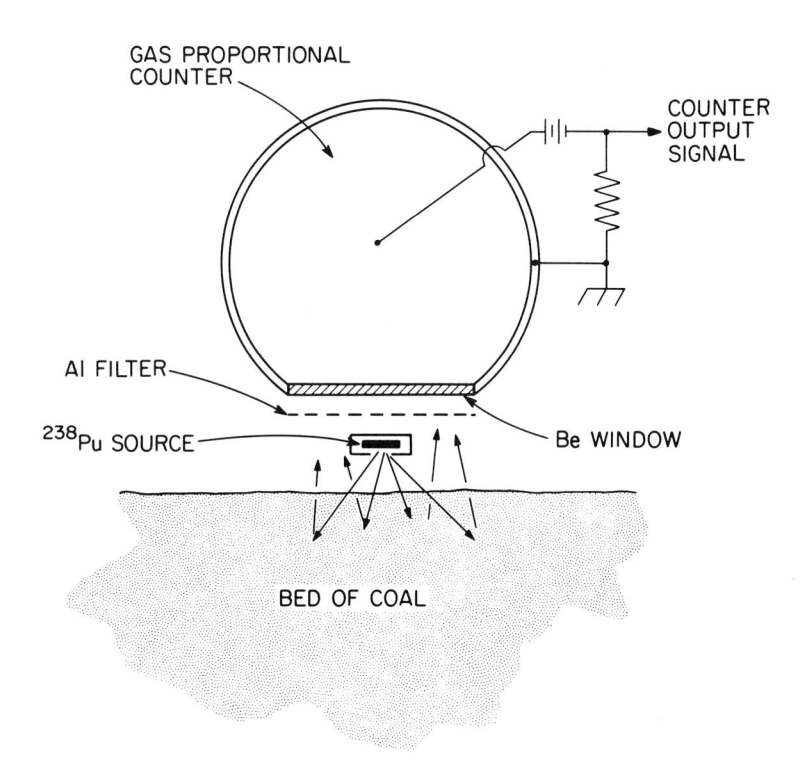

Figure 4. Schematic drawing of Sortex ash monitor sensor system.

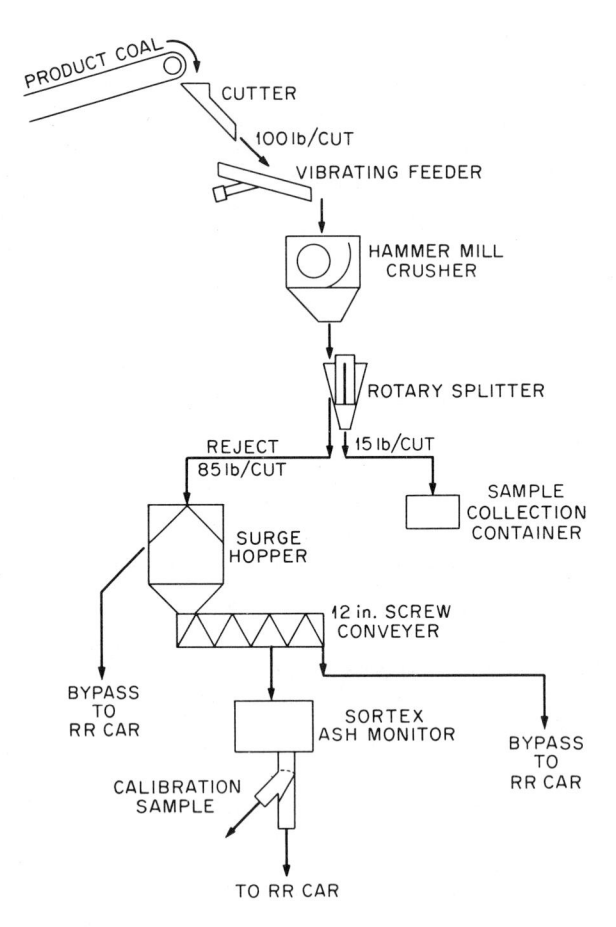

Figure 5. Typical coal sampling system and flow path.

combined with the main product stream at the railroad car loading chute.

Actual operating data indicate an excellent correlation between the ash monitor and laboratory ash analyses (6). The observed correlation equation is:

$$\text{Ash Monitor} = 0.74 \text{ (Lab Ash)} + 1.88 \qquad (2)$$

Data spanning six consecutive months of operation showed that the difference between the ash monitor results and the laboratory results was less than 1.12% ash at the 95% confidence limit. The precision of the ash monitor during the same time period averaged 0.53% ash. The failure to obtain the ideal correlation probably can be attributed to a number of factors; however, the two measurements are very different, the ash monitor provides data indicative of the elemental composition of the mineral matter in the coal while the laboratory analyses yield data reflecting the mass of metallic oxides derived from this mineral matter.

The accuracy of the ash monitor for different coal types was evaluated by Cammack and Balint (5). For fifteen different coal samples with ash content ranging from 4 to 40% ash, the accuracy of the ash measurement at the 95% confidence limit ranged from $\pm$ 0.3 to $\pm$ 2.2% ash; the average value for this confidence limit was $\pm$ 1.3% ash.

The moisture content of the coal is a major variable affecting the performance of the ash monitor. It alters the packing characteristics of the coal, and thereby, influences the bulk density of the sample. Moisture also affects the handling characteristics of the coal. Cammack and Balint (5) found that the compacted coal bed has a minimum density at 8% moisture; and for moisture levels up to 7%, the maximum error in ash analysis was 0.5% ash. Jones and Stanley (6) found that the coal handling system does not operate reliably above 8% moisture.

The operating experience at Moss No. 3 clearly indicates that the Sortex Ash Monitor is a very useful tool and that it possess sufficient reliability and accuracy to be suitable for control of plant circuitry. Through use of a correlation equation and careful calibration, the system should be able to provide the accuracy required to meet the quality control requirements of the coal preparation industry.

Nuclear density gauges are commercially available and have found use in many material handling industries. Nuclear density gauges determine the density of material by measuring the attenuation of gamma-rays passing through the material. Equation 1 provides the basis for the measurement. The gauge consists of a radioactive source, usually ^{137}Cs, a gamma-ray detector, amplifier and associated signal processing electronics.

Nuclear density gauges are typically used as process control instruments to monitor the density of water-magnetite slurries used in the heavy medium cleaning circuits. The instrument package is designed to clamp onto a process pipe. Accuracies of ± 1-2% are routinely achieved. The nuclear density gauge, configured as a line source with a linear array of detectors and coupled to a tachometer, is used as a belt scale.

The pH process monitoring and control equipment is standard industrial grade instrumentation and is available from numerous vendors. The pH measurement is commonly made in the froth flotation cells, and caustic soda or lime is added to the tailings output of the froth cells. This measurement is primarily used to control the pH of the plant process water with the set point selected to yield optimum performance from the flocculants added to the static thickener. Control of the process water acidity is also required to minimize the corrosion of process equipment.

On-line Instrumentation Applicable to the Coal Preparation Industry. Although numerous measurement concepts have been considered for on-line analysis in the coal preparation industry and several of these are being evaluated in the laboratory, e.g. Mossbauer spectroscopy (7), x-ray diffraction, and x-ray fluorescence, this section will discuss only those systems that have undergone or are scheduled for field testing and are advertised as available. These include elemental analyzer systems, a moisture monitor, and a coal-rock-water slurry meter.

Two elemental analyzer systems have been developed, the "Continuous On-line Nuclear Assay of Coal", CONAC, (Science Application, Inc., Palo Alto, CA) and "The Elemental Analyzer" (MDH Industries, Inc., Monrovia, CA). Both of these units are based upon the measurement of prompt gamma rays that are emitted from a nucleus following the capture of a neutron. This technique is commonly known as prompt neutron activation analysis, PNAA.

In a typical application a ^{252}Cf neutron source is positioned either beneath a conveyor belt or on one side of a gravity fed chute. The high energy neutrons emitted by the neutron source are thermalized by the coal sample before undergoing capture by the various sample constitutents. The resulting prompt gamma ray spectrum provides a qualitative signature for the various elements, and the number of characteristic photons, which depends upon the capture cross section and weight percent, provides a quantitative measure of each element. Most of the emitted gamma rays have energies in excess of 0.4 MeV and readily penetrate the structural material of the coal handling system as well as the mass of coal being measured. The gamma ray detector is placed outside the coal handling system. For those applications requiring analyses for major and minor elements a high resolution detector, such as germanium-lithium drifted detector, is used. If only sulfur and a few major elements in coal are desired, a medium resolution detector, such as a NaI (Tl) crystal is used. A computerized multichannel analyzer is required to process the gamma ray data. The system is calibrated using known reference samples of coal. The CONAC system can handle up to 50 ton/hr coal streams; the MDH system can handle up to 10 ton/hr.

The CONAC system can be supplied with the capability to measure a single element, e.g., sulfur, to perform a complete elemental analysis, or any intermediate level of analysis capability. The larger systems also provide an estimate of the BTU content of the coal. Generally, the results from the CONAC system are more precise than the results obtained from ASTM analytical procedures (8). Literature data indicate that the accuracy of the CONAC system surpasses that of the ASTM procedures (8); however, a careful study of its accuracy has not been completed, primarily because of the unavailability of a variety of standard samples in quantities required by the CONAC system. The heating value of the coal sample is calculated from an empirical equation (9), which requires analyses for hydrogen, oxygen, nitrogen, and sulfur. The agreement with results obtained by bomb calorimetry is excellent ($\pm$ 1.2 $\pm$ 0.3%). Coal samples with top sizes up to 8 cm per side have been successfully analyzed.

The MDH "Elemental Analyzer" performs analyses for sulfur and the major ash forming elements; an estimate of the BTU content is also provided. Computer simulations of the systems performance indicate that sulfur analyses can be performed with errors of $\pm$ 2.2% and ash analyses with average errors of $\pm$ 3.4% (10). Experimental data acquired with the system have not been reported.

A CONAC system, "Sulfcoalyzer", has been installed at Detroit Edison Company's Monroe, Michigan, power station (11). This plant is a four unit complex rated at 3000 megawatts. The "Sulfcoalyzer" cost $500,000 and will be used to control the coal blending facility. Operating data from this installation are not available. A second CONAC system will be installed at the Tennessee Valley Authority, Kingston, Tennessee, steam plant. This plant is a multi-boiler complex with 1600 megawatt capacity. The CONAC system will be used in connection with a consent agreement involving environmental regulations, compliance schedules, and methods of achieving compliance for sulfur dioxide and particulate emissions. While it seems most desirable to control coal quality at the mine or at a preparation plant, the important control point for a utility is at the delivery area at the power station. TVA's initial effort will concentrate on the use of the "Sulfcoalyzer" to monitor coal as it is delivered. A CONAC system has not been installed at any coal preparation plant.

Two installations of the MDH "Elemental Analyzer" are planned, one in a coal application and a second unit in an oil blending facility (12). The coal application is in the Homer City, Pennsylvania, coal preparation plant currently under construction. The analyzer system will provide a feedback signal that will be used to control the specific gravity of the heavy media coal cleaning operations. The installation was scheduled for Spring, 1981. The unit cost $300,000.

On-line moisture monitors, which are based on infra-red absorption, nuclear magnetic resonance, capacitance and microwave attenuation are used in numerous process industries. The applicability of these techniques to the coal preparation industry has been reviewed (13,14). The CONAC and "Elemental Analyzer" units use a microwave attenuation method for the moisture measurement. This information is used to correct the PNAA hydrogen assay data in order to calculate the hydrogen content of the coal. Kay-Ray, Inc., (Arlington Heights, Illinois) has developed an instrument that measures the moisture content of a coke stream and uses a combination of microwave and gamma ray attenuation measurements.

A schematic diagram of an on-line microwave moisture monitor is shown in Figure 6. At a frequency of 1 GHz sample thickness of approximately 15 cm are employed.

Dipole-dipole interaction between molecules placed in an oscillating electric field results in energy being transferred from the oscillating field to the sample. This phenomenon manifests itself as a frequency dependent dielectric constant and is a property common to all materials. The physical basis of the measurement is expressed by Equation 3.

$$P = P_o \exp[-2\alpha t] \qquad (3)$$

P_o is the incident power, P is the transmitted power, t is the sample thickness, and α is the attenuation factor defined by Equation 4:

$$\alpha = \omega \sqrt{\left(\frac{\Sigma_s \mu_o}{2}\right)\left[\left(1 + \tan^2\left(\Sigma_i/\Sigma_s\right)\right) - 1\right]} \qquad (4)$$

ω is the frequency, μ_o is the magnetic permeability, Σ_s is the static component of the dielectric constant, and Σ_i is the complex component of the dielectric constant, i.e., the frequency dependent term. Because water is a very polar molecule, relative to the hydrocarbon constituents of coal, it is responsible for a significant fraction of the microwave attenuation, i.e, tan $(\Sigma_i/\Sigma_s) \gg 1$, and Equation 4 reduces to

$$\alpha \simeq \omega\sqrt{\mu_o \Sigma_i/2} \qquad (5)$$

A typical response curve obtained for an individual seam of coal is shown in Figure 7. Two major aspects of the measurement are contained in this response curve. First, the curve is composed of two linear segments. The portion of the curve with the smaller slope is due to water adsorbed on the coal surface, and because of the restricted motion its loss factor, α, is less than that of free water molecules. The portion of the curve with the larger slope is due to the bulk water associated with the coal.

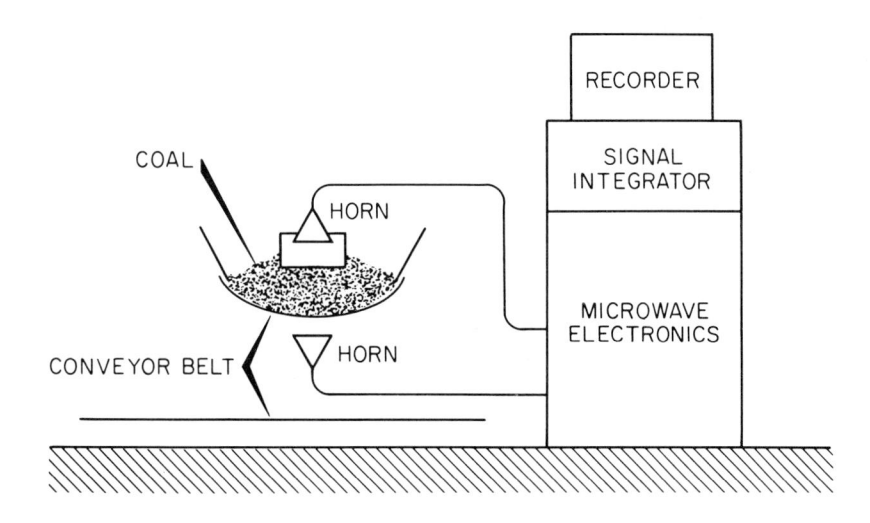

Figure 6. Schematic drawing of an on-line microwave moisture meter.

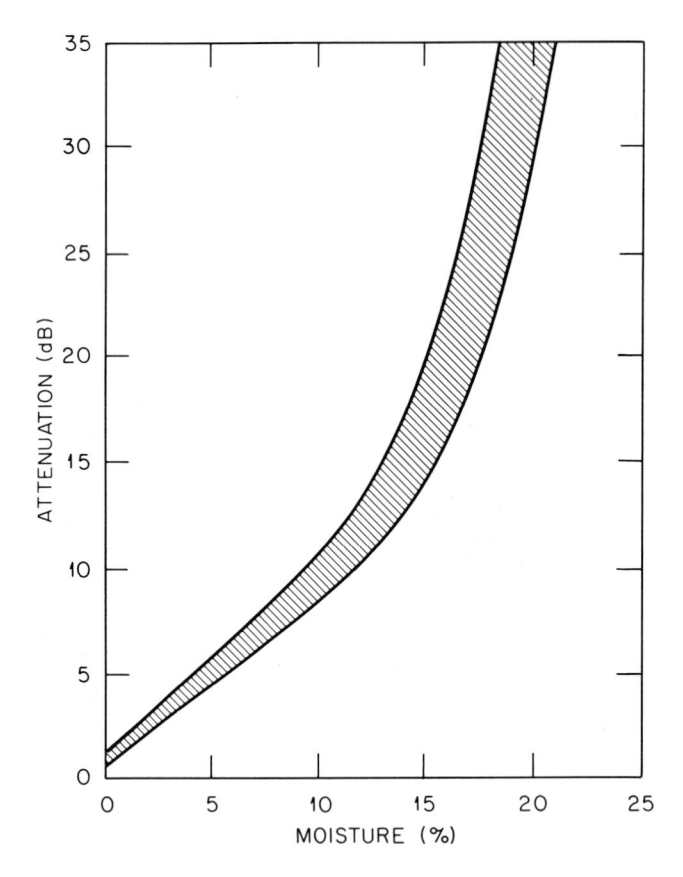

Figure 7. Response of microwave moisture meter for an individual seam of coal.

The band nature of the response curve is related to the average particle size. As the average particle size decreases the gradient of both linear segments decreases. Changes in the bulk density and the surface to volume ratio, and the resulting change in the ratio of surface to bulk water have been postulated as responsible for these changes in the slopes of the response curve (15). From the envelope of this curve and the precision of the attenuation measurement, the moisture content of a known coal can be measured with an uncertainty of $\pm$ 1% water. If the coal type is unknown the uncertainty increases to approximately $\pm$ 2.8% water (13).

Coal transport in the intermediate and fine coal processing circuits of a coal preparation plant is accomplished by means of water-coal slurries. Control of the various unit operations within these circuits requires a means of measuring the relative masses of coal, rock, and water in the feed, product and/or waste streams. Science Application, Inc., under contract from the Pittsburg Mining Technology Center, U. S. Department of Energy, has developed a slurry concentration meter (16). Although its intended application is in slurry haulage systems, it is equally applicable in the coal preparation industry.

Figure 8 contains a schematic drawing of the slurry concentration meter. The instrument package consists of three separate gauges: a gamma density gauge, a neutron gauge, and a conductivity gauge. The gamma density gauge provides a measure of the total mass, the neutron gauge provides a measure of the total hydrogen, and the conductivity gauge measures the free water in the slurry. The equations relating the observed response of each gauge to the mass of each component are:

(a) gamma gauge

$$M_c \mu_c^\gamma + M_r \mu_r^\gamma + M_w \mu_w^\gamma = -\ln(N_\gamma / \varepsilon_\gamma) \quad (6)$$

(b) neutron gauge

$$M_c \mu_c^n + M_r \mu_r^n + M_w \mu_w^n = -\ln(N_n / \varepsilon_n) \quad (7)$$

and (c) conductivity gauge

$$M_w \sigma_w^o = K \sigma_w \quad (8)$$

M_c, M_r, and M_w denote the mass per unit area of coal, rock, and water; μ_c^γ, μ_r^γ, and μ_w^γ and μ_c^n, μ_r^n, and μ_w^n are the effective gamma-ray and neutron mass absorption coefficients. ε_γ and ε_n are the efficiences of the gamma and neutron detection systems. σ_w^o and σ_w are the conductivity of the water phase measured in a reference cell and the conductivity of the slurry, respectively; K is the conductivity cell constant. Solving these

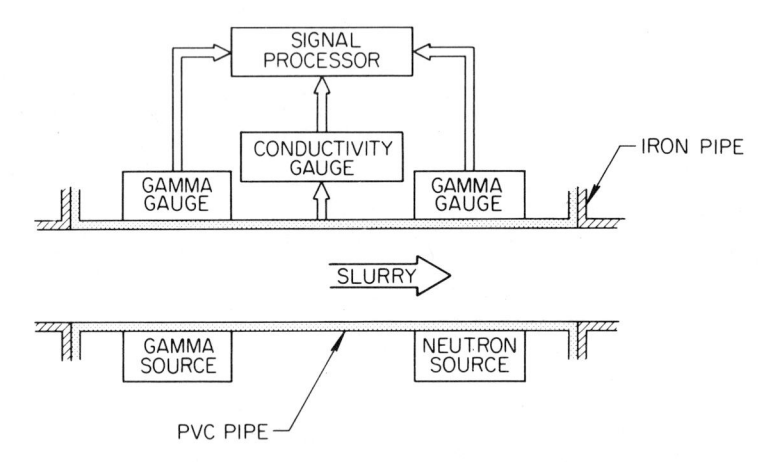

Figure 8. Schematic drawing of the coal slurry concentration meter.

equations simultaneously yields the relative proportion of coal, rock, and water in the slurry.

The slurry gauge was evaluated at the Colorado School of Mines, Research Institute. For slurries in which the particle sizes were less than 0.6 mm, the absolute difference between the measured and actual slurry composition averaged 1.7%. Slurries containing 0 to 25% coal and rock were utilized in these tests. The precision of the measurement was limited by the counting statistics. Response time is one second. For slurries containing solid particles approximately one-third the pipe diameter, large errors in the measured slurry composition were observed. These errors are probably due to the irregular movement of the large particles. The custom designed slurry concentration meter cost $150,000; mass production of the meter probably will reduce the cost to $75,000 ([17](#)).

Economic Considerations of On-Line Analytical Instruments

Most coal preparation plants use instrumentation to inform the control room operator of equipment status and guide the employees within the plant in the manual adjustment of the set points on the process equipment. Many unit operations are not equipped with any instrumentation and successful implementation of these unit operations relies totally upon the experience of an employee charged with the responsibility of monitoring specific pieces of process equipment. New plants are being designed with increased levels of automation and on-line instrumentation; however, the investment in instruments is only a small fraction of comparable investments in other process industries. For example, the project team visited two modern coal preparation plants in the Appalachian coal fields which cost $75,000,000 and the total on-line instrumentation investment was approximately $150,000. The ratio of instrumentation to plant investment in the chemical industry is significantly larger.

Several factors are responsible for the low investment in on-line instrumentation. First, coal relative to other fuels is undervalued, and as a result the coal industry is under capitalized. Second, the coal industry has been reluctant to try new technology. However, based upon visits to several coal preparation plants in the Eastern U.S., the rate of technological change within the coal preparation industry is accelerating. For example, computers are being used to automatically operate coal preparation plants. Third, an economic rationale for on-line instrumentation has not been provided. Finally, only approximately one-third of the total coal mined is cleaned, and this fraction has been declining since the mid 1960's.

The economic benefit derived from the use of on-line process control and quality control instruments ultimately must be translated into the ability to improve product quality, i.e., improve product specifications and reduce variations in product quality.

Improved on-line instrumentation should allow the coal preparation plant operator to accomplish this objective through a more efficient use of the raw material, by producing more product without increasing plant capacity, and by allowing one to operate closer to product specification limits. A reduction in operating and maintenance costs should also occur.

On-Line Instrumentation Development Needs

The on-line instrumentation development needs can be divided along the traditional lines of quality control and process control instrumentation. The interrelationship between the two and the proposed mode of interaction are shown in Figure 9. The quality control instruments, in an operational and a practical sense, are located outside the preparation plant. As a result of data obtained on product quality, the quality control instrument interacts with the plant feed and the process control instrumentation to optimize the use of raw material consistent with the desired product specifications. This feedback loop can be automatic; however, current practice places the preparation plant superintendent or engineer in the loop and the quality control measurement in a remote laboratory, resulting in manual control scheme with a long time delay. The process control instrumentation controls a single unit operation and the instrumentation is unique to each unit operation. Their set points are established by the output from the quality control instruments.

Several quality control devices are available and the development status has been summarized in a previous section. Laboratory studies indicate that the precision and accuracy of these devices are adequate to meet product certification needs. Fielding testing of these units in the coal preparation plant environment is required to demonstrate their usefulness and economic viability.

The development status of process control instrumentation lags that of the quality control instruments significantly. Nuclear density gauges function in the coal preparation plant environment. The slurry concentration meter has application in the intermediate and fine sized coal cleaning circuits and needs to be tested in a preparation plant. Other devices, such as ash monitors to control the operation of heavy media baths or jigs are not available; and instruments developed for other process industries are not suitable for use in coal preparation plants. Modeling studies of the various unit operations are required in order to ascertain the fundamental parameters required to automate the control of these systems. Primary process control instrument needs include ash, sulfur, and moisture monitors; secondary needs include an on-line washability and ash fusion measurement.

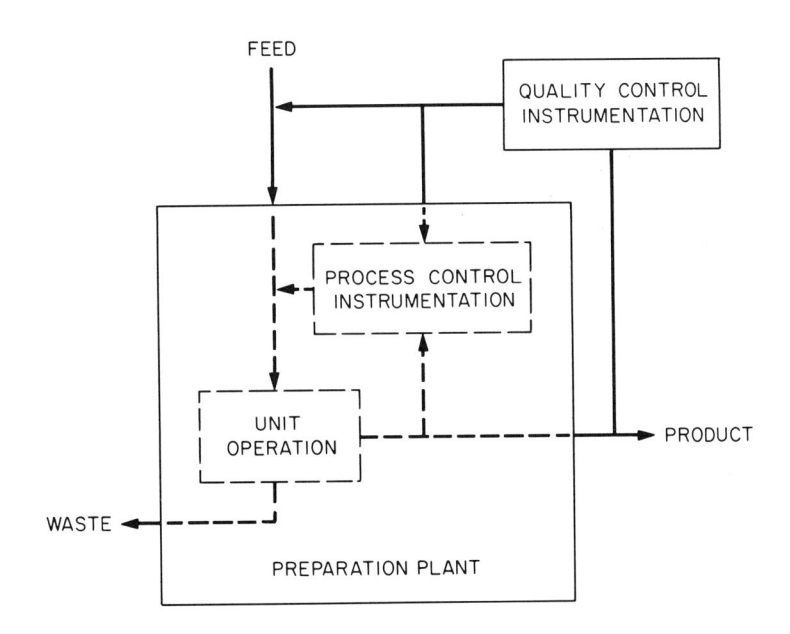

Figure 9. Interrelationship between process control and quality control on-line instrumentation.

Acknowledgement

Research sponsored by the Office of Energy Research, U. S. Department of Energy under Contract W-7405-eng-26 with the Union Carbide Corporation.

Literature Cited

1. "Engineering/Economic Analysis of Coal Preparation with SO_2 Cleanup Processes", EPA-600/7-78-002, U. S. Department of Energy, Pittsburgh, PA., 1979.
2. "Environmental Assessment of Coal Cleaning Processes: Technology Overview", EPA-600/7-79-073e, U. S. Environmental Protection Agency, 1979.
3. Vivenzio, T. A., "Impact of Cleaned Coal on Power Plant Performance and Reliability", Electric Power Research Institute, EPRI CS-1400, 1980.
4. "Gaseous Fuels; Coal and Coke; Atmospheric Analysis, Part 26", American Society for Testing and Materials, Philadelphia, PA., 1978.
5. Cammack, P.; Balint, A., Trans. SME, 1976, 260, 361.
6. Jones, D. W.; Stanley, F. L., "Ash Monitoring at Pittston Moss No. 3 Plant", Coal Convention of the American Mining Congress, St. Louis, MO, 1978.
7. Levinson, L. W., "Mossbauer Effect Spectroscopic Study of Pyritic Sulfur in Coal", EPRI FP-1228, 1979.
8. Tassicker, O. J., "Continuous Nuclear Analyzer of Coal for Improved Combustion Control", International Symposium on Pulverized Coal Firing and Its Effects, University of Newcastle, Australia, 1979.
9. Berkowitz, N., "An Introduction to Coal Technology", Academic Press, 1979, p. 35.
10. Cekorich, A.; Deich, H.; Harringon, T.; Marshall, J. H. III, "Performance of an On-line Elemental Analyzer for Coal", 2nd International Coal Utilization Conference, Houston, TX, 1979.
11. Lagarias, J.; Irminger, P.; Dodson, W., "Nuclear Assay of Coal, Vol. 8, Continuous Nuclear Assay of Coal: Progress Review and Industry Applications", EPRI FP-989, January 1979.
12. Deich, H., MDH Industries, personal communication.
13. Brown, D. R.; Gozani, T.; Elias, E.; Bozorgmanesh, H., "Nuclear Assay of Coal, Volume 4: Moisture Determination in Coal - Survey of Electromagnetic Techniques", EPRI CS-989, 1980.
14. Fauth, G., "On-Stream Determination of Ash and Moisture Content in West German Coal Preparation Plants", Proceedings of the 1980 Symposium on Instrumentation and Control for Fossil Energy Processes, Virginia Beach, VA, 1980.

15. Hall, D. A.; Sproson, J. C.; Gray, W. A., J. Institute of Fuel, 1970, 43, 350.
16. Verbinski, V. V.; Cassapakis, C. G.; deLesdernier, D. L.; Wang, R. C., "Three Component Coal Slurry Sensor for Coal, Rock, and Water Concentrations in Underground Mining Operations", 6th International Conference on the Hydraulic Transport of Solids in Pipes, University of Kent, United Kingdom, 1979.
17. Orphan, V. J., Science Applications, Inc., personal communication.

RECEIVED May 17, 1982

Coal as Energy in the Steel Industry

DAN P. MANKA

1109 Lancaster Avenue, Pittsburgh, PA 15218

>   The iron and steel industry uses up to 20 percent of the energy consumed by industry in the United States. Coal is a major source of this energy. Turning coal into coke has three major advantages to the industry: coke is essential as a source of heat in the blast furnace. One-third of the gas generated during the coking of coal is a source of heat for underfiring of the coke ovens and two-thirds is used in steel reheating furnaces. The coal-byproduct chemicals derived from the gas are increasingly valuable. Washed metallurgical bituminous coal is charged into the ovens. It must have low sulfur and low silica content; it must have proper volatile matter and expansion properties, otherwise it will stick in the ovens. Analyses of the various coals charged to the ovens will be discussed, as well as the composition of the gases liberated during the carbonization cycle. Recovery of the various chemicals from the gas will be discussed. The mixture of iron ore and limestone in the blast furnace is heated by the coke to form liquid pig iron. The gases liberated in the blast furnaces are used in stoves to preheat the air which is blown into the furnace to maintain combustion of the coke. The liquid pig iron is refined to liquid steel by direct reaction with oxygen in the Basic Oxygen Furnace. Some boiler coal is used as a source of heat in boilers to produce steam used throughout the coke and steel plants.

The iron and steel industry uses up to 20 percent of the energy consumed by all industry in the United States. Coal is a major source of this energy. Converting coal into coke serves this steel industry in three ways:

1. Coke is essential as a source of heat and as a reactant in the blast furnace to convert iron ore into pig iron.

2. One-third of the gas generated during the coking of coal is a source of heat for underfiring of the coke ovens.

3. The remaining two-thirds of the gas is used in the reheating furnaces in the steel plant.

Valuable chemicals are recovered from the gas in the coke plant. Benzene is used in the production of nylon, toluene and xylene are used as solvents and in the making of various chemicals. Phenols recovered from tar and the gas find uses in making thermosetting phenolic plastics. Ammonia is used as fertilizer in agriculture. Coumarone-indene recovered from the high boiling solvent is used in the resin industry. Naphthalene recovered from tar is used to make phthalic anhydride. There are literally hundreds of compounds in the tar and the chemicals recovered from the coke oven gas. Several of these are in sufficiently large concentrations to make recovery in pure form economically feasible.

In the production of coke, coal is heated for several hours in ovens in the absence of air at temperatures above $1000^\circ C$ to remove the volatile matter in the coal. Coal must be washed with water to remove many impurities. It is charged to the ovens in the wet or dry form.

A coke oven is constructed from high temperature, strong bricks. The oven dimensions are 15 to 18 inches in width, 15 ft. high, and 41 ft. in length. The largest width of 18 inches is on the so-called coke side where the coke is pushed out of the oven into coke cars. Subsequently, the red hot coke is cooled directly with water. The narrow, or 15 inch width, is on the other end of the oven generally called the pusher side where a large ram from a pusher car is guided into the oven to push out the coke. Between the outer walls of the ovens is an opening extending the full height and length of the oven which is the flue. Coke oven gas is burned in these flues to supply heat to the charged coal. The wall on one side of the oven is heated for a predetermined time and then the other wall is heated for the same period of time. In this manner, heat is supplied to the coal to coke the coal uniformly.

Generally, there are 59 ovens to a battery. All the volatile liquid flowing from a battery is collected in a separate collecting main before the liquids from all the batteries are mixed together ahead of the tar-liquor separating tanks. Some batteries may contain as many as 79 ovens. In this discussion we will consider a coke plant consisting of four batteries with 59 ovens each and one battery containing 79 ovens for a total of 315 ovens.

Not every coal can be charged to the ovens. The concentration of ash and sulfur must be considered to make good, strong coke. Too much sulfur in the coke will find its way into the steel. Some coals expand when they are heated, while others contract. Some coals have a high content of volatile matter, while others have a low concentration. Therefore, blends of various coal mixes are coked in small experimental ovens to determine the expansion or contraction of the coal and the physical characteristics of the final coke. If the coal expands too much, it will be detrimental to the walls of the ovens and also cause sticking of the coke, so that it cannot be pushed out of the ovens. Sticking ovens removes them from normal production, thereby lowering the daily production of coke. The sticking ovens in many cases have to be cooled down to facilitate removal of coke and again must be reheated before charging with coal. Too much contraction of the coal forms an undesirable, dense coke.

Table I lists the analysis of a typical coal mixture charged to the ovens. The lower the sulfur content, the lower the H_2S concentration in the coke oven gas that must be removed from the gas before it is burned to meet the pollution standards set by EPA. Note that the contracting coals expand somewhat during the coking process before finally contracting. The expansion and contraction of the various coals was determined on the individual coals charged to the small experimental oven. The coal mixture is determined on the quality of the final coke, which must meet certain hardness, volatile matter, ash and sulfur content to be suitable for use in the blast furnace.

The oven is charged through three openings located on top of the oven. The oven is filled to a heighth of 12 feet with 16.2 tons. A leveling bar on the pusher machine levels the coal in the oven. As noted previously, all doors and openings on the oven are closed so that the coal is heated in the absence of air, otherwise the coal would only burn and not form coke. Oxygen analysis of the gas exiting the oven during the 16-17 hour coking cycle is 0.1% or less. A total of 6950 tons of coal is charged per day, forming 4600 tons of coke per day, and 73,300,000 cubic feet of coke oven gas per day.

Coke Oven Gas Flow Diagram

The flow diagram of gas is practically the same in all coke plants. As more desulfurization processes are installed, these will vary in location depending on the type of system being installed.

In the flow diagram in Figure 1, coke oven gas rises from the coke oven, a, through standpipe, b, to gooseneck, c, where it is contacted with flushing liquor (ammonia liquor). Tar and moisture are condensed. Ammonium chloride, and a portion of the ammonia, fixed gases, hydrogen cyanide and hydrogen sulfide are dissolved by the liquor. The gas, liquor and tar enter the gas collecting main, d, which is connected to all the ovens of a battery. In some cases there may be two gas collecting mains to a battery. The gas, liquor and tar are separated in tar decanters. The tar, f, separated from the liquor, e, flows to tar storage, g. A portion of the liquor, e, is pumped to the gooseneck, c, on the top of each oven. The remainder of the liquor is pumped to the ammonia liquor still, h, where it is contacted with live steam to drive off free ammonia, fixed gases, hydrogen cyanide, and

Table I. Analysis of Typical Coal Mixture Charged to Ovens.

	% IN MIX	% VOLATILE MATTER	% FIXED CARBON	% ASH	% S	% H$_2$O	
LOW VOLATILE W. VA.	12	16.3	75.7	8.0	0.7	6.5	12.5% EXPANSION 13 psig PRESSURE
MID VOLATILE PA.	15	24.7	68.4	6.9	0.89	7.5	8.0% EXPANSION 4 psig PRESSURE
HIGH VOLATILE PA.	46	36.4	56.7	6.9	1.35	7.0	26% CONTRACTION 0.9 psig PRESSURE BEFORE CONTRACTION
HIGH VOLATILE W. VA.	27	33.6	61.4	5.0	0.64	7.4	15.5% CONTRACTION 2 psig PRESSURE BEFORE CONTRACTION
MIXTURE		31.4	62.9	5.7	1.03	6.3	10% CONTRACTION 1.7 psig PRESSURE BEFORE CONTRACTION

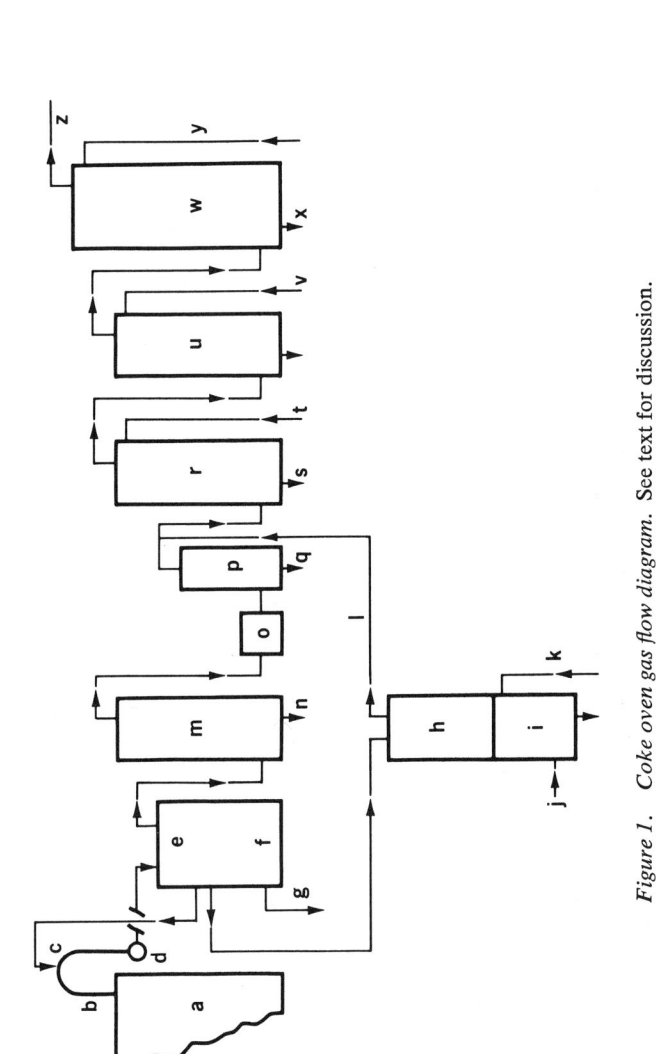

Figure 1. Coke oven gas flow diagram. See text for discussion.

hydrogen sulfide. As the liquor flows from the free still, h, to the fixed still, i, it is contacted with lime or sodium hydroxide, k, to liberate free ammonia from ammonium chloride. Live steam, admitted at j, flows up through the fixed and free stills and the ammonia, fixed gases, hydrogen cyanide and hydrogen sulfide are added through, l, to the main coke oven gas stream ahead of the ammonia saturator, r.

The coke oven gas, separated from liquor and tar in, e, is cooled indirectly with water in the primary coolers, m. The fine tar that separates from the gas is pumped through, n, to the tar storage tank. The cooled gas is pumped by exhausters, o, to the electrostatic precipitators, p, where additional fine tar is condensed and pumped through, q, to the tar storage tanks. The gas is contacted with a dilute solution of sulfuric acid in the ammonia saturator, r, to remove free ammonia. The ammonium sulfate-laden acid flows through, s, to the ammonia crystallizer (not shown) where crystals of ammonium sulfate are separated, and the remaining sulfuric acid is pumped back to the ammonia saturator through, t.

The ammonia=free gas flows to the final coolers, u, where it is further cooled by direct water contact. The water plus condensed naphthalene flows from the cooler through tar, which absorbs the naphthalene. The water is cooled and recirculated into the final cooler through, v.

The cooled gas enters the wash oil scrubbers, w, also known as benzole scrubbers, where it contacts wash oil, a petroleum oil, pumped into the scrubbers through y. The aliphatic and aromatic compounds are extracted from the gas by the wash oil. The principal components are benzene, toluene, xylenes, indene, and solvent, also known collectively as light oil. The benzolized wash oil is pumped through, x, to the wash oil still (not shown) where live steam strips out the light oil compounds. The debenzolized wash oil is cooled and returned to the wash oil scrubber. In some plants the light oil is further processed and fractionated into benzene, toluene, and xylenes and into a high-boiling solvent fraction. Naphthalene is also present in the light oil. Plants with low volumes of light oil do not have facilities for refining, therefore, the oil is sold to large refineries.

The gas from the wash oil scrubbers flows through, z, to a gas holder which tends to equalize the pressure. Booster pumps distribute one-third of the gas for underfiring of the coke ovens, and two-thirds to the steel plant where it is used as a fuel in the many furnaces.

Determining End of Coking Cycle By Gas Analysis

The length of a coking cycle for carbonizing of coal in an oven is determined from experience and the range of the flue temperature. Most coke plants also have experimental coke ovens to determine coking cycles for variable coal mixes and at different flue temperatures. Another method to determine the end of this cycle is by analysis of the gas flowing from the oven into the standpipe. This section will describe this approach.

A. Sampling System. The gas sample from the standpipe must be cooled, separated from tar and water and filtered before it is analyzed for H_2, O_2, N_2, CH_4, CO, CO_2 and illuminants.

The sampling train in Figure 2 has been used successfully for the continuous pumping of gas to a sample bottle or to a gas chromatograph. It is designed to operate for several hours so that many samples can be analyzed before there is an accumulation of tar in probe, c, or in the separators, d.

Standpipe, b, is located on the gas discharge and on top of coke oven, a. Normally, there is an opening on the side of the standpipe where live steam is admitted to remove accumulated tar in the pipe. This opening is ideally located for gas sampling, since it is generally about two to three feet above the oven.

The probe, c, extending to approximately the center of the standpipe, is a 3/4 inch or 1 inch heavy wall stainless steel pipe. The end of this pipe in the gas stream is sealed and the other end has a plug for removing accumulated tar. A 1/4-inch by approximately 3-inch slot is cut along the length of the pipe in the gas stream extending from the closed end. When the probe is inserted into the standpipe, it is important that the slot face the direction of gas flow. This position decreases the amount of tar pumped in with the sample gas. The probe is inserted into a plug similar to the one used for the steam line and pushed through the opening in the standpipe until the plug seals the opening. This insertion should be done rapidly to prevent excessive flow of coke oven gas through the opening. The probe extending outside of the pipe and the 1/4-inch pipe from the probe to the first separator, d, should be insulated.

The two separators, d, for cooling the gas and separation of tar and water, are identical. These are made from 4-inch pipe, 10- to 12-inch long, one end sealed and a flange with a gasket on the top end. The inlet pipe extends approximately half-way into the separator. Both separators are kept in a bucket of water, e. The separators should be pressure tested before they are used. The remaining piping is 1/4-inch copper or stainless tubing with Swaglock fittings. Although most of the tar is condensed in the separators, the gas is drawn through glass wool in a glass tube, f, to remove the lighter tar oil. This glass tube can be a glass bottle normally used in the laboratory to dry gases with drierite. It has an inlet connection on the bottom, an outlet connection near the top, and a metal screw cap with a seal on the top. Tygon tubing is used for the metal tube to glass connection. The gas flows through a final filter, g, that removes submicron particles, such as the Pall Trinity "Junior Size" Epocel cartridge. The gas is drawn from the standpipe by a peristaltic pump, i. Good results are obtained with this type of pump using a 1/2- or 5/8-inch diameter plastic tube for conveying the gas. The tubing is easily

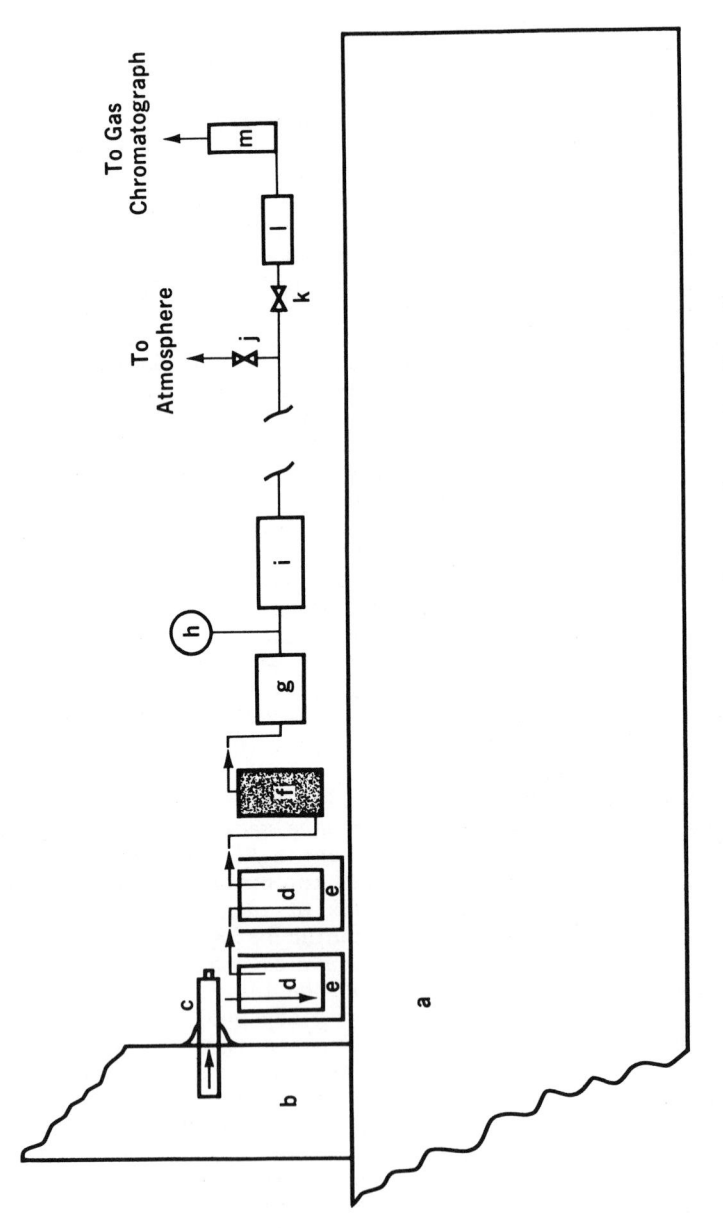

Figure 2. Sampling gas from top of coke oven. See text for discussion.

replaced if it becomes broken or contaminated. A vacuum gauge, h, ahead of the pump is valuable to detect plugging of lines or separators. When vacuum reaches 20 inches, it is most probable that tar is accumulating in the probe. A rod inserted into the probe through the plug opening of the probe will generally be sufficient to reopen the pipe.

Gas passes through valve, k, after the pump, i, and is dried in tube, l, containing drierite. The flow to the sample loop of the chromatograph is maintained at 50 cc per minute as indicated by rotometer m. Excess sample gas is vented to atmosphere through valve j.

B. Orsat Analysis. If a portable gas chromatograph is not available, a gas sampling bottle can be filled with the dried gas after rotometer, m. Analysis of hydrogen, oxygen, carbon dioxide, carbon monoxide, methane, and illuminants concentrations in the gas can be determined on an Orsat equipped with combustion apparatus to determine hydrogen and methane. However, the analysis time is lengthy so that the number of analyses made near the end of the coking cycle are too limited to obtain a true picture of the coking cycle end point.

C. Gas Chromatographic Analysis. The use of a portable gas chromatograph located in a sheltered area near the sampling system is the best method to follow the course of the coking cycle.

The curves in Figure 3 are based on the analytical results obtained on a gas chromatograph with a thermal conductivity detector (1). The separation column is a 1/8-inch by 10-foot stainless steel tube containing #5A molecular sieve and the reference column is a 1/8-inch by 67-inch stainless steel tube containing Porapak Q. The operating conditions are argon flow 12 cc per minute to each column, 100 ma cell current, column and cell temperature is 40°C. The chromatograph is standardized with a gas containing 50% H_2, 30% CH_4, 10% CO, 5% CO_2 and N_2 the balance. The standard gas is admitted into the sample line ahead of the drierite in l and the flow rate maintained through the sample loop the same as for coke oven gas. The sample loop capacity is 0.5 cc and is dependent on the sensitivity of the detector, using argon as carrier gas.

If the detector has insufficient sensitivity for CO and CH_4, a carrier gas containing 8% hydrogen and 92% helium may be used at a cell current of 150 ma.

Concentrations of CO_2 and illuminants, as ethylene, can be obtained on the Porapak column. The chromatograph must also be equipped with a sample valve for this column and a polarity switch so that the peaks are positive on the chart paper of the recorder. However, CO_2 and illuminants are not necessary to determine the end point; therefore, these can be ignored or periodically determined by the Orsat method.

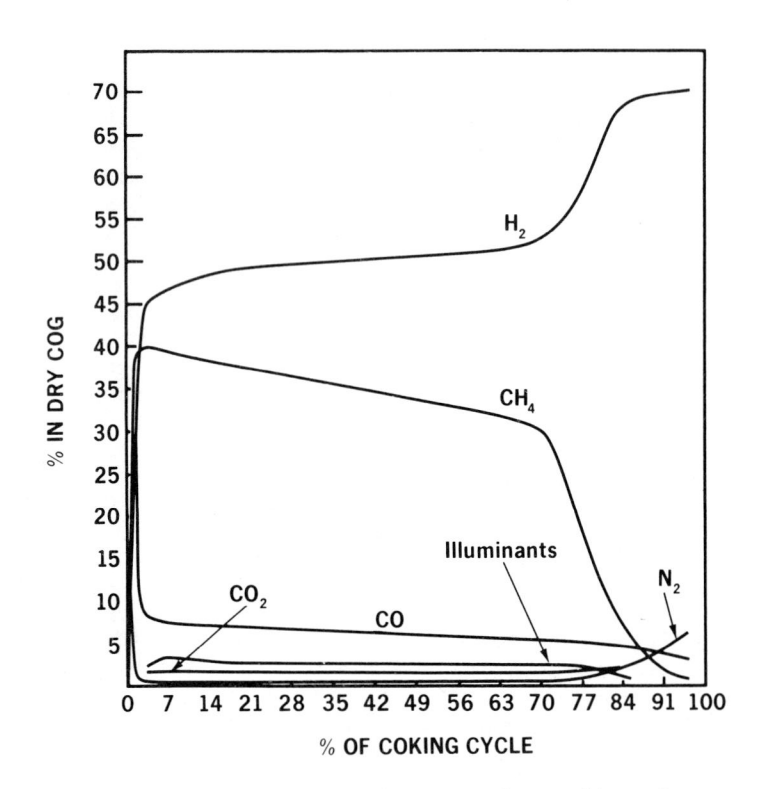

Figure 3. Composition of coke oven gas during coking cycle.

D. Results. The plot of the analytical results in Figure 3 are those for coking dry coal for 14 hours at a flue temperature of 1260°C. Concentration of hydrogen approaches a maximum of over 70% and the methane approaches the minimum of less than 1% near the end of the coking cycle. Generally, coking is continued for an additional 30 minutes after these analytical values are reached and then the coke is pushed from the oven.

Similar curves are reported (Beckman et al., 1962) for the coking of wet coals (2). However, the coking cycle is 18 to 19 hours, because a large amount of water must be evaporated from the wet coal charged to the ovens.

The major constituents of the 73,300,000 cubic feet of coke oven gas after condensation of the tar are shown in Table II.

Table II. Composition of Detarred Coke Oven Gas.

CONSTITUENT	VOLUME PERCENT
H_2	55.0
CH_4	29.0
CO	5.5
Aromatics	3.0
CO_2 and S compounds	2.6
O_2	0.9
N_2	4.0

The volume of tar in this particular coke plant is 58,900 gallons/day and the volume of light oil is 22,300 gallons/day. The light oil contains 63.0% benzene, 13.0% toluene, 5.6% xylene, and 5.1% of crude solvent. There are also produced 33.5 tons/day of ammonium sulfate and 400 gallons of tar acids, which are mainly phenols.

The coke oven gas, practically freed of tar, ammonia and light oil, is extracted with monoethanolamine to remove H_2S, HCN, and SO_2 to levels required for air pollution abatement. One third of the desulfurized gas is returned to the coke ovens as a source of heat for carbonizing the coal in the ovens. Two-thirds of the gas flows to the steel plant furnaces where it is used as the source of heat for reheating steel products.

The coke produced in the coke ovens is charged to the blast furnaces together with iron ore and limestone where, at high temperatures, the iron ore reacts with the coke and limestone to form pig iron and slag. The spent gases containing H_2, CO, and CO_2 are used to heat large volumes of air which are blown into the furnace to aid the reaction of coke, limestone, and iron ore.

Typical charges per day into a medium sized blast furnace are:

<u>Dry Coke</u>	1050 lbs./ton hot metal, also called pig iron
<u>Iron Ore</u>	3200 lbs/ton hot metal
<u>Limestone & Dolomite</u>	270 lbs/ton hot metal
<u>Wind for Combustion of the Coke</u>	51,000 to 54,000 cubic feet/ton hot metal
<u>Oil Injection</u>	100 lbs/ton hot metal
<u>Production</u>	3000 tons hot metal/day

The 3200 lbs. of ore per net ton of metal consists of 100 lbs. BOF slag, 150 lbs. of scrap and 2950 lbs. of iron ore pellets.

This illustration shows that coal is a major source of energy in the steel industry. Although the industry is in a depressed state today, it will use 75 million tons of coal in 1981.

Literature Cited

(1) Manka, D. P., "Analytical Methods for Coal and Coal Products," Vol. III, edited by C. Karr, Jr., Academic Press, New York, 1979, 6--10.

(2) Beckmann, R., Simons, W., and Weskamp, W. (1962) Brennst.-- Chem. 43 (8), 241.

RECEIVED May 17, 1982

15
Electron Optical and IR Spectroscopic Investigation of Coal Carbonization

J. J. FRIEL, S. MEHTA, and D. M. FOLLWEILER

Bethlehem Steel Corporation, Homer Research Laboratories, Bethlehem, PA 18016

The chemistry of coal carbonization was studied in an attempt to better understand the changes that take place as metallurgical coal becomes plastic. Infrared spectroscopy was used to study the transformation from coal-mesophase-semicoke and to relate the changes to those observed in the electron microscope. Spectra of semicokes produced from three different coals at various heat-treatment temperatures show changes that can be related to the fluid properties of the coals. Evidence of mesophase in the fluid range of temperature can be obtained from changes in the region of the spectra assigned to aromatic hydrogen out-of-plane bending modes. Heat-treatment temperatures above the fluid range result in condensation of the mesophase as hydrogen is lost.

Various metallurgical coals with different coking properties were examined in an attempt to learn more about the carbonization process. In our previous studies, we observed three basically different microstructures in different coals during heating stage experiments in the transmission electron microscope ([1]). A Pittsburgh No. 8 coal with good coking properties readily formed mesophase spheres that coalesced before solidification to a semicoke (Figure 1). An Illinois No. 6 sample of similar rank but little fluidity exhibited only isolated small mesophase spheres (Figure 2). A Lower Kittanning low-volatile coal formed not only spheres but rods (Figure 3).

The purpose of the present study was to examine the coal-to-coke transformation away from the heating stage of the electron microscope. We carbonized samples at various heat-treatment temperatures in a plastometer. The plastometer has the advantage over the electron microscope hot stage not only in that it is the apparatus we used in the fluidity test but also in that it minimizes thermal gradients across the sample. Semicokes thus

0097-6156/82/0205-0293$06.00/0
© 1982 American Chemical Society

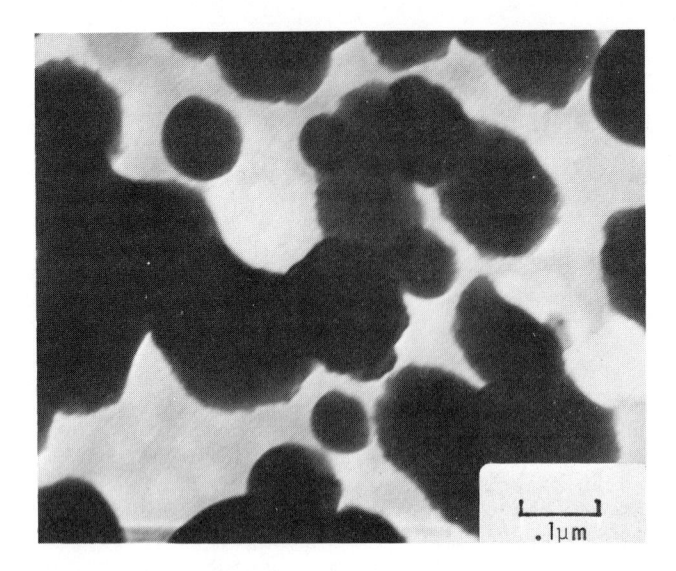

Figure 1. Electron micrograph of coalesced spheres in Pittsburgh No. 8 vitrinite on the heating stage of the TEM.

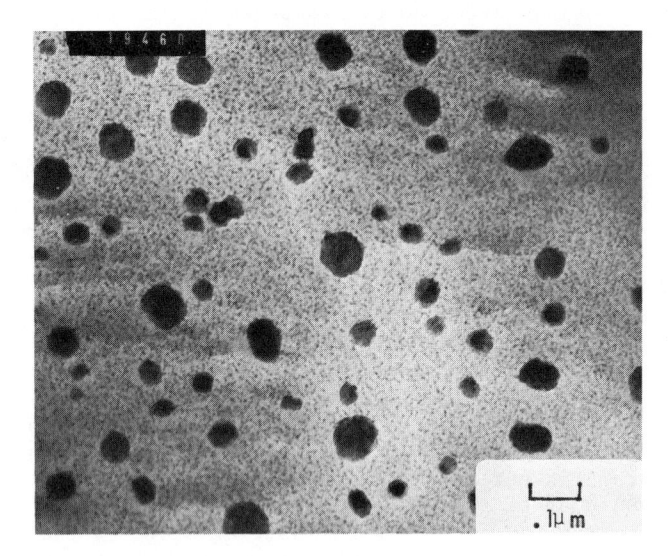

Figure 2. Electron micrograph of isolated spheres in Illinois No. 6 vitrinite.

produced were examined in the electron microscope and by infrared spectroscopy.

Studies on coal structure and its decomposition products using Fourier transform infrared spectroscopy (FTIR) were recently reported by Solomon (2). While this technique produces a greater peak-to-background ratio than is possible with the conventional infrared technique, we were still able to obtain useful information with the conventional dispersive instrument by means of multiple accummulations. Painter and Coleman also used the FTIR technique on coal and carefully assigned absorptions to specific functional groups (3). Our aim was to study the differences among coals of different coking properties and relate the growth of mesophase in these coals to the changes that take place upon carbonization. The properties of the coals studied are listed in Table I. Inasmuch as infrared spectra of coals of different rank are similar, it was necessary to select coals of markedly different coking properties to improve our chances of seeing differences.

Experimental Technique

Coal ground to <4.75 mm was placed in the plastometer equilibrated at 300°C. The coal was then heated at a standard rate of 3°C/min to the desired temperature without stirring and allowed to cool. Heat-treatment temperatures ranged in 25° increments between 300 and 550°C for the Pittsburgh No. 8 coal, between 300 and 500°C for the Illinois No. 6 coal, and between 400 and 550°C for the Lower Kittanning coal. Other samples of these coals were subjected to the standard Bethlehem plastometer test to measure the temperatures of onset of fluidity, maximum fluidity, and resolidification.

Semicokes prepared in the plastometer were ground, and observed as a dispersion in a Philips EM300 transmission electron microscope or Philips EM400 scanning transmission electron microscope (STEM). The instruments were operated at 100 kV, and the STEM was employed in both the transmitted electron and the secondary electron imaging modes. A portion of each of the samples from the plastometer was reserved for infrared spectra. These spectra were taken on KBr pellets into which a known amount of coal or semicoke was pressed. A spectrum was obtained from each of the coals and from the semicokes from each heat-treatment temperature. All spectra were acquired at room temperature. A Perkin-Elmer Model 283B automated infrared spectrophotometer was used, and each sample was scanned from 301–4000 cm^{-1} and accumulated twice. The region of the spectrum containing aromatic out-of-plane bending modes (595–900 cm^{-1}) was accumulated 25 times for each sample in order to more carefully evaluate the changes that take place in the substituted aromatic groups. All of the spectra were normalized by the computer on the basis of the weight of sample in the KBr pellet.

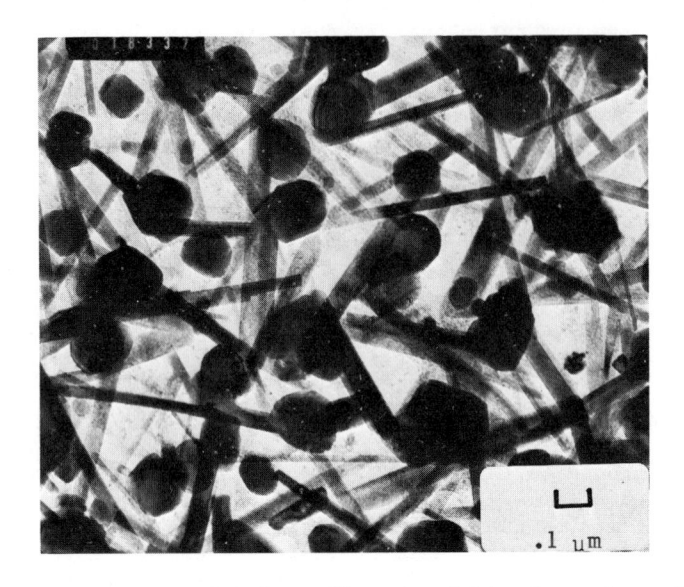

Figure 3. Spheres and rods in Lower Kittanning vitrinite shown after cooling on the heating stage of the TEM.

Table I. Some Characteristics of the Coals Used in This Study

	$Rvit^a$	Volatile matter (%)	Fluidity^b (ddpm)	Mesophase structure
Pittsburgh No. 8	0.84	38.1	5,200	coalesced spheres
Illinois No. 6	0.69	40.4	<10	isolated spheres
Lower Kittanning	1.57	19.5	<50	rods and spheres

a = vitrinite reflectance
b = dial divisions per minute

Results

Fluid properties of the coals are given in Table II. The fluid ranges of temperature of the Illinois and Lower Kittanning coals are less than that of the Pittsburgh No. 8 coal. The temperature of maximum fluidity is highest for the low-volatile Lower Kittanning coal and is lowest for the Illinois No. 6 coal. Even though the Lower Kittanning and Illinois coals have similar low fluidities, they differ significantly in rank, and the reason for the low fluidity may well be different. The ultimate analysis of the coals is shown in Table III.

Semicokes prepared at various temperatures were examined in the transmission electron microscope, but few features could be seen in the lower- and higher-temperature samples. Because of the porous and brittle nature of semicoke, it was difficult to prepare good ion-thinned sections as we did for coal. Therefore, features such as spheres and rods may have been destroyed during grinding to prepare the sample.

The Pittsburgh No. 8 semicoke that was cooled from 400°C contained remnant mesophase spheres, and these can be seen in Figure 4. This photomicrograph was taken in the secondary electron mode in the scanning transmission electron microscope. In this imaging mode, only surface features can be seen by topographic contrast. In the transmitted electron mode, most of the sample was too thick to pass electrons, and spheres were difficult to distinguish. In the Pittsburgh No. 8 sample, 400°C is the onset of fluidity, and mesophase spheres were sufficiently abundant to be observed in the STEM (Figure 4).

Figure 5 shows rod-shaped features in the 500°C Lower Kittanning semicoke seen in the transmitted electron mode. Although 500°C is beyond the temperature of resolidification measured in the plastometer, the temperatures reported for low-fluidity coals are not as meaningful as those for high-fluidity coals. In addition, fluid properties of many samples of low-fluidity coals cannot be measured in the plastometer. Accordingly, the temperatures measured for one sample may not be applicable to other samples of the same low-fluidity coal.

In the Illinois No. 6 semicokes, we were not able to detect remnants of the mesophase with the electron microscope. In each of the semicokes observed, fewer features were found than were previously seen on the heating stage. This situation might be expected because of the different sample preparation techniques. A uniform 100 nm thin section used on the heating stage is better suited for examination in the electron microscope than is a fine dispersion of particles, which may or may not have electron transparent edges.

Infrared spectra of each of the three coals are shown in Figure 6. The spectra are similar to each other, with subtle differences often being masked by the overall absorption of the coal. Although we studied coals with different carbonization

Table II. Fluid Properties from Bethlehem Plastometer Test

Coal	Initial Bonding	Maximum Fluidity	Solidification Point	Fluidity
Pittsburgh No. 8	397°C	450°C	470°C	5200 ddpm
Illinois No. 6	405	422	445	3
Lower Kittanning	457	475	480	35

Table III. Chemical Analysis and Ash Content of the Coal Studies (Wt %, Dry Basis)

Coal	C	H	N	S	O*	Ash
Pittsburgh No. 8	79.82	5.15	1.86	0.84	6.42	5.9
Illinois No. 6	76.74	5.24	1.51	1.15	9.06	6.3
Lower Kittanning	85.58	4.58	1.38	0.46	1.50	6.5

* By difference

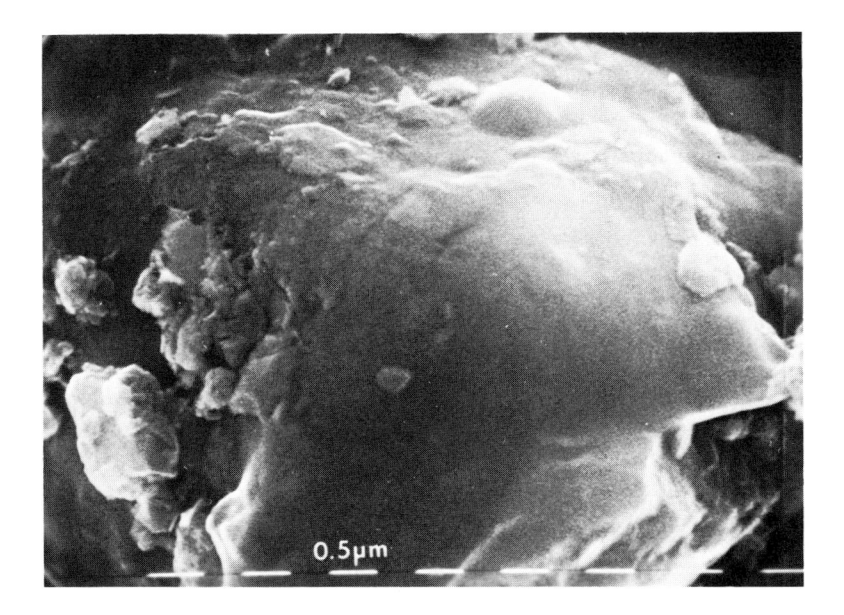

15. FRIEL ET AL. *Spectroscopic Investigation of Carbonization* 299

Figure 4. Secondary electron micrograph of semicoke from Pittsburgh No. 8 coal after heating to 400°C.

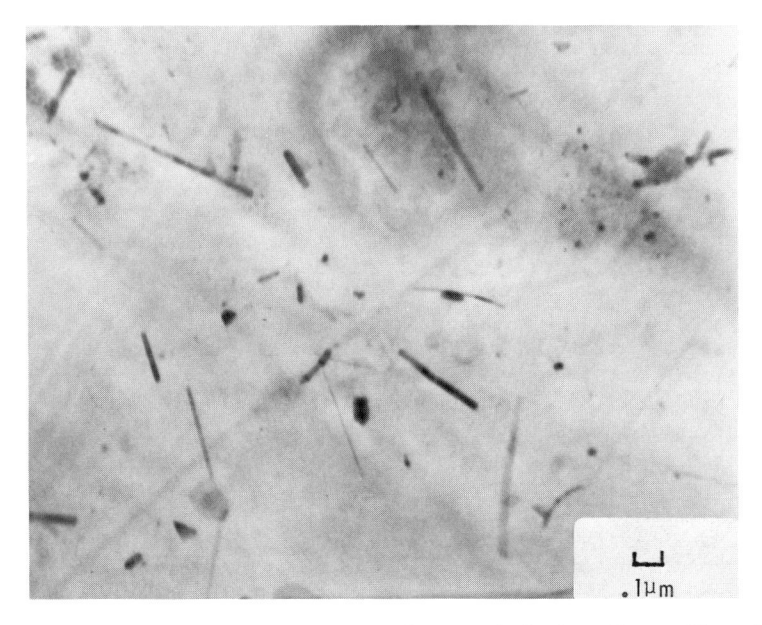

Figure 5. Transmitted electron micrograph of semicoke from Lower Kittanning coal after heating to 500°C.

properties, one must remember that these are still bituminous coals. Lowry gives many references to infrared spectroscopic studies of coals of different rank, but marked changes are not evident in the infrared spectrum (4).

In the spectra shown in Figure 6, differences can be seen principally in the 1000-1300 cm^{-1} region. There are also differences in the aromatic/aliphatic ratios, but such determinations have been made by Solomon with more sensitivity than is possible with our technique (2). The Illinois No. 6 coal shows evidence of a greater content of aromatic ethers (1200-1300 cm^{-1}). The Pittsburgh No. 8 coal absorbs throughout the 1000-1300 cm^{-1} range, owing to the presence of both aromatic and aliphatic ether groups. In this region the Lower Kittanning sample absorbs least.

Spectra of semicokes made from Pittsburgh No. 8 coal at various temperatures are shown in Figure 7. Increasing the heat-treatment temperature from 300°C causes little change in a given spectrum until 400°C. At this temperature, corresponding to the onset of fluidity, the peaks become sharper. At 450°C, corresponding to maximum fluidity, the peaks become their sharpest. Beyond the resolidification temperature of 470°C, spectral definition is lost, and by 525°C the spectrum is essentially featureless.

Spectra of semicokes from Illinois No. 6 coal are shown in Figure 8. Ether groups, and in particular aromatic ethers, are a significant feature of this coal, which is high in oxygen. Increasing temperature causes little change in this series of spectra until 400°C. Between 400 and 425°C the spectral peaks become sharpest and then decrease in the higher temperature samples. The spectra also show a probable decrease in the amount of aromatic ethers and probable increase in the amount of aliphatic ethers (1000-1200 cm^{-1}) with increasing temperature. The temperature of maximum fluidity of Illinois No. 6 coal is not easily measured in the plastometer owing to its low fluidity. However, infrared evidence indicates that the temperature of maximum fluidity is probably between 400-425°C.

Spectra of semicokes from the Lower Kittanning coal are shown in Figure 9. The spectrum with the most distinct peaks in this series is the 500°C semicoke. However, as in the case of the Illinois No. 6 series, there is not the marked improvement in spectral quality at the temperature of maximum fluidity that we observed in the Pittsburgh Seam series. There is also evidence of carbonyl groups at 1740 cm^{-1} in this series of semicokes -- possibly corresponding to esters.

The aromatic peaks between 700 and 900 cm^{-1} are shown at higher sensitivity in Figure 10. The semicokes that produced the sharpest spectrum and those from the two adjacent temperatures are shown along with the untransformed coals. In each series of semicokes the peak at 750 cm^{-1}, corresponding to 1,2 substitution, reaches a maximum near the onset of fluidity temperature. The peak at 860 cm^{-1}, corresponding to highly

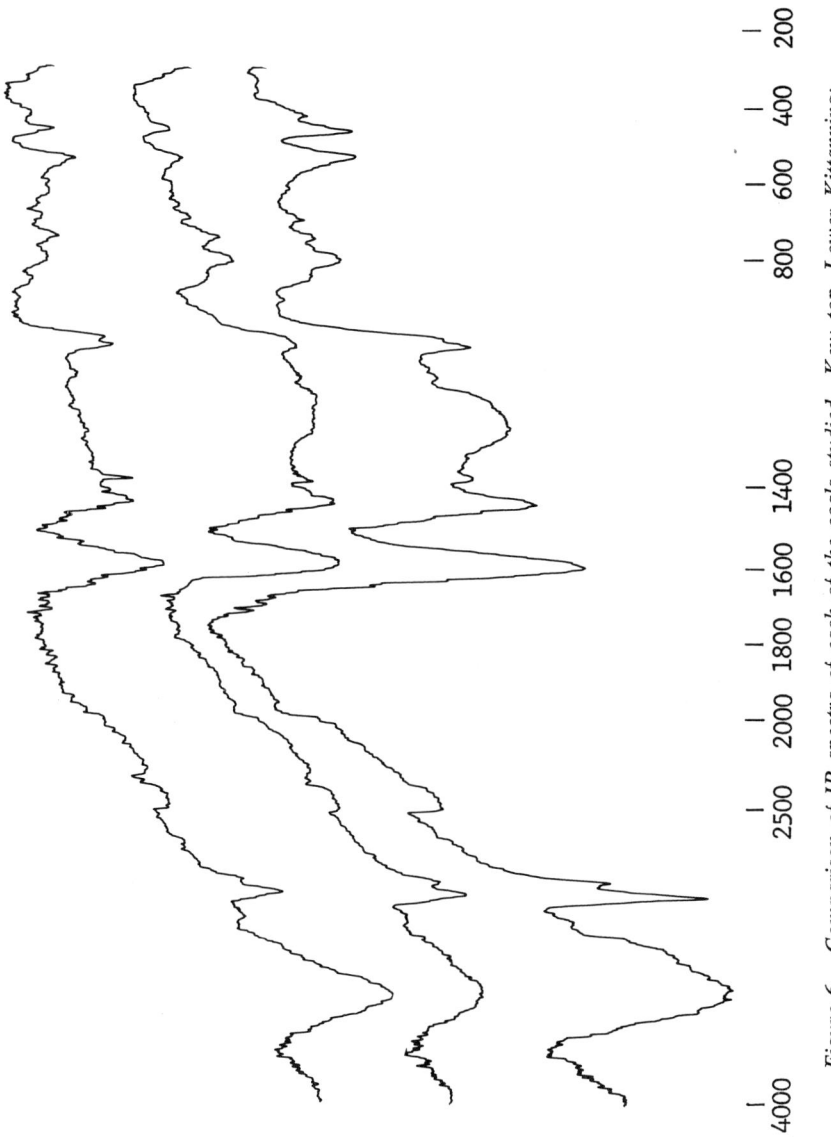

Figure 6. Comparison of IR spectra of each of the coals studied. Key: top, Lower Kittanning; middle, Pittsburgh No. 8; and bottom, Illinois No. 6.

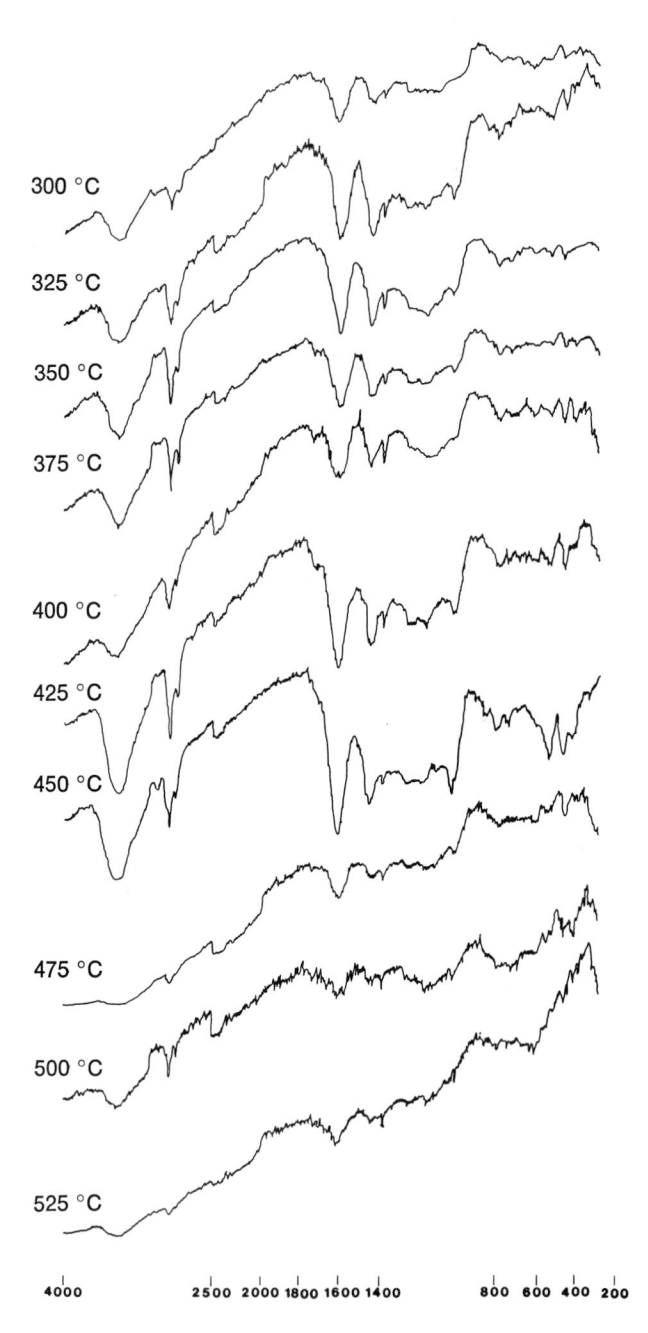

Figure 7. Spectra of semicokes prepared from Pittsburgh No. 8 coal at various heat-treatment temperatures.

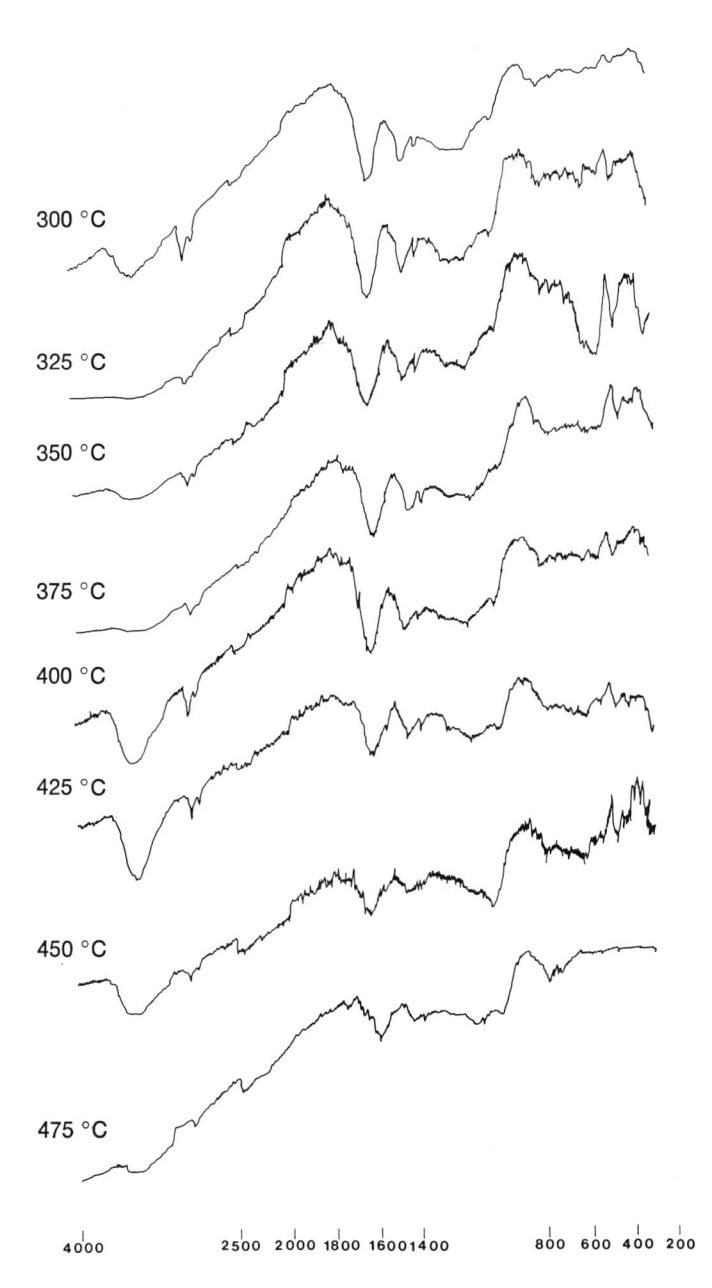

Figure 8. Spectra of semicokes prepared from Illinois No. 6 coal at various heat-treatment temperatures.

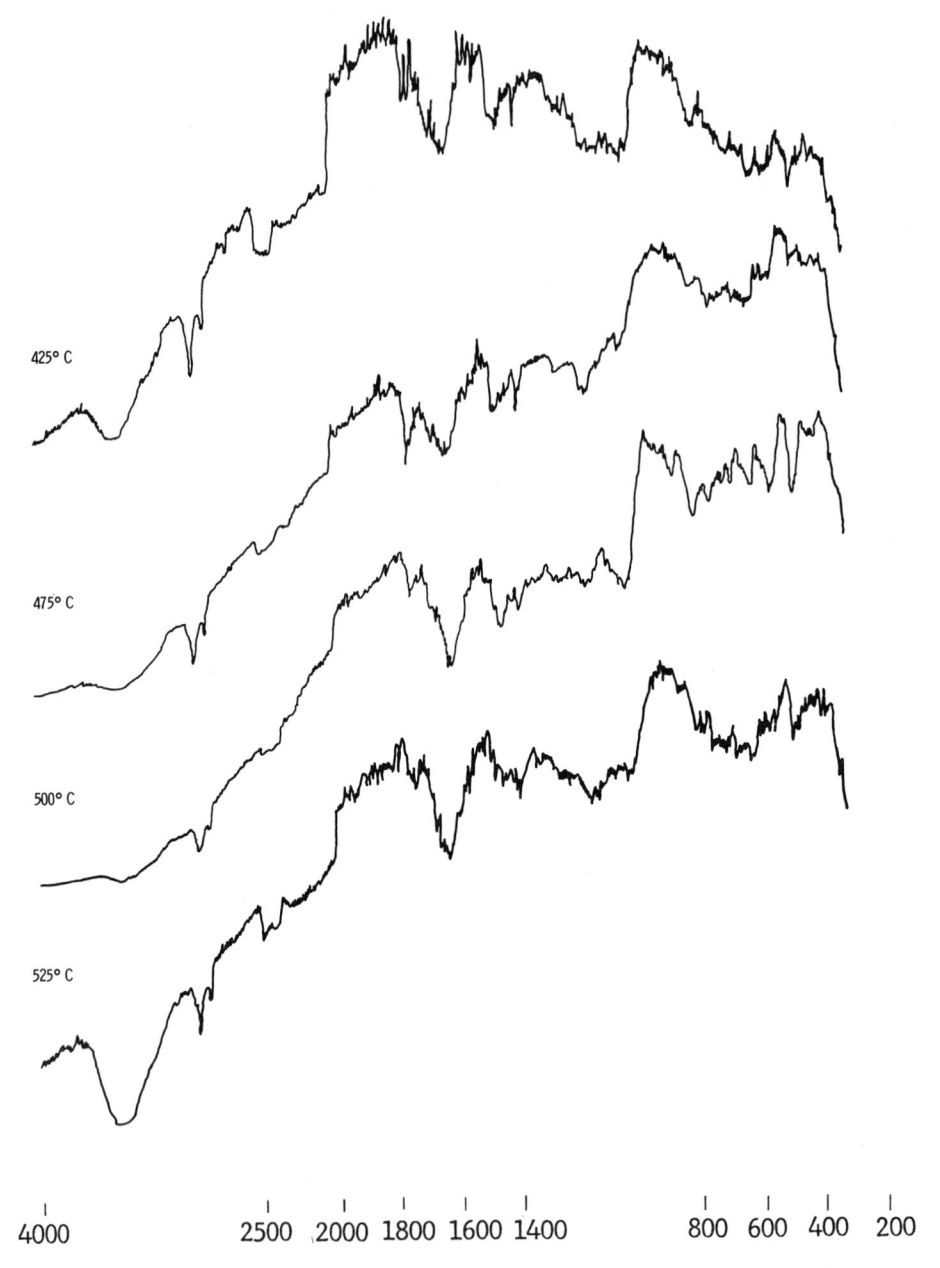

Figure 9. Spectra of semicokes prepared from Lower Kittanning coal at various heat-treatment temperatures.

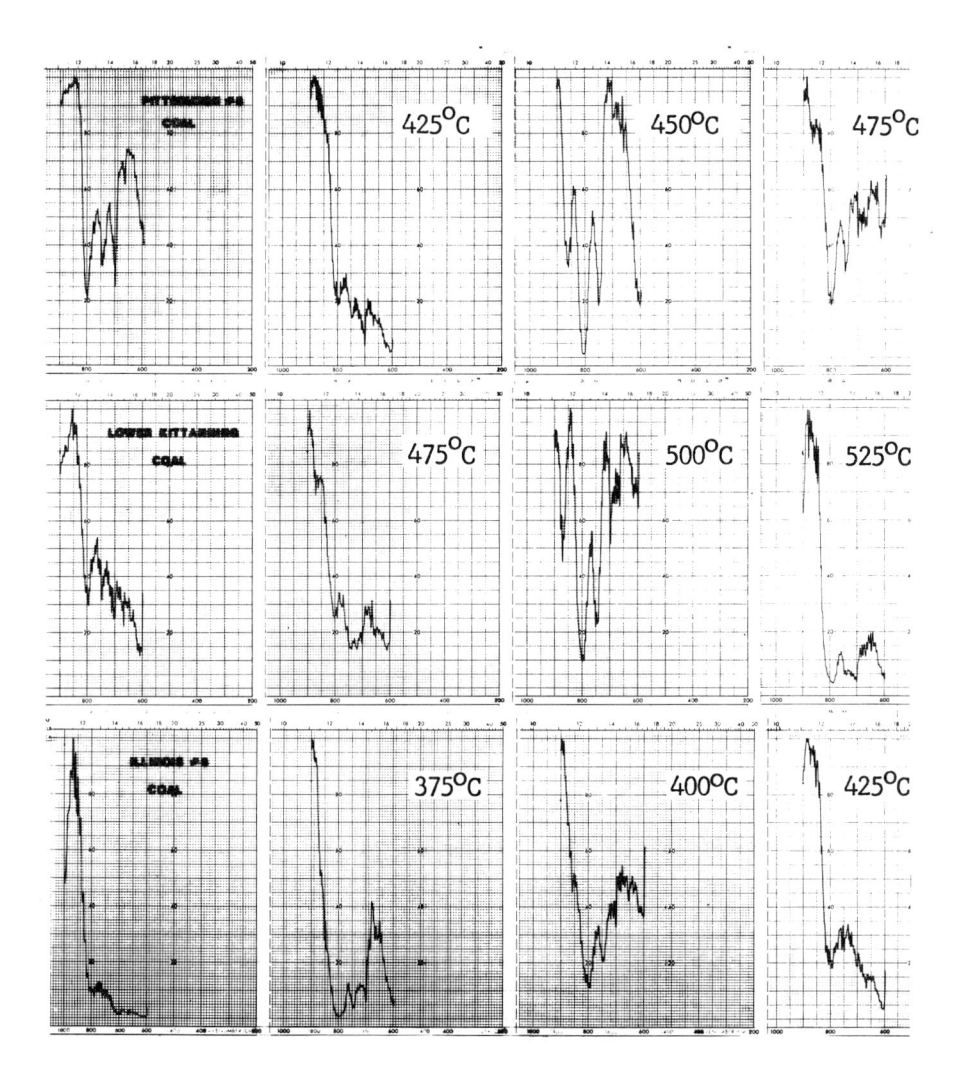

Figure 10. Spectra showing the aromatic hydrogen out-of-plane bending vibrations in different coals and semicokes—25 times accumulation. Key: top row, Pittsburgh No. 8; middle row, Lower Kittanning; and bottom row, Illinois No. 6.

substituted aromatics becomes prominent only at the next higher temperature and then diminishes greatly at the resolidification temperature.

Discussion

The data gathered from these three coals serve to provide insight into the chemical changes that take place as metallurgical coal goes through its plastic state while transforming into semicoke. A great deal of the more recent work on coal has dealt with coal conversion processes or additives to carbonization rather than with the carbonization of coal alone. Although the chemical changes in coal during its fluid range are still not well understood, some information can be derived from observations of the mesophase and from changes in functional groups as inferred from infrared spectra.

As a highly fluid coal such as Pittsburgh No. 8 is heated, the coal structure begins to break down into smaller units probably consisting of 2-5 fused aromatic rings along with the aliphatic and alicyclic component. The spectra up to 400°C show an increase in the methyl and methylene bending peaks at 1380 and 1450 cm^{-1}, possibly resulting from the cleavage of ethylene bridging units as suggested by Solomon (5). The Illinois No. 6 coal does not show this effect. In the Illinois coal the initial breakdown, which is critical to fluidity, perhaps does not occur to the same extent as it does in the Pittsburgh No. 8 coal. The increased amount of aromatic compared with aliphatic ether groups in the Illinois No. 6 sample may inhibit breakdown of the coal structure. Aromatic ether linkages are known to be strong bonds and may be at least partially responsible for the lack of fluidity of this coal. With increasing temperature up to the onset of fluidity, the number of ether linkages remains almost constant. There may be an equilibrium at any given time between cleavage of existing ether bonds and formation of new ones by dehydration.

The small aromatic units formed by breakdown of the coal structure are similar to those in tar, although their molecular weight is probably greater. The spectra of tar collected during the pilot-scale carbonization of Pittsburgh No. 8 and of Lower Kittanning coals are shown in Figure 11. It can be seen that the tars are rich in aromatic compounds and are remarkably similar to each other in spite of the differences in rank and properties of the coals. The strongest peak is the 750 cm^{-1} vibration of ortho-substituted aromatics. Naphthalene, anthracene, and similar polynuclear aromatic molecules found in these tars are the most likely contributors to such a peak.

During the early stages of heating before the onset of fluidity, little change occurs in the spectra. The broad absorption between 2400 and and 3400 cm^{-1} associated with hydrogen bonding remains relatively constant as does the ether region between 1000 and 1300 cm^{-1}. At the temperature at which the coal

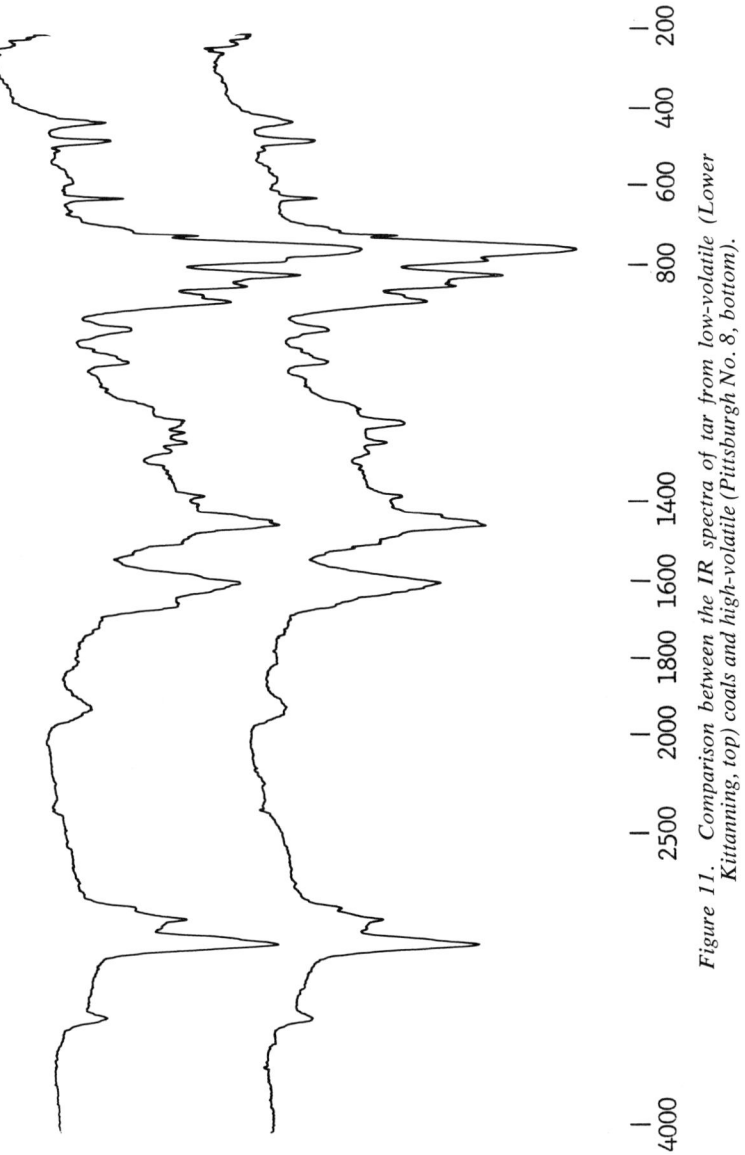

Figure 11. Comparison between the IR spectra of tar from low-volatile (Lower Kittanning, top) coals and high-volatile (Pittsburgh No. 8, bottom).

becomes plastic, the peak at 750 cm^{-1} becomes the strongest aromatic absorption and indicates that small ortho-substituted molecules such as those in tar have reached their maximum abundance. Hydrogen bonding begins to decrease at the temperature of onset of fluidity in the Pittsburgh No. 8 sample as well as in the Illinois No. 6. Apparently at this stage the aromatic molecules are highly mobile. Also, at this stage the free radical concentration in vitrinite increases sharply (6), probably as a result of cleavage of C-C and/or C-O-C bonds.

Further increase in temperature causes the small aromatic units to recombine and form aromatic sheets. It was puzzling that infrared spectra seem to give evidence for more light aromatic molecules than the tar yield of coal could account for. This inconsistency can be resolved if what occurs is that the light aromatics are formed and then most of them polymerize to form sheets. The remaining light aromatic molecules that do not polymerize are carried off as tar. The sheets can then form layered structures held together by London dispersion forces, which increase with increasing molecular size. Structures such as these are capable of diffracting electrons and are probably the liquid-crystalline mesophase seen on the hot stage of the electron microscope.

At the temperature of maximum fluidity the vibration at 860 cm^{-1} reaches its maximum. This peak, which arises from highly substituted aromatic rings, is consistent with a model of liquid crystals composed of layered aromatic sheets with lone hydrogen atoms only on the edges. The 860 cm^{-1} peak becomes significant in each coal at only one temperature. In another Pittsburgh No. 8 sample, whose fluidity was severely reduced by oxidation, the 860 cm^{-1} peak did not become significant at any temperature. Hence, fluidity and mesophase are evidently related.

Continued carbonization beyond the temperature of maximum fluidity causes resolidification. The 860 cm^{-1} peak disappears as the coalesced mesophase loses more volatile matter, mostly hydrogen, and the aromatic sheets bond to each other, but not in a well-ordered fashion. It is likely that the hydrogen lost at this stage comes from the edges of the sheets, leaving free radicals. This interpretation is consistent with the observation by Petrakis et al that the maximum number of free radicals measured by electron paramagnetic resonance occurs roughly around the resolidification temperature (6). In classical liquid-crystal systems, on heating, crystals transform to liquid crystals and then to an isotropic liquid, the transformations being reversible. In the carbonaceous mesophase, on the other hand, decomposition proceeds simultaneously with the phase transformations, so that the reactions are not reversible.

In semicokes cooled from above the resolidification point, the infrared spectra become less intense as electronic absorption becomes significant. Further heating of semicokes after solidification evolves primarily hydrogen.

Conclusions

From the foregoing data and previous heating stage experiments we conclude that the carbonization of metallurgical coal proceeds via a mesophase composed of sheets of aromatic rings possessing functional groups similar to those in the parent coal. Mesophase differs from coal in that it has a greater aromatic component and is more ordered. During carbonization, coal reaches its maximum fluidity as the mesophase coalesces. As carbonization proceeds beyond the fluid range, decomposition will have occurred and a disordered semicoke will form. The properties of this semicoke and of the resulting coke are controlled to a large extent by what happens during the plastic stage. Therefore, what we learn about the mesophase we hope will help us achieve better control of the carbonization process and improve the product.

Acknowledgments

The authors gratefully acknowledge the assistance of N. Marion for preparing the semicoke samples and W. S. Szulborski for taking the infrared spectra.

Literature Cited

1. Friel, J. J.; Mehta, S.; Mitchell, G. D.; Karpinski, J. M. Fuel 1980, 59, 610-616.
2. Solomon, P. R. Am. Chem. Soc. Div. Fuel Chem., Preprints 1979, 24 (2) 184.
3. Painter, P. C.; and Coleman, M. M. Am. Lab. 1980, 12, 19-29.
4. Lowry, H. H. "Chemistry of Coal Utilization (Supplementary Volume)"; Wiley: New York, 1963; p 61-78.
5. Solomon, P. R.; personal communication.
6. Petrakis, L.; Grandy, D. W. Fuel 1981, 60, 115-119.

RECEIVED May 17, 1982

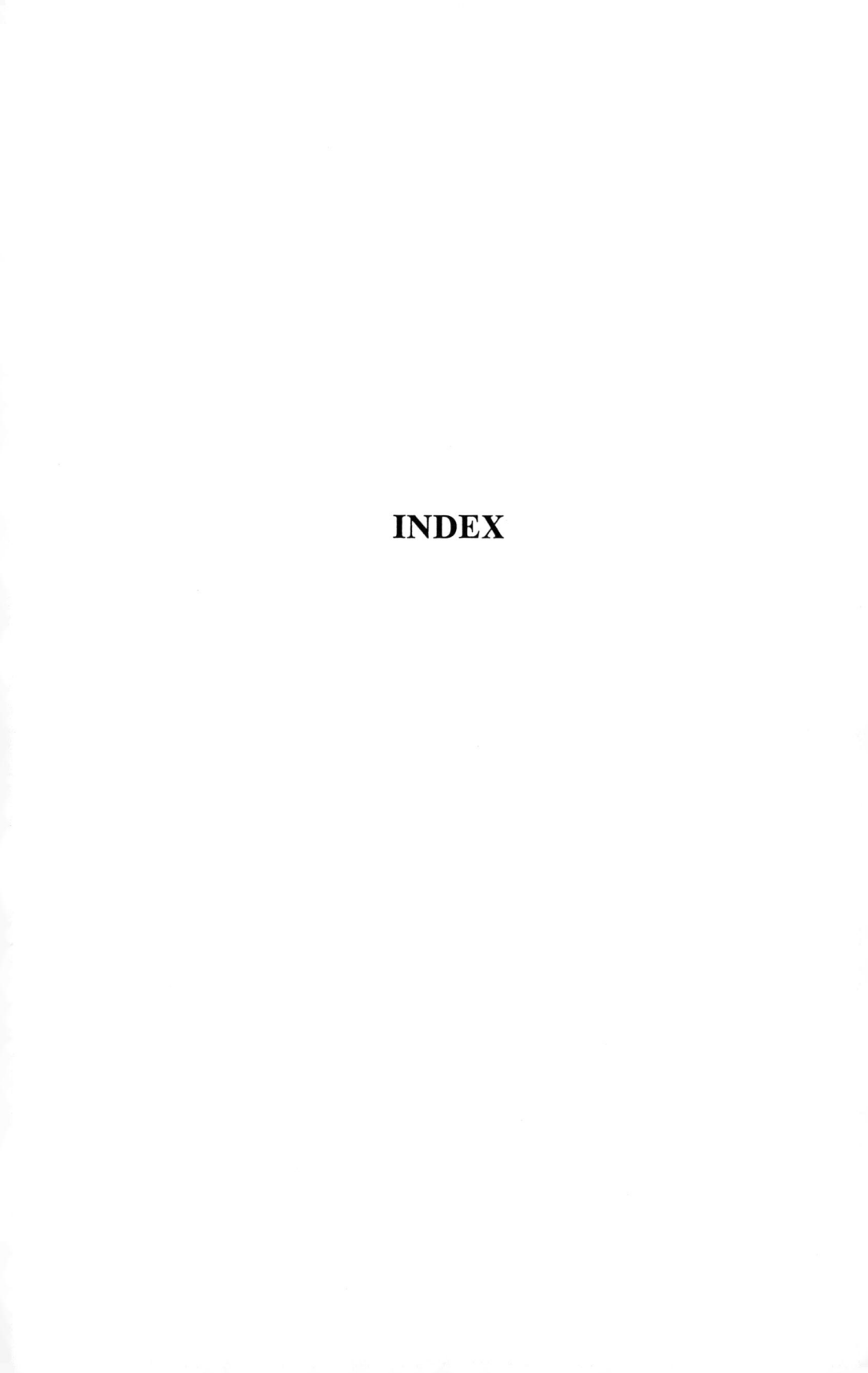

INDEX

INDEX

A

AAS—*See* Atomic Absorption Spectrometry
Absorbance variation in FTIR 94*f*
Absorption process 264
Absorptivities
   aromatic and aliphatic 97*f*
   of hydroxyl concentration, FTIR .. 95*f*
Acenaphthene and acridine, GC/MS identification in toluene subfraction of Sephadex LH-20 fractions of tar trap tar 218*t*
Acicular flow domain anisotropy 6*t*
Activated states in air oxidation 143–144
Active pitches 8
Acyclic structure 29*t*
Additives
   breeze 11
   pitch 8
AES—*See* Atomic emission spectrometry
Aliphatic absorptivities from regression analysis 97*f*
Aliphatic hydrocarbon concentration
   analysis by FTIR 61–63, 71–74
   and degree of oxidation 141
   and Fischer assay oil yield 123*f*
Aliphatic hydrogen concentration
   analysis by proton NMR 219*t*
   analysis by FTIR 93, 96–99
   correlation with tar yield 117*f*
Alkali halide matrix 49–51
Alkane content
   of fossil fuels 27–43
   of hydrocarbon minerals 28*t*
Aluminum analysis using ICP–AES .. 150*t*, 153*t*, 160*t*, 167*t*, 169*t*, 171*t*
Americana, sulfur analysis using EPM 195–202
American Society for Testing and Materials (ASTM) Standard Test Method—*See* ASTM
Analytical instruments, on-line, economic considerations 275–277
Analytical instruments in coal preparation industry, overview 259–277
Analyzer, elemental, using Californium-252 neutron source ..268–269
Anisotropic and isotropic properties of cokes and coke blends 1–7
Anistropy nomenclature 6*t*

Ankerite
   Mössbauer spectra 247*f*, 248*t*
   SEM–AIA analyses 244*t*, 250*t*
Anthracene, GC/MS identification in toluene subfraction of Sephadex LH-20 fractions of tar trap tar .. 218*t*
Anthracite, Rhode Island, Mössbauer spectra 247*f*
Antimony analyses in coal by hydride–atomic absorption spectrometry 155*t*
Apatite, SEM–AIA analyses 244*t*, 250*t*
Appalachian coals, gallium and germanium content 156*t*
Apparatus, pressure filtration and Soxhlet extraction 266–231
Applications of FTIR 77–127
Arizona coal, FTIR spectrum 66*f*
Aromatic absorptivities 97*f*
Aromatic C–H group determination by FTIR 61–63, 71–74
Aromatic carbon
   FTIR vs. NMR 100*f*
   proximate analysis fixed carbon .. 121*f*
Aromatic ethers in Illinois coal semicokes 300
Aromatic hydrogen, FTIR 93, 96–99
Aromatic hydrogen in coal gasifier tar by proton NMR 219*t*
Aromatic hydrogen out-of-plane bending vibrations shown by IR spectra 305*f*
Arsenic analysis using hydride generation–atomic absorption spectrometry 155*t*
Ash content
   of Kentucky Colonial Mine and Illinois Burning Star Mine 228*t*
   of Pittsburgh, Illinois, and Lower Kittanning coal 298*t*
   of vitrains 13*t*
   of Wyodak and Bruceton coal 136*t*
Ash monitors 263–267
Ashes, North Dakota lignite, gallium and germanium content .. 156*t*
Asphaltenes, Ashland A200 pitch analysis 13*t*
Asphaltite, Turkish
   alkane content 28*t*
   carbon-13 spin-lattice relaxation times 39*t*

ASTM for sulfur analysis 192–194, 200
ASTM/EPM comparative study 195t
Atomic absorption spectrometry,
  detection limits 153t
Atomic absorption—hydride genera-
  tion spectrometry, trace ele-
  mental analysis 155t
Atomic absorption spectrophoto-
  meter (AAS) 148, 151, 153
Atomic emission spectrometry for
  metal quantitation and
  speciation 163–182
Atomic H/C ratio of organic matter
  in hydrocarbon minerals 28t
Australian coals, gallim and
  germanium content 156t
Avgamasya-asphaltite alkanes 40f
Azapyrene, GC/MS identification
  in tar trap tar 218t

B

Backscattered electron (BSE)
  radiation in SEM–AIA 241
Bale correction, FTIR 51–52
Barite, SEM–AIA analyses 250t
Barium analysis using ICP–AES 150t,
  153t, 160t, 167t, 169t, 171t
Barium chloride formation in chlorine
  determination 186–190
Beer's law in FTIR 81–83
Benzene-insolubles in Ashland
  A200 pitch 13t
Benzo[a]fluorene and benzo[a]pyrene,
  GC/MS identification in toluene
  subfraction of tar trap tar 218
Benzo(f)quinoline and benzo(h)-
  quinoline, GC/MS identification
  in the basic subfraction of Sepha-
  dex LH-20 fraction 5 of tar
  trap tar 218t
Beryllium analysis in coal and fly ash
  using ICP–AES 150t, 153t, 160t
Bethlehem plastometer test, fluid
  properties 298t
Beulah North Dakota lignite, carbon
  dioxide yield vs. time from
  pyrolysis 112f
Bevier, sulfur analysis by EPM 198t
Bituminous coal
  alkane content 28t
  DRIFT analysis 133–143
  FTIR 67–68, 97f
  secondary electron image and
    inverted backscattered elec-
    tron (BSE) image analysis 243f
Blaenhirwaun Pumpquart 13t
Blast furnace, typical charges per day   292
Blending procedures 7–8, 11–12
Blind Canyon, sulfur analysis
  using EPM 195–202
Bonding of Pittsburgh, Illinois, and
  Lower Kittanning coal 298t
Boron analysis
  using ICP–AES 167t, 169t, 171t
  using Parr oxygen bomb elemental
    analysis 152t, 153t, 160t
Bottom ash, gasifier cleanup device
  samples 215t
Breeze additives 11
Bromide analysis 149
Bromine analysis using Parr oxygen
  bomb 152t, 153t, 160t
Bruceton chemical analysis 136t
Btu, low, sasification 205–222

C

^{13}C NMR—See Carbon-13 NMR
Cadmium, detection limits and analysis using ICP–AES 167t, 169t, 171t
Caking coal, 2–3
Calcite, SEM–AIA analyses 244t,
  250t, 255t
Calcium, detection limits and analysis
  by ICP–AES 150t, 153t, 160t,
  167t, 169t, 171t
Californium-252 neutron source for
  elemental analyzer system 268–269
Capacitance, on-line moisture
  monitors 270–273
Carbon
  anisotropic 2–3
  aromatic, C(ar), correlation of
    proximate analysis fixed
    carbon 121f
  aromatic, FTIR vs. NMR 100f
  content in coal 298t
  functional group concentrations,
    determination 94–99
  Kentucky Colonial Mine and Illinois
    Burning Star Mine 228t
  Pittsburgh, Illinois, and Lower
    Kittanning coal 289t
  Six Bells, Cortonwood Silkstone,
    and Maltby Swallow Wood
    coal 12t
  vitrains 13t
  Wyodak and Bruceton coal 136t
Carbon-13 NMR of coal-derived
  products and spin-lattice relaxation
  time measurements 34–41
Carbon dioxide
  concentration determined by
    GC 289–292
  gasification 11
  yield vs. time from pyrolysis of
    Beulah North Dakota lignite 121f

INDEX

Carbon monoxide spectrum 114f
Carbonaceous system and liquid
  crystals 3–4
Carbonization process 2–3
  electron optical and IR spectro-
    scopic investigation 293–309
  gas analysis to determine length
    of coking cycle 287
Categories, SEM–AIA mineral 244t
Cesium iodide matrix 49–51
Chars, FTIR spectra during
  thermal decomposition 119f
Cherokee, sulfur analysis by EPM 198t
Chloride analysis 149
Chlorine analysis 185–190
  of Indian coals 190t
  using Parr oxygen bomb
    elemental analysis ..152t, 153t, 160t
Chlorite, Mössbauer and SEM–
  AIA analyses 244t, 248t, 250t
Chloroform, solvent matrix for
  ICP–AES 167t
Chromatogram, reconstructed ion,
  toluene subfraction from silica
  gel column of Sephadex
  LH-20 column 216f
Chromatograms, reconstructed ion,
  from GC/MS analysis of vapor
  phase organic extractable 212f
Chromatography
  gas—See GC
  liquid—See LC
  size exclusion, ICP–AES
    detection 172–180
  solvent, metal detection in coal-
    derived process 172
Chromium
  analysis in coal and fly ash using
    ICP–AES 150t, 153t, 160t
  detection
    limits and analysis by
      ICP–AES 167t, 169t, 171t
    separation of process solvent 181f
Clarion, sulfur analysis by EPM 195–202
Classes size in SEM–AIA of coal 244t
Classification by solvent solubility ..225–236
Cleanup device samples, dichloro-
  methane extractables from bottom
  ash, cyclone ash, tar trap tar,
  scrubber water, scrubber tar 21
Cleanup system, METC low Btu
  gasifier, schematic diagram 207f
Clintwood, sulfur analysis by
  EPM 195–202
Clyde Ironworks coke, optical and
  SEM micrograph 15f, 16f
Coal-derived process solvent
  chromatography, metal
  detection 172

Coal-derived products solvent
  solubility analysis 225–236
Coal gasifier tar, proton NMR, mole
  percent of hydrogens 219t
Coal–oil slurry, high conversion, and
  solvent solubility 229–236
Coal ranking systems, comparison of
  U.K. and U.S. 12t
Coal structure and proximate analysis
  fixed carbon relation 118, 121–122
Coal tar, alkane characterization 27–43
Coal-tar branched/chain alkanes,
  Rexco, proton-decoupled
  carbon-13 .... 36t, 35–38
Cobalt analysis in coal and fly ash
  using ICP–AES 150t, 153t, 160t
Cokes
  nomenclature, optical texture in
    polished surfaces 6t
  optical properties 4–7
  optical and SEM micrograph of
    Spencer Wharf and Clyde
    Ironworks 15f, 16f
  properties and carbonization 2–7
  strength 5, 8
Coke oven, construction 282
Coke oven gas
  composition during coking
    cycle 290f, 291t
  flow diagram 283–286
Coking cycle, gas analysis to determine
  length 286–292
Coking potential of coal blends 7
Colormetric procedure for the determination of germanium 158
Column, elution profiles of compounds
  from Sephadex LH-20 .... 209f
Column, silica gel, reconstructed ion
  chromatogram of toluene
  subfraction 216f
Column packing, separation of silicon
  model compounds with silicon
  detection 178f
Complex of gallium—malachite green   156
Composition of coke oven gas during
  coking cycle 290f, 291t
Computer applications, FTIR 52–58
Computer programs, library search
  routines 102, 104
Concentration meter, slurry 273–275
Conductivity gauge for slurry
  concentration meter 273–275
Continuous on-line nuclear assay of
  coal (CONAC) 268–269
Copper, detection limits and analysis
  by ICP–AES 150t, 153t, 160t,
                                    167t, 169t, 171t
Coquimbite, Mössbauer and SEM–AIA
  analyses 248t, 250t

Corrections for FTIR mineral analysis	85
Cortonwood Silkstone coal, analysis	12*t*
Cost of elemental analyzers	269
Creosote oil, pressure filtration	236*t*
Cross-linking in liquid crystals	3–4
Crude oils, alkane content	28*t*
Crystals, liquid, and mesophase	3–4
Curve, coal washability	262*f*
Curve resolving, FTIR	59, 61
Cyclic monoterpane, diterpane, sesquiterpane, sterane, and triterpane, structure	29*t*
Cyclohexane solubility in coal-derived products	225
Cyclone ash, gasifier cleanup device samples	215*t*

D

Data analysis operations, FTIR	52–58
Density gauges, nuclear	267–268, 273
Derivative spectroscopy, FTIR	59, 60*f*, 68*f*, 72*f*
Detarred coke oven gas, composition	291*t*
Detection gamma and neutron	273–275
ICP–AES for size exclusion chromatography	172–180
iron, separation of process solvent	179*f*
metal, in coal-derived process solvent chromatography	172
silicon, iron, and chromium, separation of process solvent	177–182
Detection limits ICP–AES	166
spectroscopic procedures	153*t*
Deuterium-exchanged Pittsburgh seam coal, FTIR spectra	86*f*
Dibenzothophene, GC/MS identification in toluene subfraction of tar trap tar	218*t*
Dichloromethane extractables from glass fiber Tenax filter, reconstructed ion chromatograms from GC/MS analysis and mass distribution	212*f*, 215*f*
Dicyclic hydrocarbons, structure	29*t*
Dietz, sulfur analysis by EPM	195–202
Difference methods, FTIR	54–55
Difference spectrum of acetylated lignite	68*f*
using DRIFT	138*f*, 139*f*, 140*f*
Diffuse reflectance IR Fourier transform spectroscopy—*See* DRIFT	
Diffuse reflectance spectroscopy	82, 84–85
Dimethylbenzoquinoline, dimethylnaphthalene, dimethylphenanathrene, and dimethylquinolines, GC/MS identification in tar trap tar	212*t*, 218*t*

Diterpane, cyclic structure	29*t*
Dolomite, SEM–AIA analyses	244*t*, 250*t*
Domains	6*t*
DRIFT (diffuse reflectance IR Fourier transform spectroscopy)	133–143
DRIFT spectra of Wyodak coal	136*f*, 138*f*, 140*f*, 142*f*

E

Eastern coals, gallium and germanium content	156*t*
Economic considerations of on-line analytical instruments	275–277
EDX—*See* Energy dispersive x-ray	
Electron image inverted backscattered, of Pust seam Montana lignite and energy-dispersive x-ray spectrum of a maceral	253*f*
secondary, of bituminous coal	243*f*
Electron micrograph of coalesced and isolated spheres in Illinois No. 6 vitrinite and Pittsburgh No. 8 vitrinite	294*f*
Electron micrograph transmitted of semicoke from Lower Kittanning coal	301*f*
Electron optical IR spectroscopic investigation of coal carbonization	293–309
Electron probe microanalysis for direct determination of organic sulfur	191
The Elemental Analyzer	268–269
Elemental detection limits monitored in various solvents for ICP–AES	167*t*
Elkhorn, sulfur analysis by EPM	195–202
Elution profiles of compounds from Sephadex LH-20 column	209*f*
Emission lines of elements in ICP–AES	166–172
Energy dispersive x-ray (EDX)–scanning electron microscope (SEM) vs. electron probe microanalysis for sulfur analysis	192–194, 200
Energy-dispersive x-ray spectrum of a maceral of Pust seam Montana lignite	253*f*
EPM—*See* Electron probe microanalysis	
Eschka method, chlorine determination	190
Ethers, aromatic, in Illinois coal semicokes	300
Experimental apparatus for determination of chlorine in organic combination	188*f*

INDEX

Extinction coefficients, mean, for absorbing groups of 156 model compounds ... 101t
Extracts, fossil fuel, structural spectroscopic analyses ... 28
Exxon–Baytown analytical methods for coal and fly ash analysis ... 161f

F

Factor analysis, FTIR ... 55–58
Fast Fourier transformation of the free-induction decay of coal-derived products ... 34
Ferric oxide, Mössbauer spectra ... 247f, 248t
Field desorption MS ... 31, 43
Fies coal, elemental analysis using ICP–AES ... 168–172
Filter, glass fiber Tenax, reconstructed ion chromatograms from GC/MS analysis of vapor phase organic extractable ... 212f
Filter sample, raw gas, mass distribution of dichloromethane extract ... 215t
Fischer assay oil yield correlation with aliphatic hydrocarbon concentration ... 123f
Fissure formation, coal blends ... 5, 7
Fixed carbon, proximate analysis ... 118, 121–122
Flow domain anisotropy ... 6t
Fluid melts and carbonization process ... 2–3
Fluid phases, reactions with coal ... 133
Fluid properties from Bethlehem plastometer test ... 298t
Fluidity of Pittsburgh, Illinois, and Lower Kittanning coal ... 296t, 298t
Fluidized-bed coal tar, carbon-13 spectrum ... 35
Fluorene and fluoranthene, GC/MS identification in toluene subfraction of tar trap tar ... 218t
Fluoride analysis ... 149
Fluorine analysis using Parr oxygen bomb elemental analysis ... 152t, 153t, 160t
Fly ash, elemental analysis ... 147
Fossil fuels
 alkanes in ... 27–43
 spectroscopy ... 28, 29–43
Freeport, sulfur analysis by EPM ... 195–202
FTIR ... 47–74
 applications ... 77–127, 113
 determination of hydroxyl groups ... 61–71
 free-induction decay ... 34
 of gases ... 109
 inversion recovery technique ... 37
 of liquids ... 104–108
 model compounds ... 99–104

FTIR—
Continued
 vs. NMR for aromatic carbon ... 100f
 of solids ... 79–99
 thermal decomposition study ... 113, 116–118
 spectra
  of chars during thermal decomposition ... 119f
  of coals and their vacuum distilled tars ... 117f
  of 4-hydroxyl quinoline and 1-napthol, comparison ... 103f
  of lignite ... 80f, 83f
  of liquefaction feedstock and products ... 124f
  of napthalene and quinoline, comparison ... 103f
  of raw vs. chemically cleaned coal ... 125f, 126f
  of tars from a Pittsburgh seam bituminous coal ... 120f
  of vitrinite concentrate ... 53f, 72f
  scattering and mineral correction ... 87f
  vs. wet chemistry ... 100f
FTIR, diffuse reflectance—See DRIFT
Fuels, fossil
 alkanes in ... 27–43
 spectroscopy ... 28, 29–43
Functional group, aliphatic C-H, and degree of oxidation ... 141
Functional group, carbon, determination ... 98–99
Functional group, quantitive determination using FTIR ... 47–74
Furnace, blast, typical charges per day ... 292
Fusinite, organic sulfur analysis by EPM ... 194–202

G

Gallium, spectrophotometric determination ... 154–161
Gallium (III)—malachite green reaction ... 156
Gamma density gauge for slurry concentration meter ... 273–275
Gas analysis to determine length of coking cycle ... 286–292
Gas, coke oven, composition during coking cycle ... 290f, 291t
Gases from pyrolyzed coal, high resolution spectra ... 110f, 111f, 114f
Gases, analysis using FTIR ... 109
Gasification
 in carbon dioxide ... 11
 and heat treatment of metallurgical coke ... 10–11, 14–17
 low Btu ... 205–222

Gasifier cleanup device samples, percent dichloromethane extractables from bottom ash, cyclone ash, tar trap tar, scrubber water, and scrubber tar 212t
Gasifier, METC low Btu, schematic diagram 207f
Gauss-Lorentz sum and product functions, FTIR spectral analysis 61
GC analysis 31, 34, 210, 289–292
GC/FTIR analysis 113
GC/MS analysis
　of coal-derived products 31, 41, 43
　of gasifier process streams 210–222
Gel permeation chromatography 208–218
Germanium, spectrophotometric determination 154–161
Gas flow diagram, coke oven 283–286
Glass fiber Tenax filter, reconstructed ion chromatograms from GC/MS analysis of dichloromethane extractables 212f
Goethite, particle-size parameter distribution in middle Kittanning coal 249t, 250t, 257f
Gypsum, SEM–AIA analyses 244t, 250t

H

^{1}H NMR—See Proton NMR
Halite, SEM–AIA analyses 244t, 250t
Halogen analysis using ion analyzer .. 149
Hartshorne, sulfur analysis using EPM 195–202
Hazard, sulfur analysis using EPM 195–202
Hematite, Mössbauer and SEM–AIA analyses 249t, 250t
Heptane, solvent matrix for ICP–AES 167t
Heptane-insolubles in Ashland A200 pitch 13t
Heat treatment and gasification of metallurgical coke 10–11, 14–17
High resolution FTIR spectra of gases from pyrolyzed coal 110f, 111f, 114f
High-conversion coal–oil slurry solubility 229–236
Hydrocarbon, aliphatic
　correlation of Fischer assay oil yield 123f
　structures 29t
Hydrocarbon minerals, alkane content and H/C ratio of organic matter 28t
Hydrodesulfurization effect on coal structure 122, 124–127
Hydrogen
　aliphatic, correlation with tar yield 117f

Hydrogen—Continued
　aromatic, out-of-plane bending vibrations shown by IR spectra 305f
　FTIR analysis 93, 96–99
　proton NMR analysis, mole percent 219t
Hydrogen bonding O–H absorption 102
Hydrogen/carbon ratio of organic matter in hydrocarbon minerals 28t
Hydrogen chloride formation in chlorine determination 186–190
Hydrogen content
　of Kentucky Colonial Mine and Illinois Burning Star Mine coal 228t
　of Pittsburgh, Illinois, and Lower Kittanning coal 298t
　of Six Bells, Cottonwood Silkstone, and Maltby Swallow Wood coals 12t
　using slurry concentration meter 273–275
　of vitrains 13t
　of Wyodak and Bruceton coal 136t
Hydrogen distribution in fossil fuel alkane extracts from proton NMR spectra 33t
Hydrogen transfer 9
Hydroxyl group determination by FTIR 61–71, 92, 95f
4-Hydroxyl quinoline and 1-napthol, comparison of the FTIR spectra 103f

I

ICP–AES 148–151, 163–182
ICP–AES, detection limits for different spectroscopic procedures 153t
Illinois coal
　Burning Star Mine, analysis 228t
　gallium and germanium content 156t
　IR spectra 301f, 300–307
　No. 6 vitrinite, electron micrograph of isolated spheres 294f
　sulfur analysis by EPM 198t
Illite, Mössbauer and SEM–AIA analyses 244t, 248t, 250t, 255t
Indian coals
　chlorine contents 190t
　gallium and germanium content 156t
Indian fly ash, gallium and germanium content 156t
Inductively coupled plasma–atomic emission spectrometry—See ICP–AES
Industry
　coal preparation, analytical instruments 259–277
　steel, coal as energy 281–292
Inertinite macerals, sulfur analysis by EPM 195–202

Infrared spectroscopy—*See* IR
Inorganic elements in macerals, characterization 252–254
Instrumentation in coal preparation industry 259–277
Integration limits for absorbing groups of 156 model compounds 101*t*
Interferences in gallium and germanium determination 157*t*
Intermediate states in air oxidation 143–144
Inversion-recovery Fourier transform (IRFT) technique of coal-derived products 37
Inverted backscattered electron (BSE) image of bituminous coal 243*f*
Inverted backscattered electron (BSE) image of Pust seam Montana lignite 253*f*
Ion analyzer for halogen analysis 149
Ion chromatogram, reconstructed from GC/MS analysis 212*f*, 216*f*
IR absorption and on-line moisture monitors 270–273
IR spectroscopy 31
of coal carbonization 293–309
Fourier transform—*See* FTIR
Fourier transform, diffuse reflectance—*See* DRIFT spectra of Pittsburgh coal and semicokes, Illinois coal and semicokes, and Lower Kittanning coal and semicokes ... 300–307
Iron-57 Mössbauer spectroscopy of coal minerals 241, 245–252
Iron analysis
in coal and fly ash using ICPES 150*t*, 153*t*, 160*t*
in clays, Mössbauer spectra 247*f*, 248*t*
using ICP–AES 150*t*, 153*t*, 160*t*, 167*t*, 169*t*, 171*t*
Iron detection, separation of process solvent 179*f*
Iron model compounds, molecular structures 176*f*
Iron–sulfate, SEM–AIA analyses 244*t*, 250*t*, 255*t*
Isocratic separation, organometallic systems 174, 175
Isoprenoid, acyclic structure 29*t*
Isotropic properties of cokes and coke blends 1–7

J

Jarosite, Mössbauer spectra 247*f*, 248*t*
Jarosite, SEM–AIA analyses 244*t*, 250*t*

K

Kaolin, DRIFT spectra 138*f*, 139*f*, 140*f*
Kentucky coal, sulfur analysis by EPM 198*t*
Kentucky 9/14 coal, pressure filtration 232*t*
Kentucky Colonial Mine, analysis 228*t*
Kentucky solvent refined coal processing conditions 170*t*
Kinetic measurements of gases using FTIR 109
Kittanning, sulfur analysis by EPM 195–202
Kuwait crude branched/cyclics, proton NMR spectra 33*t*

L

Laboratory instrumentation 260, 263
Lafayette coal, elemental analysis using ICP–AES 168–172
LC/FTIR 104
LC/ICP–AES 163–182
Lead determination
using hydride-atomic absorption spectrometry 155*t*
using ICP–AES detection limits 167*t*, 169*t*, 171*t*
using Parr oxygen bomb elemental analysis 152*t*, 153*t*, 160*t*
Least squares optimization procedure, FTIR spectral analysis 61
Leaves, orchard, elemental analysis .. 152*t*
Lepidocrocite, Mössbauer and SEM–AIA analyses 249*t*, 250*t*
Library search routines, computer programs 102, 104
Lignite ashes, North Dakota, gallium and germanium content 156*t*
Lignite
alkane content 28*t*
Beulah North Dakota, carbon dioxide yield vs. time from pyrolysis 112*f*
FTIR spectra 80*f*, 83*f*, 97*f*
Pust seam Montana, inverted backscattered electron image 253*f*
Turkish, alkane characterization 27–43
Linearity, gallium and germanium determination in coal 157*t*
Liptinite macerals, sulfur analysis by EPM 195–202
Liquefaction feedstock and products, FTIR 122, 124*f*
Liquid cell spectrum of recycle solvent 107*f*
Liquid chromatography—*See* LC
Liquid crystals and mesophase 3–4
Liquids, FTIR 104–108

Lorentzian band shapes, FTIR
  spectral analysis 61
Low Btu gasification 205–222
Low temperature ash spectrum by
  addition of spectra from mineral
  library 88f
Lower Kittanning coal and semicokes,
  IR spectra 300–307
Lower Kittanning vitrinite, spheres
  and rods 296f

M

Macerals
  aromatic C–H out-of-plane bend-
    ing mode, factor analysis 58f
  characterization of inorganic
    elements 252–254
  organic sulfur analysis by EPM ...194–202
  of Pust seam Montana lignite,
    energy-dispersive x-ray
    spectrum 253f
Magnesium analysis detection limits
  and analysis by ICP–AES ..150t, 153t,
    160t, 167t, 169t, 171t
Magnetite, Mössbauer and SEM–AIA
  analyses 249t, 250t
Maltby Swallow Wood coal analysis .. 12t
Manganese, detection limits and analysis by ICP–AES 167t, 169t, 171t
Manton Main Parkgate 13t
Marcasite, Mössbauer and SEM–AIA
  analyses 248t, 250t
Mass distribution of dichloromethane
  extract of a raw gas filter sample 215t
Mass distribution
  of 0°C condenser sample of gasifier
    process streams 214t
  of Sephadex LH-20 fractions of tar
    samples from Venturi scrubber
    decanter outlet and the tar
    trap 217t
Mass spectroscopy—See MS
Mechanistic studies of air
  oxidation 141–144
Mean extinction coefficients for
  absorbing groups of 156 model
  compounds, FTIR 101t
Melanterite, Mössbauer and SEM–
  AIA analyses 248t, 250t
Mercury analysis using Parr oxygen
  bomb elemental analysis 152t,
    153t, 160t
Mesophase, definition 3–4
Mesophase structure of Pittsburgh,
  Illinois, and Lower Kittanning
  coal 296t
Metal quantitation in synfuels,
  using ICP–AES 163–182
Metallic materials in process
  solvents 172–180

Metallograms 172–173
Metallurgical coke 123
Metallurgical coke, point-
  counting 10, 21–24
Metallurgical cokes, Spencer Works
  Wharf coke and Clyde Iron-
  works coke 10
METC low Btu gasifier and cleanup
  system, schematic diagram 207f
Methods
  Exxon–Baytown analytical, for coal
    and fly ash analysis 161f
  Soxhlet-extraction vs. pressure-
    filtration 225
Methylacridine, Methylbenzoquino-
  line, Methyldibenzothiophene,
  Methylphenanthridines, and
  Methylquinoline, GC/MS iden-
  tification in tar trap tar 212t, 218t
Micrinite, organic surfur analysis
  by EPM 194–202
Micrograph—See Optical and second-
  ary electron (SEM) micrograph
Micrograph, electron
  of coalesced spheres in Pittsburgh
    No. 8 vitrinite 294f
  of isolated spheres in Illinois No.
    6 vitrinite 294f
  of semicoke from Lower
    Kittanning coal 301f
  of semicoke from Pittsburgh No.
    8 coal 299f
Microscopy, optical 9–23
Microwave moisture meter, on-line,
  and Microwave attenuation 270–273
Middle Kittanning coal, particle-size
  parameter distribution for
  goethite 249t, 250t, 257f
Middle Kittanning seam mineral
  analysis 255t
Mineral characterization
  using FTIR, spectral
    corrections 51, 55, 85
  using SEM–AIA and Mossbauer
    spectra 239–256
Mineral content blends 8
Mineral library, synthesis of low
  temperature ash spectrum by
  addition of spectra 88f
Mineral mixtures, factor analysis 56f
Minerals, hydrocarbon, alkane con-
  tent and atomic H/C ratio of
  organic matter 28t
Minicomputer, FTIR 52–58
Model compounds for quantitative
  FTIR 99–104
Moisture content
  of vitrains 13t
  of Wyodak and Bruceton coal 136t
Moisture meter, on-line microwave 270–273

INDEX

Molecular structures of iron
  model compounds 176f
Molybdenum, detection limits and
  analysis by ICP–AES 167t, 169t, 171t
Monocyclic hydrocarbons, structure .. 29t
Monoterpane, cyclic structure 29t
Montana lignite, Pust seam, inverted
  backscattered electron image 253f
Montmorillonite, Mössbauer and SEM–
  AIA analyses 244t, 248t, 250t
Montan wax total alkanes
  proton NMR spectra 33t
  total ion current constructed
    from GC/MS 44f
Mosaics 6
Mössbauer spectroscopy, iron-57,
  of coal minerals 245–252
MS, field desorption 43

N

Nailstone Yard 13t
Napthalene and quinoline, comparison
  of the FTIR spectra 103f
1-Napthol and 4-hydroxyl quinoline,
  comparison of the FTIR spectra .. 103f
Nematic liquid crystals 3–4
Neutron gauge for slurry concen-
  tration meter 273–275
Nickel analysis using ICP–AES
  150t, 153t, 160t, 167t, 169t, 171t
Nitrogen analysis
  of Kentucky Colonial Mine and
    Illinois Burning Star Mine 228t
  using Parr oxygen bomb elemental
    analysis 152t, 153t, 160
  of Pittsburgh, Illinois, and Lower
    Kittanning coal 298t
  of vitrains 13t
  of Wyodak and Bruceton coal 136t
Nitrogen compounds, ring 102
NMR spectroscopy
  carbon-13, of coal-derived
    products 34–41
  vs. FTIR for aromatic carbon 100f
  on-line moisture meters 270–273
  proton, mole percent of hydrogens
    in coal gasifier tar 219t
  spectra, proton of Montan wax
    total alkanes 33t
  spectra, photon of Kuwait crude
    branched/cyclics 33t
Nomenclature, optical texture in
  polished surfaces of cokes 6t
North Ceylon Meadow Vein 13t
North Dakota lignite ashes, gallium
  and germanium content 156t
Nuclear density gauges .... · 267

O

Ohio coal, sulfur analysis by EPM ..195–202
Oil, creosote, pressure filtration 236t
Oil shales
  alkane content 28t
  characterization using FTIR 122
On-line analytical instrumentation. 259–277
On line microwave moisture meter..270–273
On-line microwave moisture
  meter 270–273
Optical micrograph
  of Clyde Ironworks coke 16f
  of pitch-coke breeze in coke from
    Six Bells Coal 18f, 21f
  of Spencer Wharf coke 15f
Optical microscopy 9–10
Optical properties of cokes 4–7
Orchard leaves, elemental analysis 152t
Organic chlorine determination 185–190
Organic sulfur analysis by EPM 191
Organometallic systems, isocratic
  separation 174, 175
Orsat analysis 289–292
Oven, coke, construction 282
Overhauser enhancement 35
Oxidation, air, studies of coal using
  DRIFT 133–143
Oxidation product, detection using
  FTIR 54–56
Oxygen content
  of Kentucky Colonial Mine and
    Illinois Burning Star Mine
    coal 228t
  of Pittsburgh, Illinois, and Lower
    Kittanning coal 298t
  of vitrains 13t
  of Wyodak and Bruceton coal 136t

P

Parr oxygen bomb 148–149, 151, 152t
Particle size and tensile strength 8
Particle-size analysis by
  SEM–AIA 254, 256
Passive pitches 8
Pentacyclic hydrocarbons, structure .. 29t
Petroleum coke breeze and tensile
  strength 8
Petroleum crude, alkane
  characterization 27–43
pH monitors 268
Phase transition, liquid crystal 4–7
Phenanthrene and phenanthridine,
  GC/MS identification in tar
  trap tar 218t
Phenolic hydrogen in coal gasifier
  tar by proton NMR 219t

Phenylnaphthalene, GC/MS identification in tar trap tar 218*t*
Phosphorus analysis using Parr oxygen bomb elemental analysis 152*t*, 153*t*, 160*t*
Photoacoustic spectroscopy 82
Physical properties of nematic liquid crystals 3–4
Pitch, Ashland A200 ....10, 13*t*
Pitch additives 8
Pitch and coals, blending 11
Pitch–coke breeze in coke from Six Bells coal, optical micrograph ..18*f*, 21*f*
Pittsburgh coal
  electron micrograph of coalesced spheres in vitrinite 294*f*
  FTIR spectra 85*f*, 86*f*, 89*f*, 120*f*
  IR spectra 301*f*, 300–307
  SEM of semicoke from 299*f*
  sulfur analysis by EPM 195–202
Plant, coal preparation 260–262
Plastic and post-plastic zone of blends 7
Point-counting, metallurgical coke 10, 21–24
Poly-branched hydrocarbons, structure 29*t*
Porosity and coke strength 5
Post-plastic and plastic zone of blends 7
Potassium analysis in coal and fly ash using ICP–AES 150*t*, 153*t*, 160*t*
Potassium bromide matrix 49–51
Potassium bromide pellets, preparation 79–81
Potassium bromide spectrum of recycle solvent 108*f*
Powdered coal, analysis using DRIFT 135, 136*f*
Pressure filtration for solvent solubility analysis of coal-derived products 225–236
Pristane, carbon-13 spin-lattice relaxation times 39*t*
Process control on-line instrumentation 275–277
Process solvents containing metallic materials 172–182
Processing conditions for Kentucky solvent refined coal 170*t*
Product quality and on-line analytical instruments 275–277
Properties
  coke, and carbonization 2–7
  fluid, from Bethlehem plastometer test 298*t*
  nematic liquid crystals 3–4
Proton-decoupled carbon-13 NMR of Rexco coal–tar branched/chain alkanes 35–38

Proton NMR 33
  spectra of Kuwait crude branched/ cycles and Montan wax total alkanes 33*t*
  mole percent of hydrogens in coal gasifier tar 219*t*
Proximate analysis fixed carbon relation and coal structure ..118, 121–122
Pseudovitrinite, organic sulfur analysis by EPM 194–202
Pust seam coal, inverted backscattered electron image 253*f*
Pyrene, GC/MS identification in tar trap tar 218*t*
Pyridine, solvent matrix for ICP–AES 167*t*
Pyrite
  Mössbauer spectra 247*f*, 248*t*
  SEM–AIA analyses 244*t*, 250*t*, 255*t*
Pyritic sulfur analysis 194–195
Pyrolysis analysis 113
Pyrolysis of Beulah North Dakota lignite, carbon dioxide yield vs. time 112*f*
Pyrolyzed coal, high resolution spectra of gases 110*f*, 111*f*, 114*f*

Q

Quality control on-line instrumentation 275–277
Quantitative analysis
  of functional groups by FTIR 47–74
  of gases by FTIR 109
  of liquids by FTIR 104
  of minerals by SEM–AIA and Mössbauer spectra 239–256
  of model compounds by FTIR 99–104
Quartz, SEM–AIA analyses 244*t*, 250*t*, 255*t*
Quinoline-insolubles in Ashland A200 pitch 13*t*

R

Range for AAS determination of hydride-forming elements 153*t*
Ranking systems of coal, comparison of U.K. and U.S. 12*t*
Rapid EPM method for organic sulfur content 200
Raw vs. chemically cleaned coal, FTIR spectra 125*f*, 126*f*
Reaction of gallium (III)–malachite green 156
Reactions, coal with fluid phases 133
Reactivity of cokes and optical properties 5–7

INDEX

Reconstructed ion chromatogram—
  See Chromatogram, reconstructed ion
Reflectance, diffuse, IR Fourier transform spectroscopy—*See* DRIFT
Regression analysis to determine aromatic and aliphatic absorptivities 97f
Resinite 195, 201t
Resolution
  of difference spectrum obtained from acetylated coal 68f, 69f
  of FTIR spectra 59, 61
Rexco coal-tar
  branched/chain alkanes, proton-decoupled carbon-13 35–38
  carbon-13 spin-lattice relaxation times 39t
  GC 42f
Rhode Island anthracite, Mössbauer spectra 247f
Ribbons 6t
Ring nitrogen compounds 102
Roddymoor Ballarat 13t
Rods in Lower Kittanning vitrinite 296f
Rutile, SEM–AIA analyses 244t, 250t

S

Sample preparation for FTIR 49–51
Saturated hydrocarbons, structures 29t
Scanning electron microscope (SEM)–energy dispersive x-ray (EDX) vs. electron probe microanalysis for sulfur analysis 192–194, 200
Scanning electron microscopy 9–23
  and automated image analysis (AIA) 241–245, 250–256
  and micrographs of Clyde Ironworks and Spencer Wharf cokes 15f, 16f
Scattering and mineral correction of FTIR spectrum 87f
Scattering process 264
Scrubber tar and scrubber water, gasifier cleanup device samples .. 215t
SEC—*See* Size exclusion chromatography
Second derivative spectra 60f
Secondary electron image of bituminous coal 243f
Secondary electron micrograph of semicoke from Pittsburgh No. 8 coal 299f
Selenium determination in coal by hydride generation–atomic absorption spectrometry 155t
SEM—*See* Scanning electron microscopy

Semicoke
  from Lower Kittanning coal, TEM 301f
  from Pittsburgh No. 8 coal, SEM .. 299f
  from Pittsburgh, Illinois, and Lower Kittanning coal IR spectra 301f, 300–307
Semifusinite, organic sulfur analysis by EPM 194–202
Sensitivity
  for AAS determination of hydride forming elements 153t
  for gallium and germanium determination in coal 157t
Separation of process solvent 177–182
Sephadex LH-20 fractions of tar samples from Venturi scrubber decanter outlet and tar trap, mass distribution 217t
Sesquiterpane, cyclic structure 29t
Sewell, sulfur analysis by EPM ... 195–202
Siderite, clays, Mössbauer spectra 247f, 248t
Silicate, SEM–AIA
  analyses 244t, 250t, 255t
Silicon, detection limits and analysis by ICP–AES 150t, 153t, 160t, 167t, 169t, 171t
Silicon model compounds, separation with silicon detection as a function of column packing 178f
Silver, detection limits and analysis by ICP–AES 167t, 169t, 171t
Singly-branched hydrocarbons, structure 29t
Six Bells coal
  analysis 12t
  optical micrograph of pitch-coke breeze in coke 18f, 21f
Size exclusion chromatography with ICP–AES detection 172–180
Slit width for AAS determination of hydride forming elements 153t
Slurry concentration meter 273–275
Slurry, high-conversion coal–oil, and solvent solubility 229–236
Sodium analysis in coal and fly ash using ICP–AES 150t, 153t, 160t
Solid sample analysis using FTIR 79–99
Solidification of Pittsburgh, Illinois, and Lower Kittanning coal 298t
Solubility analysis using pressure filtration 225–236
Solvent
  elemental detection limits and emission lines monitored using ICP–AES 167t
  process, detection of metals 172–182
  recycle, liquid cell and potassium bromide spectrum 107f, 108f

Solvent chromatography and metal detection in coal-derived process 172
Solvent refined coal, metal analysis 166–172
Solvent solubility analysis using pressure filtration 225–236
Sorption-desorption monitoring by DRIFT 135–143
Sortex Ash Monitor sensor system 265f
Source, Californium-252 neutron, for elemental analyzer system 268–269
South African coals, gallium and germanium content 156t
Soxhlet extraction, apparatus and procedure 230–231
Soxhlet-extraction method vs. pressure-filtration method 225
Speciation in synfuels, ICP–AES 163–182
DRIFT, of Wyodak coal 136f, 138, 140f, 142f
energy dispersive x-ray 253f
FTIR
of carbon monoxide 114f
of chars during thermal decomposition 119f
of coals and their vacuum distilled tars 117f
of 4-hydroxyl quinoline and 1-napthol, comparison 103f
of lignite 80f, 83f
of liquefaction feedstock and products 124f
liquid cell, of a recycle solvent 107f
of naphthalene and quinoline, comparison 103f
of a raw vs. chemically cleaned coal 125f, 126f
recycle solvent, potassium bromide 108f
scattering and mineral correction 87f
synthesis of low temperature ash by addition of spectra from mineral library 88f
of tars from a Pittsburgh seam bituminous coal 120f
of vitrinite concentrate 53f, 72f
high resolution of gases from pyrolyzed coal 110f, 111f, 114f
IR for Pittsburgh, Illinois, and Lower Kittanning coal and semicokes 300–307
Mössbauer, Rhode Island anthracite 247f
proton NMR
of Kuwait crude branched/ cyclics and Montan wax total alkanes 33t
of tar trap tar 220f

Spectral corrections and mineral analysis, FTIR 85
Spectral subtraction, FTIR 54–55
Spectral synthesis 85, 88–92
Spectrometric determination of germanium 154–161
Spectrometry, inductively coupled plasma emission (ICPES) 148–151
Spectrophotometer, atomic absorption (AAS) 148, 151, 153
Spectrophotometric determination of gallium 154–161
Spectroscopic analyses, structural, of fossil fuel extracts 28
Spectroscopy of fossil fuels 28, 29–43
Spectroscopy
derivative, FTIR 59, 60f, 68f, 72f
detection limits 153t
diffuse reflectance 82, 84–85
IR, electron optical investigation of coal carbonization 293–309
Mössbauer, of coal minerals, iron-57 241, 245–252
photoacoustic 82
Spencer Wharf coke, optical and SEM micrograph 15f
Spheres in Lower Kittanning, Pittsburgh No. 8, and Illinois No. 6 vitrinite, electron micrograph 294f, 296f
Spin-lattice relaxation time measurements, carbon-13 37–41
Sporinite, organic sulfur analysis by EPM 194–202
Steel industry, coal as energy 281–292
Sterane, cyclic structure 29t
Strength of coke 5, 8
Strength testing 10
Structural spectroscopic analyses of fossil fuel extracts 28
Structure
analysis using FTIR 113
mesophase, of Pittsburgh, Illinois, and Lower Kittanning coal 296t
molecular, of iron model compounds 176f
and proximate analysis fixed carbon relation 118, 121–122
of saturated hydrocarbons 29t
Subbituminous coal
analysis using DRIFT 133–143
analysis using FTIR 97f
Sulfcoalyzer, CONAC system 269
Sulfur analysis
of Ashland A200 pitch 13t
of Kentucky Colonial Mine and Illinois Burning Star Mine 228t
organic, by EPM 191–202

INDEX

Sulfur analysis—
Continued
 using Parr oxygen bomb 152t, 153t, 160t
  of Pittsburgh, Illinois, and Lower
   Kittanning coal 298t
 SEM–AIA analyses 244t, 250t
 of vitrains 13t
 of Wyodak and Bruceton coal 136t
Sulfur contents
Sunnyside, sulfur analysis by EPM 195–202
Superactive pitches 8
Surface complexes in air oxidation 143–144
Sylvite, SEM–AIA analyses 244t, 250t
Symmetry, liquid crystals 3–4
Synfuels, metal quantitation and
 speciation using ICP–AES 163–182
Synthesis of low temperature ash
 spectrum by addition of spectra
 from mineral library 88f
Synthesis, spectral 85, 88–92
System
 coal sampling 266f
 CONAC, solfcoanalyzer 269
 Sortex Ash Monitor sensor 265f
Szomolnokite, Mössbauer and SEM–
 AIA analyses 248t, 250t

T

Tar
 alkane characterization 27–43
 coal gasifier, proton NMR, mole
  percent of hydrogens 219t
 Pittsburgh seam bituminous coal,
  FTIR spectra 120f
 Rexco, GC 42f
 yield correlation with aliphatic
  hydrogen 117f
Tar trap tar
 gasifier cleanup device samples 215t
 mass distribution of Sephadex
  LH-20 fractions of tar samples
  from Venturi scrubber decan-
  ter outlet 217t
Tin determination in coal by hydride–
 atomic absorption spectrometry .. 155t
Tebo, sulfur analysis by EPM 198t
Tellurium determination in coal by
 hydride–atomic absorption
 spectrometry 155t
Tensile strength, cokes 5, 8
Tetracyclic hydrocarbons, structure .. 29t
Tetrahydrofuran solubility, coal-
 derived products 225
Thermal decomposition study
 using FTIR 113, 116–118
TIC—*See* Total ion current
Tilmanstone 13t

Tin, detection limits and analysis
 by ICP–AES 167t, 169t, 171t
Titanium, detection limits and analysis
 by ICP–AES 150t, 153t, 160t,
                       167t, 169t, 171t
Toluene subfraction from silica gel
 column of Sephadex LH-20
 column, reconstructed ion
 chromatogram 216f
Toluene, solvent matrix for
 ICP–AES 167t
Total ion current constructed from
 GC/MS of Montan wax total
 alkane fraction 44f
Trace element analysis
 of a solvent refined coal in
  pyridine 167t
 in coal by hydride generation–
  atomic absorption
  spectrometry 155t
Transmitted electron micrograph of
 semicoke from Lower
 Kittanning coal 301f
2,4,6-Trichlorophenol in chlorine
 determination 187
Tricyclic hydrocarbons, structure 29t
Trimethylnaphthalene, trimethyl-
 phenathrene, and trimethyl-
 quinolines, GC/MS identification
 of tar trap tar 218t
Triterpane, cyclic structure 29t
Turkish asphaltite, carbon-13 spin-
 lattice relaxation times 39t
Turkish asphaltites, alkane content ... 28t
Turkish lignites, alkane
 characterization 27–43
Turkish Montan wax, alkane content .. 28t

V

Vacuum distilled tars, FTIR
 spectra of coals 117f
Vanadium, detection limits and anal-
 ysis by ICP–AES 150t, 153t, 160t,
                         167t, 169t, 171t
Vapor generation–atomic absorption
 spectrometry 151
Vapor phase organic extractable from
 Tenax, reconstructed ion
 chromatogram 212f
Venturi scrubber decanter outlet and
 the tar trap, mass distribution of
 Sephadex LH-20 fraction of tar
 samples 217t
Viscosity and fluid melts 2
Vitrain analysis 13t
Vitrinite
 concentrate, FTIR spectrum 53f, 72f